Sustainable Development Goals Series

The **Sustainable Development Goals Series** is Springer Nature's inaugural cross-imprint book series that addresses and supports the United Nations' seventeen Sustainable Development Goals. The series fosters comprehensive research focused on these global targets and endeavours to address some of society's greatest grand challenges. The SDGs are inherently multidisciplinary, and they bring people working across different fields together and working towards a common goal. In this spirit, the Sustainable Development Goals series is the first at Springer Nature to publish books under both the Springer and Palgrave Macmillan imprints, bringing the strengths of our imprints together.

The Sustainable Development Goals Series is organized into eighteen subseries: one subseries based around each of the seventeen respective Sustainable Development Goals, and an eighteenth subseries, "Connecting the Goals", which serves as a home for volumes addressing multiple goals or studying the SDGs as a whole. Each subseries is guided by an expert Subseries Advisor with years or decades of experience studying and addressing core components of their respective Goal.

The SDG Series has a remit as broad as the SDGs themselves, and contributions are welcome from scientists, academics, policymakers, and researchers working in fields related to any of the seventeen goals. If you are interested in contributing a monograph or curated volume to the series, please contact the Publishers: Zachary Romano [Springer; zachary.romano@springer.com] and Rachael Ballard [Palgrave Macmillan; rachael.ballard@palgrave.com].

Rajah Rasiah · Anter Venus ·
Nur Annizah Ishak
Editors

Sustainable Agriculture and Food Security in Southeast Asia

Towards Climate Resilience and Digitalization

Springer

Editors
Rajah Rasiah
Asia-Europe Institute
Universiti Malaya
Kuala Lumpur, Malaysia

Anter Venus
Universitas Pembangunan National
Veteran Jakarta
Jakarta Selatan, Indonesia

Nur Annizah Ishak
Faculty of Business and Administration,
Universiti Malaya
Kuala Lumpur, Malaysia

ISSN 2523-3084 ISSN 2523-3092 (electronic)
Sustainable Development Goals Series
ISBN 978-981-95-0476-3 ISBN 978-981-95-0477-0 (eBook)
https://doi.org/10.1007/978-981-95-0477-0

This Springer imprint is published by the registered company Springer Nature Singapore Pte Ltd.
The registered company address is: 152 Beach Road, #21-01/04 Gateway East, Singapore 189721, Singapore

Foreword

The societies and economies of Southeast Asia have never received the attention and scholarship that their size and importance warranted. Their total population of nearly 700 million is far larger than that of North America or the European Union and is even slightly larger than the population of Latin America and the Caribbean. In recent years, several nations in the region have received attention as countries experiencing rapid economic development but the focus of this attention has been on the role of the export of manufactures. In a similar vein much of the historical interest in these countries was primarily focused on their plantation crops and minerals (rubber, tin oil, etc.). Large components of the economy in the region, particularly the agricultural sector other than plantation crops, was left out. This book, *SustainableAgriculture and Food Securityin Southeast Asia* edited by Rajah Rasiah, Anter Venus, and Nur Annizah Ishak, is a major effort to fill that gap.

The formation of the Association of Southeast Asian Nations (ASEAN) did increase the visibility of the region, but the great diversity of the countries that make up ASEAN makes analysis of the region, taken as a whole, challenging. At one end of the spectrum is Singapore, a wholly urban society with one of the highest per capita incomes in the world. Then there are the rapidly industrializing and urbanizing economies of large countries such as Vietnam and Malaysia. At the other end of the spectrum is Myanmar that remains poor and is currently wracked by a civil war.

Less well known is that the rural population of these nations at 333 million is roughly half of the total population of 685 million (in 2023). Most of this rural population is directly or indirectly involved in agriculture and in most of Southeast Asia the agricultural sector of each country is also undergoing significant changes.

The great achievement of the editors and chapter authors of *SustainableAgriculture and Food Securityin Southeast Asia* is that they have produced a volume that covers, in depth, the entire region. Many of the chapters focus on the experience of a specific country in the region—the diversity of the individual national experiences makes that necessary. Each country chapter, however, deals with a similar set of questions. The overall result thus brings out both the similarities and differences of the Southeast Asian region.

The title of the book and many of the chapters include the terms sustainability and food security. These are major themes of the book, but the book also deals with many other dimensions of agricultural development in the region as well. The particular emphasis on food security results from the fact that many countries in Southeast Asia have become dependent to an important degree on food imports that at times, such as during the COVID-19 epidemic, have been disrupted. Sustainability was a characteristic of much of the region's traditional agriculture, but the methods used in the subsequent development of the sector were not sustainable. That, as the book makes clear, has begun to change, and the newer sustainable techniques are playing an increasing role.

Readers of this book, therefore, will get a comprehensive view of how the different nations of Southeast Asia have dealt with the challenges facing their agricultural development efforts. The book is an important complement to the larger literature on the development of urban manufacturing in Southeast Asia.

February 2025

Dwight H. Perkins
Harvard University
Cambridge, Massachusetts, USA

Preface

The Asia-Europe Institute with the Centre for ASEAN Regionalism (CARUM) as the initiating anchor, successfully obtained funding from the Open Society Foundation (OSF) to publish four books on Southeast Asia highlighting critical aspects of the economies of each of the countries that constitute the subcontinent. With a focus on understanding the dynamics of development on Southeast Asia, this is the first of a series of research-based books launched by the Centre for ASEAN Regionalism that are targeted at understanding the key pillars of the subcontinent. This is the first of these books that focuses on agriculture and food security. In addition, to examining the state of agriculture and food security in Southeast Asia's constituent nations, the chapters seek to locate policies the governments in these countries have put in place to strive for achieving the United Nations' Sustainable Development Goals (SDGs).

Southeast Asia has its own unique indigenous peoples that make the region distinctive. The historical roots of Southeast Asia can be traced to sedentary agriculture with the early Indianization that still shows the remains of magnificent temples still visible in the region. Indian and Southeast Asian relations began in prehistoric times and continued until Western colonialism was established from the sixteenth to the nineteenth centuries. The contemporary nations of Brunei, Cambodia, Indonesia, Laos, Malaysia, Myanmar (Burma), Singapore, Thailand, Timor-Leste (East Timor), the Philippines and Vietnam historically encompass a region strongly influenced by Indian, Chinese, Islamic and European civilizations. The Portuguese and Spanish were the first to colonize Southeast Asia as spices and Christianity dominated their early focus. With the exception of Thailand, all other Southeast Asian nations came under colonial rule from the sixteenth century starting with the Philippines by the Spanish. Thailand is the only Southeast Asian country that was not colonized by the European powers. Nevertheless, sedentary agriculture and fishing dominated the economies of these countries until colonial interest resulted in the introduction of mining, cash crop production and plantation agriculture. In fact, at the time of independence after the Second World War, Southeast Asia was ideologically divided with the allies seeking to retain the individual nations within the free world. Following independence from France in 1954, Vietnam, Cambodia and Laos, and following independence from Britain in 1948, were increasingly disengaged from the free world. While American attempts to prevent the spread of communism in Indonesia, and the military Junta in Myanmar staying closed, ASEAN existed in its early years until 1992 but was limited to the free world economies of Indonesia, Malaysia, the Philippines, Thailand and Brunei.

The Association of Southeast Asian Nations (ASEAN) was founded when the foreign ministers of Indonesia, Malaysia, the Philippines, Singapore and Thailand signed the ASEAN declaration on 8 August 1967 in Bangkok. The initiatives towards the formation of ASEAN started originally when the Southeast Asian Treaty Organization (SEATO) was founded in 1954 and subsequently the Association of Southeast Asia in 1961. The early engagement of ASEAN gained importance owing to the Vietnam War, which lasted from 1955 until 1975. ASEAN's first summit meeting took place in Bali in 1976, which resulted in the signing of the Treaty of Amity and Cooperation. Subsequent initiatives include the triangular economic development initiated by Singapore, and the founding of the ASEAN Free Trade Area (AFTA) in 1992. The adoption of the ASEAN Charter in 2007 provided the legal and institutional framework for its three core pillars, i.e. the ASEAN Economic Community,the ASEAN Political-Security Community and the ASEAN Socio-Cultural Community. The end of the Cold War as President Gorbachev brought about the dismantling of the Soviet Union and the Warsaw Pact in 1991 attracted stronger initiatives to raise regional integration among the ASEAN nations with Vietnam, Laos, Myanmar and Cambodia transiting to join ASEAN. AFTA subsequently evolved into the implementation of the ASEAN Economic Community (AEC) in 2015. In this context, Timor-Leste is still seeking political stability to pursue similar development goals. We are expecting the entry of Timor-Leste as the 11th member soon. The country was part of the ASEAN when the country was a part of Indonesia earlier.

Key topics covered in this book are significance of agriculture to the national economies, poverty, standards of living, labour productivity, export significance, the importance of organic and traditional farming and climate resilience, and the proliferation of smart technologies in agriculture. Given the critical importance of food production to reduce dependence on imports, food security is a major topic addressed by all the chapters. The evidence is clear that small city states and nations, such as Singapore and Brunei, and countries still grappling with low incomes, such as Cambodia, Lao PDR and Myanmar have increased emphasis on smart farming to raise productivity and reduce reliance on imports.

As medium and large economies, Indonesia, Myanmar, the Philippines, Thailand, Vietnam and Malaysia exhibit a wide range of agriculture, sedentary, small-scale and large-scale farming. Indonesia, Malaysia and Thailand show heavy dominance of large-scale plantation agriculture with strong specialization in oil palm and rubber cultivation. Agricultural parastatals have been important in Indonesia, Malaysia and Thailand to support small farmers, though small-scale farming is important, especially to support rural households in all countries, except for Singapore where farming is focused increasingly on high yielding crops that show high returns. All countries in Southeast

Asia have supplied to the United Nations Convention for Climate Change (UNFCCC) developed Intended National Determined Contribution (INDC) disclosing their timelines to achieve carbon net zero emissions. Initiatives are also being taken by these countries to strengthen climate resilience and digitalization to meet the United Nations Sustainable Development Goals (SDGs).

Given the infancy of the new approaches undertaken by the Southeast Asian economies, there is still much room to achieve food self-sufficiency and overall agricultural sustainability. All countries have sought smartification initiatives to raise agricultural productivity, but the emphasis has been stronger in Singapore and Brunei, both of which face serious shortage in farm acreage and facing massive imports. Timor-Leste is also a small country but is presently lacking a focused approach to establishing economic strategies.

While bringing together the state of agriculture in Southeast Asia by individual countries and focused on screening the strategies taken by governments to raise agricultural productivity, exports, and self-sufficiency rates, the book examines the efforts taken by these countries to achieve the United Nations' SDGs in general, and to alleviate poverty and promote climate resilience in particular. In doing so, the chapters deal with the reality facing agriculture in these countries, i.e. urbanization and industrialization have increasingly squeezed agriculture thereby rapidly lowering the acreage enjoyed by the sector. While the intensity has varied, all countries in the subcontinent have responded in varying scales to promote small-scale, intensive farming through the introduction of smart technologies.

Although Teng (2024) edited an impressive book on food security in Asia, the chapters focused on particular aspects of food security rather as systematics assessment of individual countries in Southeast Asia. This book for the first time brings together chapters on agricultural policies, and agricultural development of the eleven countries that constitute Southeast Asia, including the non-Association of Southeast Asian (ASEAN) member, Timor-Leste. As a developmental feature of the book, nationals have been critical contributors of each of the chapters. Jerry Courvisanos is the principal author of Timor-Leste, but has prepared the chapter with two Timor-Leste citizens, i.e. Apolinário Magno and Elsa Pinto. I take this opportunity to thank all the important contributors to make this initiative possible. First and foremost, I would like to thank the Open Society Foundation (OSF) for generously funding the flights of all participants to the book workshop in Jakarta on 15–17 November 2023. I would in particular thank Premesh Chandran, Agantaranansa Juanda and Sophia Hurtado Epstein from OSF for taking the lead to support the workshop. I would like to thank the Universitas Pembangunan Nasional Veteran Jakarta for footing the domestic costs involved to host the workshop, accommodation at La Premiere Hotel in Jakarta, all meals and domestic transport. A Thanks also go to Anter Venus, Dianwicaksih Arief and Fachru Nofrian for being wonderful hosts at the workshop. A special thanks to Zahadin Omar who solved numerous technical problems I faced while preparing this manuscript. I wish to also thank Afiqah Rosden, Arief Izzuadin, Marsha, Noraerna Abdul

Rahim, Ummi Afiqah and Diana Ab Aziz of the Asia-Europe Institute for helping out in various ways to support the workshop. Finally, I wish to thank my colleagues Kevin Fernandez, Adil Johan, Johan Saravanamuttu and Roy Anthony Rogers for conversations on the topics addressed in this book. The errors that remain are mine.

Kuala Lumpur, Malaysia Rajah Rasiah

Praise for *Sustainable Agriculture and Food Security in Southeast Asia*

"Food security is national security. The recent COVID-19 pandemic crisis has unmasked the vulnerabilities in ASEAN food systems. The crisis exposed a complex matrix of challenges involving climate change, geo-political dynamics, depleting resources, technological disruption, changing population demography are among them. The future will be a multiplication of the current mix. This book provides an excellent mosaic of policies and future strategies of ASEAN countries toward climate-smart and technology-driven agriculture."

—Professor Fatimah Mohamed Arshad, *Consultant Fellow and Chair of Food Security, Institute of Tropical Agriculture and Food Security, Universiti Putra Malaysia*

Contents

Editors and Contributors

About the Editors

Rajah Rasiah is the Executive Director of the Asia-Europe Institute (Universiti Malaya) since 2022. He obtained his Ph.D. (economics) from Cambridge University in 1992 and was a Rajawali Fellow at Harvard University in 2014. He was awarded the Celso Furtado Prize in 2015 for advancing the frontiers of social science (economics) thought by the World Academy of Sciences, the Merdeka Prize in 2018 for scholastic achievement by the Merdeka Prize Committee, and the Distinguished Professorship in Economics in 2017 by the Ministry of Higher Education.

Anter Venus is Associate Professor of Strategic Communication with Specialization in Sustainable Communication at the Faculty of Social and Political Sciences, Universitas Pembangunan National Veteran Jakarta. He is also the Rector of the university. He teaches and conducts research in the fields of strategic communication, sustainable communication, campaign management, and Nusantara interethnic communication. His latest research conducted with Trisia Megawati is entitled: identifying and addressing the right to burn by indigenous people. He has also led fire stewardship practices in South Sumatra in 2024.

Nur Annizah Ishak is a Senior Lecturer in the Department of Political Science, Public Administration and Development Studies, Faculty of Business and Economics, Universiti Malaya. She holds a Ph.D. (Economics) from University Malaya, MA (Economics) from University of Missouri, USA, MEC (Economics Development) and BEC (Economics) from Universiti Kebangsaan Malaysia. Her research interests include development economics, sociology, poverty, and community development. She has published numerous articles in academic journals and books.

Contributors

Rose Abdullah Faculty of Agriculture, Universiti Islam Sultan Sharif Ali, Bandar Seri Begawan, Brunei

Osumanu Haruna Ahmed Faculty of Agriculture, Universiti Islam Sultan Sharif Ali, Bandar Seri Begawan, Brunei

Yot Amornkitvikai College of Population Studies, Chulalongkorn University, Bangkok, Thailand

Dianwicaksih Arieftiara Universitas Pembangunan Nasional Veteran Jakarta, Depok, Indonesia

Khammerng Bannalath Ministry of Finance, Lao Statistics Bureau, Vientiane, Laos

Nguyen Thanh Bình Can Tho University, Can Tho, Vietnam

Teerawat Charoenrat Faculty of Economics, Khon Kaen University, Khon Kaen, Thailand

Jerry Courvisanos Federation University Australia, Melbourne, Australia

Mohamad Fadhil Hasan Universitas Pembangunan Nasional Veteran Jakarta, Depok, Indonesia

Nur Annizah Ishak Faculty of Business and Economics, Universiti Malaya, Kuala Lumpur, Malaysia

Dang Kim Khôi vietnam national university of agriculture, Hanoi, Vietnam

Kum Kim KOTRA Phnom Penh, Phnom Penh, Cambodia

Elyssa Kaur Ludher ISEAS-Yusof Ishak Institute, Singapore, Singapore

Apolinário Magno Universidade Nacional Timor Lorosa'e, Dili, Timor-Leste

Fachru Nofrian Universitas Pembangunan Nasional Veteran Jakarta, Depok, Indonesia

Rene E. Ofreneo School of Labor and Industrial Relations, University of Philippines, Diliman, Quezon City, Philippines

Elsa Joaquina Araújo Pinto Universidade Nacional Timor Lorosa'e, Dili, Timor-Leste

Nguyen Hong Quan Vietnam National University, Ho Chi Mnh, Vietnam

Rajah Rasiah Asia-Europe Institute, Universiti Malaya, Kuala Lumpur, Malaysia

Naw Shareen Kasetsart University, Bangkok, Thailand

Bach Tan Sinh Vietnam National University, Ho Chi Mnh, Vietnam

Latdavanh Songvilay Lao Academy of Social and Economic Sciences, Macroeconomic Research Institute, Vientiane, Laos

Paul Teng ISEAS-Yusof Ishak Institute, Singapore, Singapore

Le Van Thuy Tien Can Tho University, Can Tho, Vietnam

Nguyen Minh Tu Vietnam National University, Ho Chi Mnh, Vietnam

Abbreviations

ACEF	Agricultural Competitiveness Enhancement Program
ACT	Agrifood Cluster Transformation
ADB	Asian Development Bank
ADPC	Asian Disaster Preparedness Center
ADS	Agriculture Development Strategy
AEC	ASEAN Economic Community
AFIP	Agrifood Innovation Park
AFMA	Agricultural and Fisheries Modernization Act
AFTA	Asean Free Trade Area
AI	Artificial Intelligence
AoA	Agreement on Agriculture
ARBs	Agrarian Reform Beneficiaries
ASEAN	Association of Southeast Asian Nations
ATOs	Agricultural Technical Officers
AVA	Agrifood and Veterinary Authority
B2B	Business-to-Business
B2C	Business-to-Consumer
BAAC	Bank for Agriculture and Agricultural Cooperatives
BAPANAS	Badan Pangan Nasional (National Food Agency)
BERNAS	Padiberas Nasional Berhad
BIMAS	Bimbingan Masyarakat (Society Training)
BMPT	Badan Perusahaan Bahan Makanan dan Pembuka Tanah
BPMD	Badan Pendidikan Masyarakat Desa (Village Society Education Agency)
BULOG	Badan Urusan Logistik (National Logistic Agency)
BUMN	Badan Usaha Milik Negara (National State Enterprise)
CADT	Cambodia Academy of Digital Technology
CAGR	Cumulative Average Nominal Growth Rate
CARL	Comprehensive Agrarian Reform Law
CARUM	Centre for ASEAN Regionalism
CCPAP	Climate Change Priority Action Plan
CDRI	Cambodia Development Resource Institute
CFW	Cash for Work
CLMV	Cambodia, Lao PDR, Myanmar, and Vietnam
COP	Conference of Parties
CPI	Consumer Price Index
CRC	Center For Research and Communication

CSA	Climate Smart Agriculture
DA	Department of Agriculture
DAE	Department of Agricultural Extension
DAR	Department of Agrarian Reform
DEMAS	Dewan Masyarakat (Conseils of Society)
DEPA	Digital Economy Promotion Agency
DGA	Digital Government Development Agency
DMH	Department of Meteorology and Hydrology
DOSM	Department of Statistics Malaysia
DTAC	Total Access Communication Public Company Limited
EFAP	Emergency Food Assistance Project
ELCs	Economic Land Concessions
EMS	Early Mortality Syndrome
EPS	Earning per Share
ESG	Environmental, Social and Governance
EU	European Union
EUDR	EU Deforestation Regulation
EU-FF	Food Facility from the European Union
EWEAS	Early Warning Early Actions
FAO	Food and Agriculture Organization
FDA	Food and Drug Administration
FELCRA	Federal Land Consolidation and Rehabilitation Authority
FELDA	Federal Land Development Authority
FFW	Food for Work
FRESH	Future Ready Food Safety Hub
G2G	Government-to-Government
GAP	Good Agricultural Practice
GDP	Gross Domestic Product
GFSI	Global Food Security Index
GHG	Greenhouse Gas
GIS	Geographic Information System
GIZ	German Agency for International Cooperation
GMP	Good Manufacturing Practice
GPS	Global Positioning System
GSA	Green and Sustainable Agriculture
HACCP	Hazard Analysis Critical Control Point System
HVC	Grow High-Value Crops
HYV	High-Yielding Variety
IDE	International Development Enterprises
IDR	Import Dependency Ratio
IDR	Indonesian Rupiah
IFPRI	International Food Policy Research Institute
ILO	International Labour Organization
IMF	International Monetary Fund
INDC	Intended Nationally Determinant Contributions
IoT	Internet of Things
IPOA	Indonesian Palm Oil Association
IR	Industrial Revolution

IRRI	International Rice Research Institute
ISPO	Indonesia Sustainable Palm Oil
ITA	International Trade Administration
KOGM	Komando Operasi Gerakan Makmur
KOPERTA	Koperasi Produksi Pertanian (Agriculture Cooperation of Production)
KOTOE	Komando Otoritas Ekonomi
KUD	Koperasi Unit Desa (Village Unit of Cooperation)
KUT	Koperasi Unit Tani (Agriculture Association of Cooperation)
LASES	Lao Academy of Social and Economic Sciences
LED	Light Emitting Diode
LGUs	Local Government Units
LIFT	Livelihoods and Food Security Trust Fund
LPP	Farmers' Organisation Authority
LSB	Lao Statistic Bureau
LTS4CN	Long-Term Strategy for Carbon Neutrality
M99	Masagana 99
MAF	Ministry of Agriculture and Forestry
MAFF	Ministry of Agriculture, Forestry and Fisheries
MAPSA	Myanmar Agriculture Policy Support Activity
MARDI	Malaysian Agricultural Research and Development Institute
MCO	Movement Control Order
MDGs	Millennium Development Goals
MEF	Ministry of Economy and Finance
MES	Myanmar Engineering Society
MGS	Myanmar Geosciences Society
MIMU	Myanmar Information Management Unit
MLFRD	Ministry of Livestock, Fisheries and Rural Development
MND	Ministry of National Development
MOAC	Ministry of Agriculture and Cooperatives
MOAI	Ministry of Agriculture and Irrigation, former MOALI
MOALI	Ministry of Agriculture, Livestock and Irrigation
MOC	Ministry of Commerce
MoH	Ministry of Health
MONREC	Ministry of Natural resources and Environmental Conservation
MPC	Malaysian Productivity Corporation
MPI	Ministry of Planning and Investment
MPOB	Malaysian Palm Oil Board
MRI	Macroeconomic Research Institute
MSMEs	Micro, Small and Medium Enterprises
NABC	National Agricultural Big Data Center
NADP	National Agricultural Development Policy
NAFP	National Agrofood Policy
NAP	National Agricultural Policies
NAPA	National Adaptation Programme of Action

NbS	Nature-based Solutions
NCCAP	National Climate Change Action Plan
NCDM	National Committee for Disaster Management
NCSD	National Council for Sustainable Development
NEA	National Environment Agency
NEDA	National Economic and Development Authority
NEP	New Economic Policy
NESDB	Office of the National Economic and Social Development Board
NESDC	Office of the National Economic and Social Development Council
NFA	National Food Authority
NGOs	Non-government Organizations
NiDA	National Information Communication Technology Development Authority
NIPTICT	National Institute of Posts, Telecommunications, Information and Communication Technology
NIS	National Institute of Statistics
NOCs	Notices of Coverage
NREP	National Renewable Energy Program
NRF	National Research Foundation
NSDP	National Strategic Development Plan
NSPARD	National Strategy on Poverty Alleviation and Rural Development
NTB	Nusa Tenggara Barat
NTT	Nusa Tenggara Timor
OAE	Office of Agricultural Economics
OCED	Organisation for Economic Co-operation and Development
OFWs	Overseas Filipino Workers
OJK	Otoritas Jasa Keuangan (Indonesian Financial Services Authority)
OPSSB	Organisasi Pelaksana Swasembada Beras (Rice Self-Sufficiency Organization)
OSF	Open Society Foundation
PAGASA	Philippine Atmospheric, Geophysical and Astronomical Services Administration
PBSN	Perkebunan Besar Swasta Nasional (National Private Large Plantation)
PGK	Poverty Line
PhilMech	Philippine Center for Postharvest Development and Mechanization
PhilRice	Philippine Rice Research Institute
PNK	Area Fishermen's Organization
PNSB	Pembangunan Nasional Semesta Berencana
PPD	Primary Production Department
PPK	Area Farmers Organization
PPPs	Public-Private Partnerships
PTD	Pamong Tani Desa

R&D	Research and Development
RCEP	Regional Economic Partnership Agreement
REPELITA	Rencana Pembangunan Lima Tahun (Five Year Development Plan)
RISDA	Rubber Industry Smallholders Development Authority
RRD	Red River Delta
RSPO	Roundtable Sustainable Palm Oil
RTL	Rice Tariffication Law
SAT	Singularity Aerotech
SDGs	Sustainable Development Goals
SEA	South-East Asia
SEATO	Southeast Asian Treaty Organization
SFA	Singapore Food Agency
SGD	Singapore Dollar
SME	Small and Medium-sized Enterprises
SPCR	Strategic Program for Climate Resilience
SRI	System for Rice Intensification
SSM	Special Safeguard Measure
SSR	Self-Sufficiency Rate
SWOT	Strengths, Weaknesses, Opportunities and Threats
TFP	Total Factor Productivity
TRC	Telecommunication Regulator of Cambodia
UAV	Uncrewed Aerial Vehicles
UNAPCAEM	United Nations Asian and Pacific Center for Agricultural Engineering and Machinery
UNDP	United Nations Development Programme
UNFCCC	United Nations Framework Convention for Climate Change
USDA	United States Department of Agriculture
UUPA	Undang-Undang Penamanan Agraria (Agrarian Law)
VAC	Integrating Farming *(Vuon-Ao- Chuong)*
VAT	Value Added Tax
VMD	Vietnam Mekong Delta
VOA	Voice of America
VOC	*Vereenigde Oostindische Compagnie*
VPHC	Veterinary Public Health Centre
WFP	World Food Programme
WMT	Weather Modification Technology
WTO	World Trade Organization

List of Figures

List of Tables

1 Strengthening Agricultural Sustainability and Food Security in Southeast Asia

Rajah Rasiah

Abstract

Southeast Asia has been one of focus regions in international efforts to stimulate digitalization and climate resilience, including in the agricultural sector. This chapter introduces the whole region historically to document the trend changes in the contribution of the agricultural sector to the national economies of the region, including government policies to promote the sector. Of particular importance since the 1940s, 1950s and 1960s, and in the case of Timor-Leste, since independence in 2002 is government focus on strengthening food security in general and self-sufficiency in rice cultivation in particular. The chapter also examines initiatives of the individual countries in the region to smartify their economies following the launching of the sustainable development goals to stimulate digitalization towards raising productivity and strengthening climate resilience.

Keyword

Agricultural sustainability · Food security · Digitalization · Climate resilience · Southeast Asia

The old focus on agriculture in overall economic development is that it will continue to remain as a vital sector, especially in large and medium economies but its share in GDP would gradually fall. The Lewisian argument of the interrelationship between agriculture and manufacturing evolves to reach an equilibrium as wages between the two sectors equalize (Lewis, 1955). The establishment of such an equilibrium is driven by both technical change in agriculture and the out-migration of surplus labour (disguised unemployment) from agriculture to manufacturing and services. However, with the rising significance of food as an essential good and with countries efforts to build food production capacity to ensure food security, the share of agriculture in GDP appears to show a plateauing. The experience of the Britain, France, Germany, Japan, and the United States shows a gradual fall and

R. Rasiah (✉)
Asia-Europe Institute, Universiti Malaya, Kuala Lumpur, Malaysia
e-mail: rajah@um.edu.my

R. Rasiah et al. (eds.), *Sustainable Agriculture and Food Security in Southeast Asia*, Sustainable Development Goals Series, https://doi.org/10.1007/978-981-95-0477-0_1

eventually a plateauing.[1] Nevertheless, while the share of agriculture in GDP will gradually fall to be overtaken by manufacturing first, and then eventually services, successful developers seek to stimulate technological upgrading so that agricultural productivity continues to rise. The same can be expected of manufacturing where the sector's productivity continues to rise even after its share in GDP starts to fall (Rowthorn & Wells, 1987).

Interesting accounts on agriculture among the market economies include that on Indonesia by Hill (2000), on Malaysia by Rasiah et al. (2022a), Rasiah et al. (2022b)), on the Philippines by Balisacan and Hill (2003), on Thailand by Pongpaichit and Baker (2002), and on Vietnam by Ljunggren and Perkins (2023) offer helpful backdrop on the national economies in general but also touching among other things on agriculture. Much have happened since and there is little account of the development of agriculture since the millennium in the other Southeast Asian economies. This book seeks to capture the overall development of agriculture in these economies but specifically focusing on the period after the turn of the millennium. A major point to address in all economies regardless of whether they are large or small is the changing pressure on land prices and land availability as rapid economic growth and urbanization squeeze agriculture to the point that only high productivity farming based on little land use becomes important in small countries (such as Singapore and Dili), large cities in medium (e.g. Kuala Lumpur, Bangkok, Vientiane, Phom Penh, Manila, Hanoi, and Yangon), and large countries (such as Jakarta and Bandung). Consequently, as manufacturing and construction expands, the growing shortage in land leads to a shift from extensive to intensive farming, and the extensive use of mechanization and automation. Southeast Asia is no different in facing increasing pressures to deal with such realities.

In addition, two major technological developments have gained significance since the millennium that have driven massive transformation in the global economy and have also been stimulating technical change in Southeast Asia. The first is the global push for containing climate change and global warming (Nordhaus, 2008, 2018; Stern, 2007). While the initiative gained currency well before the millennium, the scale of a collaborative global effort was triggered by the 21st conference of parties (COP) of the United Nations Framework Convention for Climate Change (UNFCCC) that culminated in the Paris Accord when 192 countries pledging to contribute to capping temperature rise to 1.5 degrees Celsius over the next 100 years with 2005 taken as the base year (Rasiah et al., 2017). Although manufacturing is the main polluting sector, all sectors have come under scrutiny to reduce hazardous emissions, and liquid and solid waste with the emphasis shifting to the circular economy. Since the environment is a global common (global public good), and it flags the need to check negative externalities from destroying planet Earth (Baumol & Oates, 1988; Hardin, 1968; Marshall, 1890; Pigou, 1932; Rasiah et al., 2022a, 2022b). Such concerns have driven the world to come together (albeit one can question the conviction of some countries to do so) to mitigate climate change and global warming. This has been aided by the rising importance of traditional farming methods in countries (Teng, 2024), such as Thailand. Consequently, traditional and organic farming has risen in importance.

The second major technological transformation is the transition from IR3.0 to IR4.0. While the former compliments the latter, the former is focused on automation and computerization while the latter is driven by the transfer of artificial intelligence that has resulted the development of autonomous machines, such as drones and robots. On the one hand, the proliferation of IR 4.0 technologies has provided the potential for a quantum

[1] The share of agriculture in the GDP of the Britain, France, Germany, Japan, and the United States fell in trend terms from 1960 to 2010. From latest data supplied by the World Bank (2022), the highest share of agriculture value added in GDP in the periods 1990–2021 for Britain (1.4%) and France (3.1%) was recorded in 1990, 1991–2021 for Germany (1.1%) was recorded in 1991, for Japan (1.9%) in the period 1994–2021, and the United States (1.3%) in the period 1997–2021 was recorded in 1997 (World Bank, 2022). The lowest share of agriculture in GDP in the same commensurate periods for Britain (0.6%) and France (1.3%) was recorded in 2005 and 2009, respectively, for Germany (0.7%) in 2015, for Japan (1.0%) in 2020, and for the United States (0.8%) in 2019 (World Bank, 2022).

leap in productivity; its success also depends on the ability to evolve the embedding ecosystem instruments to support it, such as, broadband fibre optic cables, internet of things and cloud computing, and big data analytics. Importantly, the ability of IR4.0 to support productive transformation in agriculture will require a sea change in the way education and training is set up to produce the labour force to support digitalization and digitization.

While the developments noted above are already emerging to usher growth and economic development, especially since the turn of the millennium, given that these technologies are associated with both public goods and public utilities (in which social returns will exceed private returns),[2] the role of government will be critical. Hence, the transition to sustainable and digitalized agriculture requires a massive focus by governments. However, as the experience of past successful technological advances experienced by countries such as South Korea, Japan, Singapore, and Taiwan shows, robust screening mechanisms tied to stringent performance standards are vital to ensure that the rents distributed to promote such a transition are not dissipated (Johnson, 1982; Amsden, 1989; Wade, 1990; Rasiah, 2020). This is the carrots and sticks approach advanced by Sen (1981). Whereas Japan, South Korea, and Taiwan focused on national capital, Singapore used the leveraging approach to condition the participation of foreign multinationals to stimulate upgrading though its size meant that the focus only took place in manufacturing and services with agriculture largely insignificant.

Changing Share of Agriculture in the National Economies

Since the transition economies of Cambodia, Lao PDR, Myanmar, and Vietnam (CLMV) were involved in civil war with cold war interference until the 1980s, 1990s, and early 2000s, we analyse the contribution of agriculture to the national economies separate from the Southeast Asian market economies. Not only were these former economies plagued by war and international restrictions they were also little integrated in global trade and investment.

Southeast Asia is also characterized by the city state of Singapore, the small petroleum rich Brunei that is largely natural resource-driven, the small and still largely rural Timor-Leste (which became an independent country in 2002), and the isolated country of Myanmar that has faced economic sanctions from the United States and Europe since 2003. These specificities have had an impact on the nature of structural change experienced by these countries. Also, while Burma (now Myanmar) declared independence in 1948, the country has faced several military coups that has resulted in military dictatorships since.

In addition, although Vietnam, Lao PDR, Cambodia, and Myanmar declared independence from France in 1945, and 1953 and 1953, respectively, all three were embroiled in a crippling civil war from 1955 till 1975. The brutal Khmer Rouge took control of Cambodia in 1975 before being overthrown by the invading Vietnamese forces in 1979. However, the unrest in Cambodia continued till the end of 1997 (BBC, 2018).

The Association of Southeast Asian Nations (ASEAN) was formed in 1967 with Indonesia, Malaysia, Philippines, Singapore, and Thailand as founding members. Brunei became an ASEAN member in 1984. Vietnam (1995), Lao PDR (1997), Myanmar (1997), and Cambodia (1999) subsequently joined the ASEAN in the 1990s to enjoy the trade and privileges accorded to the association, which in 2015 were grouped under the ASEAN Economic Community (AEC) in 2015 (ASEAN, 2015). However, the military coup in Myanmar of 2021 and the continued genocide in the country has driven the country to further isolation since. Timor-Leste has been granted observer status in 2022 by the ASEAN members. Consequently, it is best to examine the five founding members and Brunei separately from the CLMV countries and Timor-Leste.

[2] However, while public goods are neither excludable nor rivalrous, public utilities are excludable and rivalrous (Samuelson, 1954).

Agricultural Share in GDP

Among the market economies except for Indonesia where data from the World Bank (2022) is only available from 1983, the market economies of Southeast Asia show a trend fall in GDP share of agriculture in GDP 1960 till 2021 with somewhat of a plateauing effect from the early 1990s (Fig. 1.1). Indonesia also shows the same trend from the available 1984–2021 data. Commercial agriculture, especially into oil palm and rubber cultivation, has stimulated some relative but slow and fluctuating expansion in agriculture from around the second half of the 1990s.

Although the data for the transition economies and Timor-Leste is only available from the late 1980s (Vietnam), 1990s (Cambodia and Lao PDR), and early 2000 (Myanmar and Timor-Leste), except for an early rise in the agricultural share in GDP in Timor-Leste, a trend fall has set in since (Fig. 1.2).

Agricultural Exports

Since agro-processing beyond off-state activity is best included in manufacturing, agricultural exports here refer to agricultural raw material exports. Figures 1.3 and 1.4 show agricultural raw materials exports in total merchandize exports of the Southeast Asian market economies and the transition economies and Timor-Leste.

Agricultural raw material exports in merchandise exports of Brunei, Indonesia, Philippines, Singapore, and Thailand fell in trend terms from 1.7, 45.6, 25.6, 34.4, and 34.9% in 1962 to almost zero, 4.6, 1.3, 0.4, and 4.2% in 2021 (Fig. 1.4). The commensurate shares for Malaysia were 51.8% in 1964 and 1.4% in 2021, respectively. The sharp fall in agricultural raw material exports was driven by both a rapid expansion in manufacturing and services, and downstream processing of agricultural goods.

There were no data available for the CLMV countries and Timor-Leste for a significant part of the 1970s, 1980s, and 1990s because of war-time disruptions and the trade embargo on Vietnam until 1994. Nevertheless, these countries also show a trend decline in the share of agricultural raw material exports in merchandise exports (Fig. 1.4). Cambodia's agricultural raw material exports fell in trend terms from 44.7% in 1962 to 2.6% in 2021. Meanwhile, agricultural raw material exports from Lao PDR and Myanmar rose from 6.9 and 18.3% in 1962 to 81.5 and 47.7%, respectively, in 1973. The last shares in total merchandise exports of Timor-Leste and Vietnam were 25.1% in 2017 for the former and 1.9% for the latter in 2021. Vietnam in particularly has experienced a significant fall in agricultural raw exports as domestic consumption grew with rapid economic growth.

Employment and Labour Productivity

As with value added, the role of agriculture in employment creation also gradually falls as the manufacturing and service sectors start growing, though the shrinking agricultural sector tends to show a trend rise in productivity in successful economies. In small economies, the contribution of agriculture to national employment falls quickly as these economies grow.

Agricultural employment in national employment fell in trend terms in the larger economies of Indonesia, Malaysia, Philippines, and Thailand from 55.5, 22.5, 44.9, and 60.3%, respectively, in 1991 to 28.5, 10.3, 22.9, and 31.4% in 2019 (Fig. 1.5). Employment contribution in national employment in the smallest economies among the small Southeast Asian market economies, i.e. Singapore and Brunei. Singapore's agricultural employment in total employment fell gradually from 0.3% in 1991 to almost 0%, respectively, in 2019 while that of Brunei fell from 2.0% in 1991 to 0.6% in 2014 before rising again to 2.0%.

The contribution of agriculture in national employment has also fallen in trend terms among the transition economies and Timor-Leste (Fig. 1.6). The agricultural employment in total employment fell in trend terms in Cambodia, Lao PDR, Myanmar, Timor-Leste, and Vietnam from 78.1, 86.8, 61.0, and 70.9%, respectively, in 1991

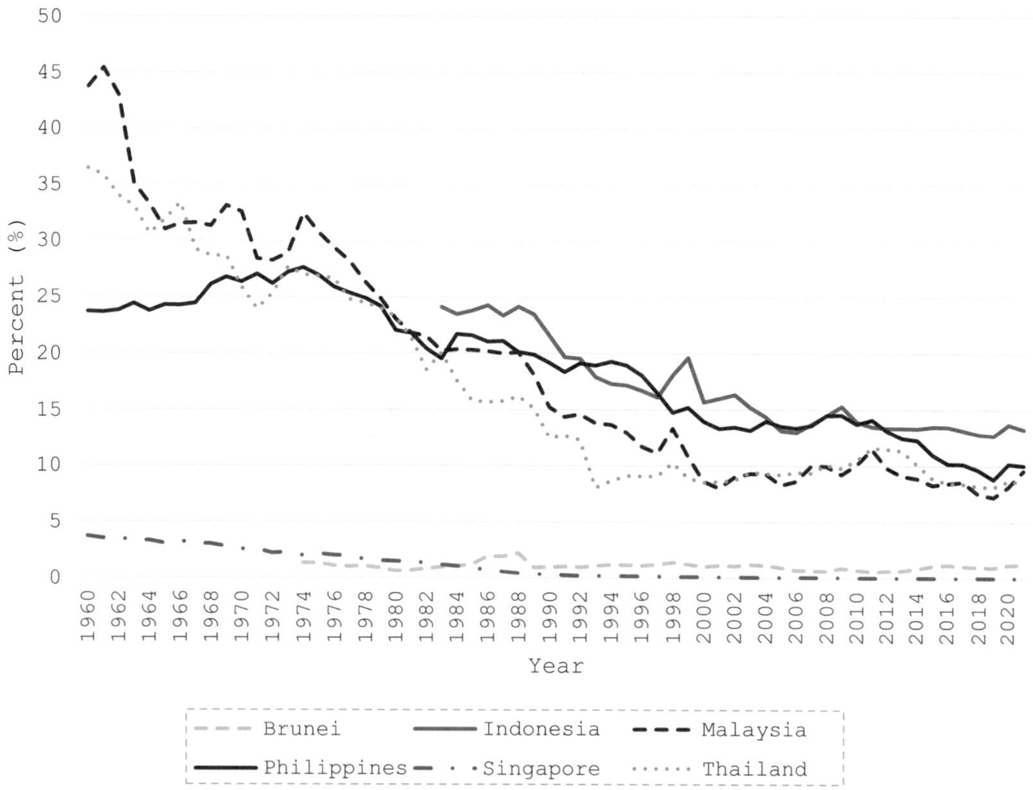

Fig. 1.1 Share of agriculture in GDP, market economies, Southeast Asia, 1960–2021. *Source*: World Bank (2022)

to 34.5, 61.4, 53.2, 39.4, and 36.2%, respectively, in 2019 (Fig. 1.4).

Although labour productivity does not show the direct effects of capital inputs on productivity, it is preferred over total factor productivity (TFP) owing to the problems associated with how technology is endogenized in the models used to measure it, which may explain why Young's (1994, 1995) estimations showed extremely low annual average growth TFP growth recorded by Singapore, South Korea, and Taiwan over the period 1970–1985. TFP estimations of especially poorer economies tend to understate the importance of higher thresholds of capital expenditure owing to the infancy of infrastructure development.

Among the Southeast Asian market economies, Brunei enjoyed the highest annual average labour productivity growth between 1960 to reach its peak of US$127.7 thousand in 2014 before falling in trends terms and US$33.3 thousand in 2019 (Fig. 1.7).

All CLMV countries show a rise in labour productivity (Fig. 1.8). Vietnam shows the fastest rise in labour productivity among the CLMV countries and Timor-Leste. Timor-Leste shows a relative stagnation in labour productivity, though its labour productivity rate was the highest among these countries over the period 2001–2016. Vietnam and Cambodia overtook Timor-Leste in 2017.

Food Import Dependence in Southeast Asia

Food security in the face of population growth has been a perennial problem. Indeed, the green revolution experiments and exercises had proliferated across Southeast Asia with the International Rice Research Institute (IRRI) an independent organization, was founded by the Ford and Rockefeller Foundations and the Philippines government in 1960 with its headquarters located

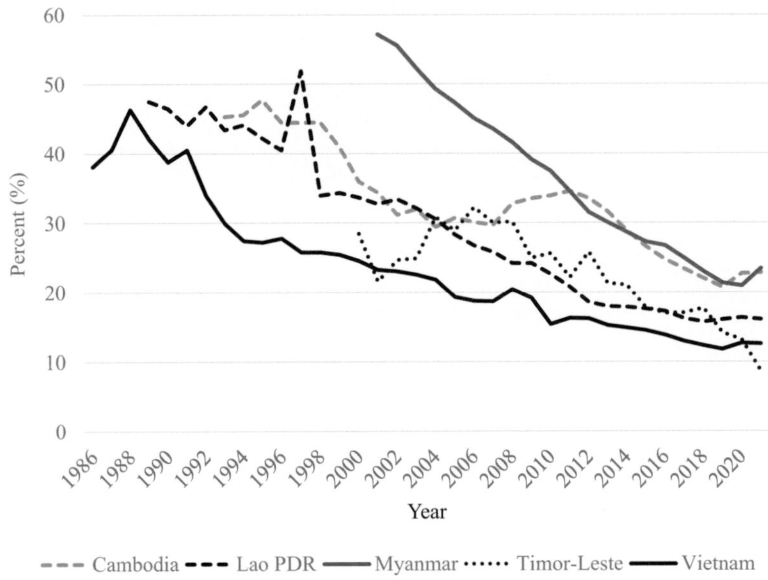

Fig. 1.2 Share of agriculture in GDP, CLMV countries and Timor-Leste, 1986–2021. *Source*: World Bank (2022)

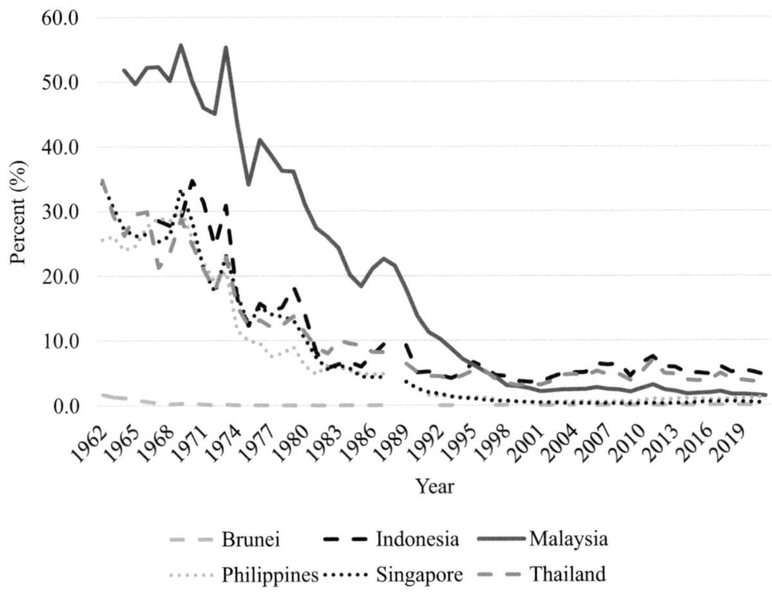

Fig. 1.3 Agricultural raw material exports in merchandise exports, Southeast Asian market economies, 1962–2021. *Source*: World Bank (2022)

in Los Banos, Laguna, the Philippines heralded a major push to develop high yielding varieties of rice that supported multiple crop production annually (Mackill & Khush, 2018). While George (1977, 1988) and Lappe and Collins () have argued succinctly that distribution is the biggest problem why scarcity and starvation exist over the growing concerns over overpopulation (Malthus, 1798; Meadows et al, 1972), overdependence on food imports to feed the domestic population has become a serious issue gripping several economies when exporting countries reduce food exports, which often happens during bad seasons and rising domestic demand. The COVID-19

Fig. 1.4 Agricultural raw material exports in merchandise exports, CLMV countries and Timor-Leste, 1962–2021. *Source*: World Bank (2022)

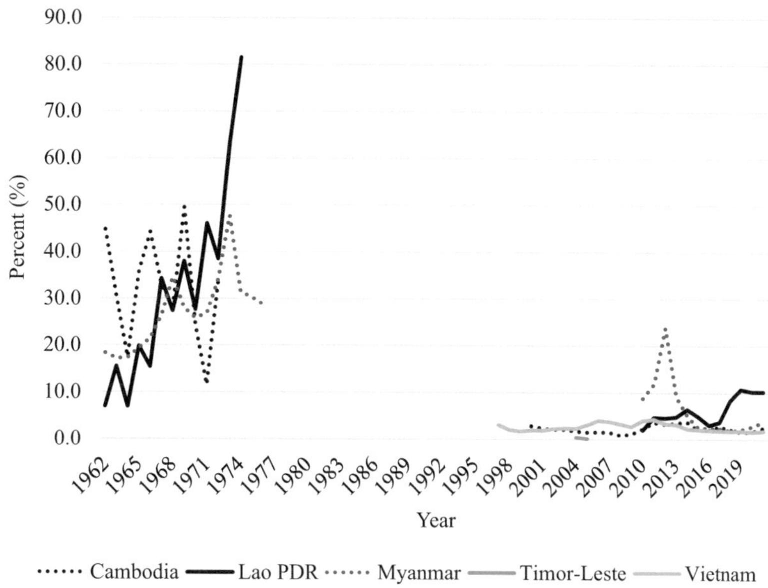

Fig. 1.5 Agricultural employment in national employment, Southeast Asian market economies, 1991–2019. *Note*: Estimations by the International Labour Organization (ILO). *Source*: World Bank (2022)

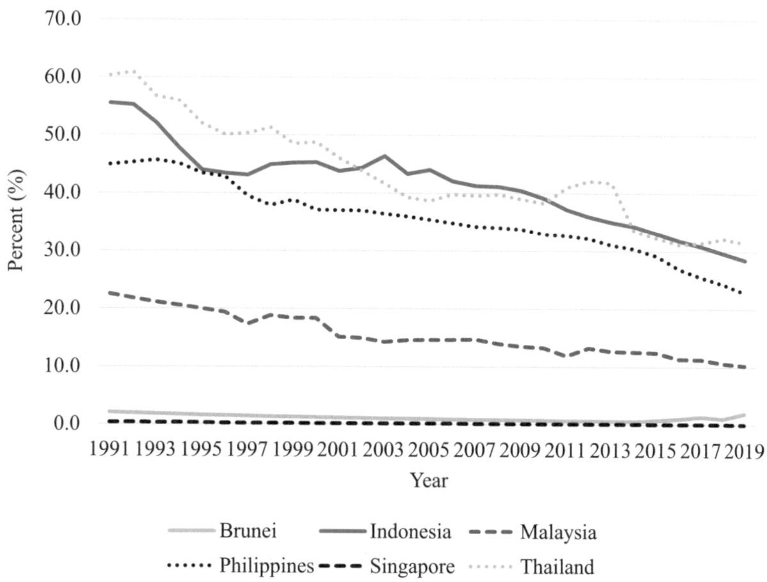

pandemic outbreak since 2020 and the Russia-Ukraine war have once again raised serious food scarcity owing to supply chain disruptions and economic sanctions imposed on Russia. Hence, this book seeks to examine the food trade balance of the Southeast Asian countries to see how dependent these countries are.

Figures 1.9 and 1.10 show the food trade balance [(X − M)/(X + M)] of the Southeast Asian countries. The food trade balance of Thailand, Malaysia, and the Philippines shows a trend fall from their peaks in 1976 (0.836), 1998 (0.377), and 1974 (0.606) to 0.354 in 2020, 0.162 in 2018, and -0.423 in 2021 (Fig. 1.9). Brunei has been a highly food import-dependent country throughout the period 1962–2021 with a marginal improvement in 2021 (-0.880). From its peak of 0.450 in 1971, the food trade balance of Indonesia fell

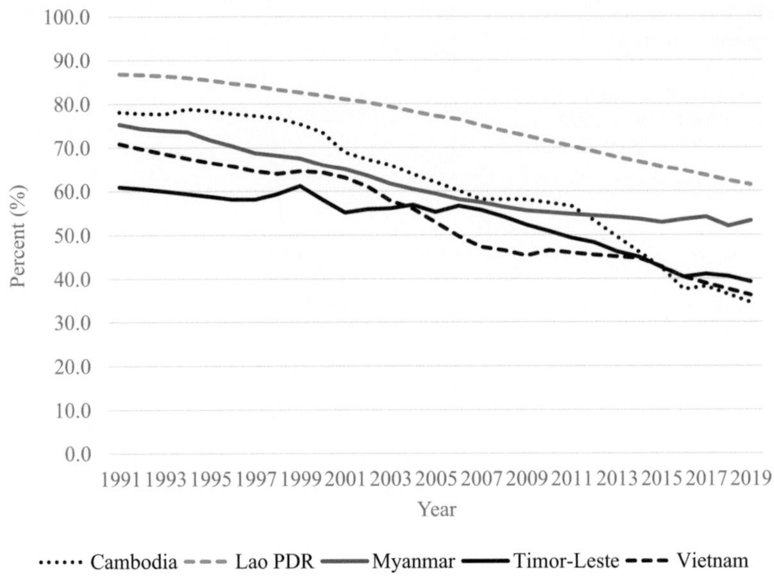

Fig. 1.6 Agricultural employment in national employment, CLMV countries and Timor-Leste, 1991–2019. Note: Estimations by the International Labour Organization. *Source*: World Bank (2022)

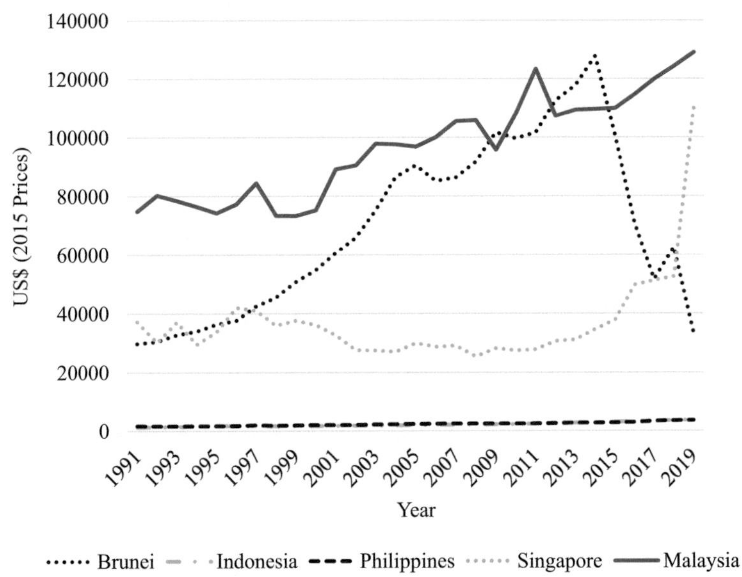

Fig. 1.7 Labour productivity, Southeast Asian market economies, 1960–2019

sharply from 1972 to slightly above zero or negative from then on until 1983 (0.017). It reached close to zero again in 1995. It has since recorded higher balances in the range of 0.205–0.447 in the remaining years of 1984–2020. Singapore's food trade balance improved in trend terms over since 1962 before enjoying a marginal surplus in 2019 (0.018) before turning negative again in 2020 (−0.042).

With the exception of Brunei, while the overall food trade balance of the Southeast Asian market economies is not heavily import-dependent, individual food items in countries, such as Malaysia and Singapore, show heavy reliance on imports. The individual chapters in the book shall pick up these imbalances.

The World Bank data on the CLMV economies of Cambodia, Lao PDR, Myanmar, and Vietnam

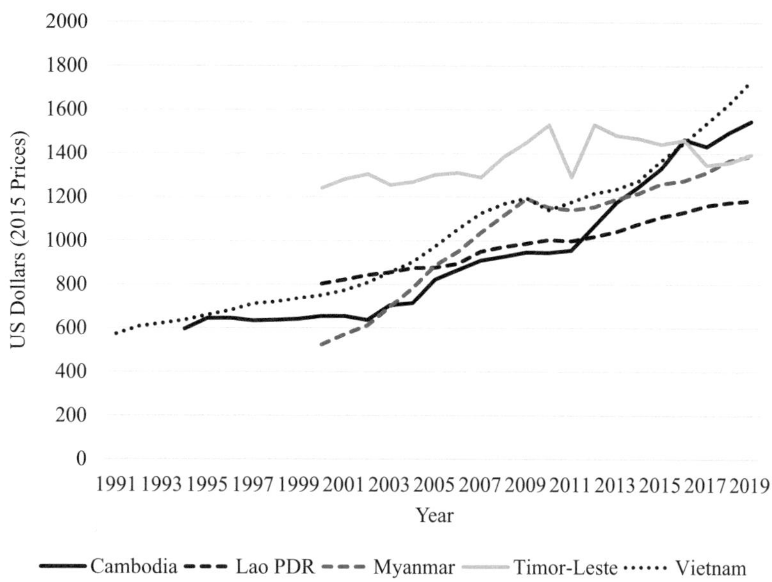

Fig. 1.8 Labour productivity, CLMV countries and Timor-Leste, 1991–2019. *Source*: Compiled from the World Bank (2022)

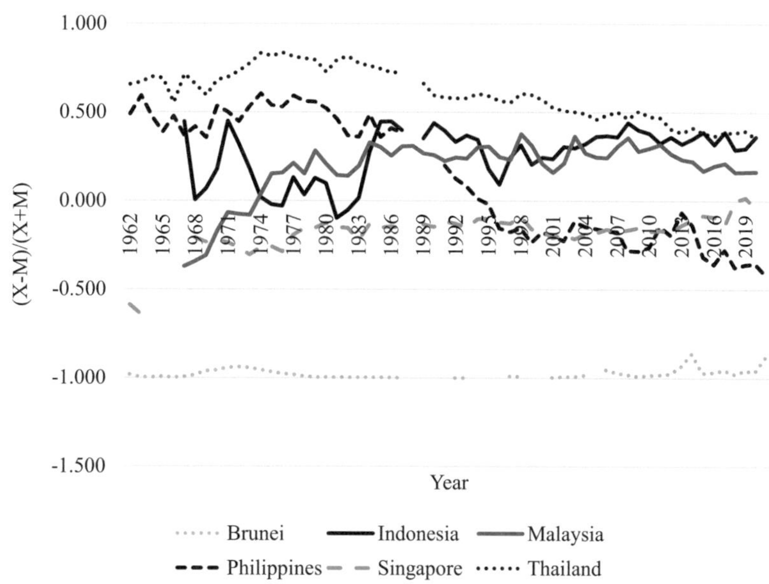

Fig. 1.9 Food trade balance, Southeast Asian market economies, 1962–2021. *Note*: X—exports; M—imports. *Source*: Computed from World Bank (2022)

and Timor-Leste does not cover the period 1971–99, 1975–2009, 1977–91 and 1993–2009, 1962–96, and 1962–2003, 2006–12, 2014–2016, and 2018–2021 (Figure 1.10). There has been a trend fall in the food trade balance of Lao PDR and Vietnam, while Myanmar's fell in the period 2010–2016 before improving thereafter. From a positive balance in the period 1962–68, Cambodia's food trade balance became negative in the period 2000–2020, though it has shown trend improvements in that period.

Southeast Asian Initiatives on Greening

Initiatives to green the Southeast Asian economies gained wind from the 1990s and especially since the turn of the millennium both through international and national efforts, which were accelerated by the launching of the millennium development goals (MDGs) and its successor, sustainable development goals (SDGs) by the United Nations,

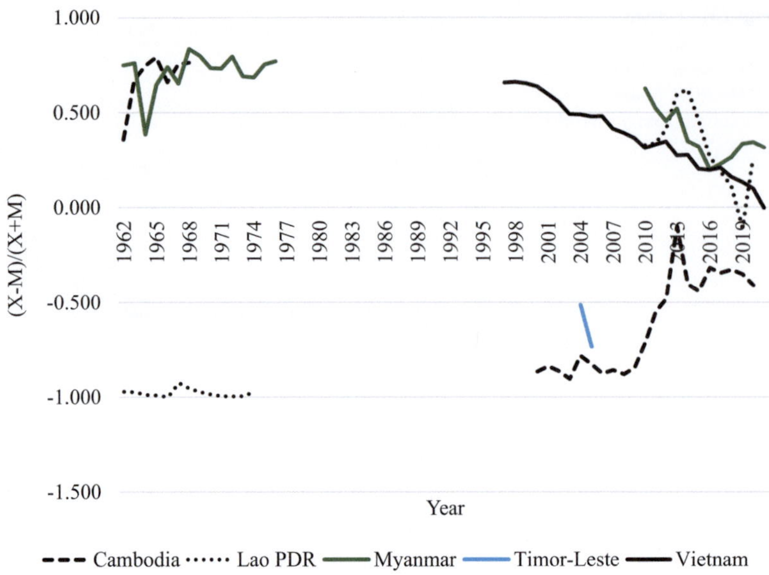

Fig. 1.10 Food trade balance, CLMV countries and Timor-Leste, 1962–2021. *Note*: X—exports; M—imports. *Source*: Computed from World Bank (2022)

and efforts of the United Nations Convention for Climate Change seeking to mitigate climate change and global warming. At the 2015 Conference of Parties (COP)21 195 countries committed to capping temperature rise. Carbon pricing and carbon taxes have increasingly been adopted since as individual countries established carbon net zero targets (Nordhaus, 2008, 2018; UNFCCC, 2021).

In addition to hazardous emissions, effluents, and the dumping of solid waste, there has also been the progressive promotion of traditional (including medicinal) and organic farming, including through certifications of organically grown crops, such as wild coffee and the harvesting of coffee seeds from civet droppings. While markets offer premier prices for organic output, the market for such produce is still small in Southeast Asia. Nevertheless, the individual country chapters that follow this chapter will address the state of such developments under the rubric of sustainable development.

Digitalization Efforts in Southeast Asia

While scholars are still grappling with the potential explosion of unemployment with the proliferation of path-breaking disruptive technologies associated with Industrial Revolution 4.0, artificial intelligence, robots, drones have increased the use autonomous navigation. While businesses and countries catching the wave of change are expected to appropriate soaring economic synergies from digitalization, the enormous costs to install the digital infrastructure to facilitate big data analytics, internet of things, cloud computing, as well as the fabrication of robots and drones threaten to create a digital divide (Rasiah et al., 2023). Fortunately, governments in Southeast Asia have chosen to catch the waves of change to strive for its expansion.

However, the poorer economies in Southeast Asia, such as Myanmar, Timor-Leste, Cambodia, and Lao PDR, are struggling to accumulate the funds to lay the digital infrastructure. Also, rural and the poor households in urban areas are also unable to attract adequate installations for achieving an egalitarian access to the digital infrastructure. Both robots and drones have raised agricultural productivity significantly in the developed and rapidly growing emerging economies. The pressure to raise agricultural productivity has become even more urgent as governments in Southeast Asia are addressing food security since serious shortages emerged owing to disruptions in supply chains and major supply sources, especially the Russia-Ukraine war that has restricted the global supply of rape seed and sunflower seed oil, wheat, and fertilizers to grow them.

Overall, the European Union has sanctioned a number of policy studies and policy approvals on the broad generic industrial policy approach drawn from the mission-oriented approach pioneered by Mazzucato (2013, 2018). The policy guide for greening and digitalization thus applies to all sectors and is easier to formulate, regulate, and calibrate (see Foray, 2018). Southeast Asian nations can look to adapt such a framework but would be better comprehended with a profound understanding all sectors as distinctly unique that also differs by location and their conditions at timing of the policy formulation (see Nazia & Rasiah, forthcoming).

Trajectory of Farming and Embedding Ecosystem Instruments

Southeast Asia offers a unique platform to examine farms in a trajectory towards upgrading as Indonesia, Thailand, and Vietnam are quite advanced in small-scale food farming while Indonesia, Malaysia, and Thailand support large-scale plantation agriculture. Meanwhile, agriculture in Brunei, Cambodia, Laos PDR, Myanmar, and Timor-Leste is still nascent. A profound assessment of agriculture in these countries can provide the phases that differentiate developmental phases and by national specificities the state of agriculture in these countries. The findings can not only offer learning experiences from within Southeast Asia but also to locate them against best practices by extensive and intensive farming trajectories.

Indonesia and Malaysia were the world's chief exporters of palm oil, Thailand a major exporter of rubber, while Indonesia and Thailand are become major exporters of horticultural products and fruits. In addition, Thailand is a major exporter of poultry and aqua-products. Meanwhile, small land-based economies of Brunei, Singapore, and Timor-Leste are heavily reliant on food imports. Malaysia has faced chronic food trade deficits over the period 1989–2021.

Book Outline

Following this introductory chapter, in chapter two, Ludher and Teng (2024) discuss the Singapore chapter by tracing the history of its agriculture and food security issues. The evidence they amass show that agricultural production and efficiency improved in the first decade since independence but declined over the next few decades before resurging in importance following concerns over food security. Food policy has consequently become important in government planning with a strong focus on food safety and emphasis on novel foods. The chapter draws on secondary data and information to provide an overview of Singapore's food and agriculture policies over the decades, which includes a profound assessment of the opportunities and challenges faced by Singapore's food and agriculture sector with focus on the adoption of novel high-tech practices in food and agriculture. The chapter finishes with recommendations for improving the food and agriculture sector in Singapore.

Ahmed and Rose (2025) examine in chapter three the development of agriculture in Brunei Darussalam. Petroleum and natural gas have long been the backbone of the country's economic growth, accounting for over half of the country's GDP. Given the vicissitudes of volatile price fluctuations of oil and gas, its government has increasingly attempted to reduce dependence on this finite export resource. The country has since 2021 launched major economic diversification initiatives to support sustainable economic growth and development with a focus on downstream oil and gas processing, food, tourism, information and communications technology, and services. The prioritization of smart agriculture has been extended into agriculture and fisheries. Although the contribution of the agricultural and fisheries sectors to the economic growth of the country is still small, the authors argue that Brunei is poised to achieve self-sufficiency in broiler chicken, eggs production, day old chicks, fisheries, fertilized eggs, and fruit production in the near future. The emphasis on agriculture has also been extended into the production of cattle, goats, sheep, buffalos, vegetables, and paddy, which is targeted to raise labour productivity by enhancing human capital, attracting investment, supporting resource optimization and value addition, reducing postharvest losses, and through extensions into agrofood processing. This chapter also discusses the economic sustainability of the agricultural and fisheries sectors in Brunei, especially since the introduction of leveraging strategies to develop agriculture and strengthen food security.

Annizah and Rasiah (2024) examine in chapter four the relationship between agriculture, food security, and poverty in Malaysia. Especially among the rural communities in the country, agriculture is not just an economic sector providing a means of income but also is a way of life. Agriculture remains major source of livelihood for the majority of the population, making it a key player in the ongoing fight against poverty. In fact, food security and poverty are interrelated issues that impact the well-being and quality of life of individuals and communities around the rural community in Malaysia. Although Malaysia has made significant progress in poverty alleviation since 1971, it has since 1990 faced food insecurity problems. Despite the implementation of various poverty reduction programmes and social safety nets to support the vulnerable communities, including promoting structural change into manufacturing and service sectors to create job opportunities, it is still grappling with low wage and food insecurity problems. The chapter argues that the momentum achieved in both absolute and relative poverty reduction in the 1970s needs revisiting to address the country's main social problems. Its focus on climate smart agriculture targeted at raising productivity growth and building climate resilience in sync with the United Nations sustainable development goals (SDGs) is impressive but requires effective appraisal to ensure its success.

In chapter five, Dianwicaksih, Mohamad Fadhil, and Fachru examine the contribution of agriculture to GDP in Indonesia as the other sectors have increasingly become more prominent. Nevertheless, agricultural value added in the country has continued to grow steadily. Following the end of colonial rule, the early governments in Indonesia began to focus on food in general and rice cultivation in particular. Although cash crops, such as tea and rubber, were important earlier, oil palm gained prominence especially in the Islands outside Java from the 1990s. The mechanization and the deployment of high yielding seeds, fertilizer, and pesticides in agriculture increased in significance from the time they were introduced in the 1960s. A major expansion that has made Indonesia prominent is the country's rise to become the world's largest lander of tuna fish since 2006. Although Indonesia still showed net imports of rice in 2023, its self-sufficiency rate, exceeding 95% over the period 2018–2023, is still better than some other countries in Southeast Asia, such as Malaysia. However, as climate change issues and agricultural acreage fell following urbanization and industrialization, government promotion of climate resilience from the late 1990s and digitalization from 2014 began to expand strongly. The government's focus on environmental, social, and governance (ESG) practices has especially risen strongly since the introduction of the United Nations sustainable development goals.

The sixth chapter on Thailand by Teerawat and Yot (2024) begins by discussing the growing importance of agriculture to the Thailand economy, including the policies the government launched to promote the sector. While Thailand enjoys food security if the measure is confined to trade measurements, it does raise a number of security issues facing especially small rural farmers. Thailand's agriculture sector has often sheltered the national economy whenever external shocks destabilized economic development, but it is now facing serious challenges. While historically a major contributor to exports and rural livelihoods, the sector has been grappling with price volatility, resource depletion, and farmer indebtedness. Government policies following growing emphasis on sustainability and commercialization initiatives of the large farmers have often resulted in inadequate support, leading to declining economic contribution and persistence of poverty among especially the small farmers. Recent initiatives, such as digital agriculture and climate resilience, hold promise but have yet to fully address the sector's deep-rooted issues. Overcoming challenges like small farm size, landlessness, ageing farmers, and limited market access requires a comprehensive approach focused on sustainable practices, value chain development, and farmer empowerment.

Ofreneo (2024) examines the state of agriculture in the Philippines in chapter seven. The Philippines, host to the International Rice Research Institute (IRRI), is the birthplace of Asia's green revolution in rice production. In the second half of the 1970s, the country, a rice importer for a century, attained self-sufficiency in rice production. However, in the 1980s, it reverted to the status of a major rice importer in which the agricultural sector has declined continuously, with the Philippines becoming more and more dependent on imported agricultural products ranging from rice and corn to vegetables and fruits and milk and meat. The sector's share in GDP has plunged from 25% in 1980 to less than 10% in 2023. The chapter argues that this dismal development in Philippine agriculture was due primarily to policy errors, starting with the adoption by the government of a disastrous "agricultural deregulation" programmes imposed by the World Bank. This deregulation programme has been compounded by the failure of the government to provide a holistic and pro-farmer development blueprint for the sector, on the one hand, and the persistence of the neo-liberal economic technocrats in pursuing a rudderless liberalization programme, on the other.

In chapter eight, Bach et al. (2024) present commendable account of the significant progress Vietnam has enjoyed in agricultural production since the 1990s, which has helped strengthen food security in the country, as well as enable strong exports. However, they also acknowledge significant challenges the country is facing ecologically and socially. In doing so, this chapter analyses agricultural development and food security in Vietnam with a focus on agricultural development and food security using indicators (such as changes in share of agriculture in GDP and employment in overall employment, farm acreage, production and yields of key crops and livestock; food farming; food security and consumption; and rural livelihood). Subsequently, it discusses government policies to promote sustainable agriculture. The chapter then assesses climate-resilient smart agricultural practices, such as organic and traditional farming, circular agriculture, digital transformation, and the impact of science, technology, and innovation policies in agriculture. The chapter finishes with an analytical proposal towards accelerating green transition in agriculture.

Chapter nine by Kum Kim (2024) presents the history, as well as the development of agriculture in Cambodia with a focus on its significance to the national economy, policies that the government has implemented and their impact on the livelihood of the people of Cambodia. Important issues addressed in the chapter include food security and rice self-sufficiency, and the effects of policies, including climate change. While the chapter points to improvements in the agricultural sector, it also addresses several challenges, the poor households have faced in the country.

Latdhavanh and Bannalath (2024) start chapter ten by stating that the agricultural sector in Lao PDR has been targeted by the government for development and to drive socio-economic growth since a majority of the country's population rely on agriculture for their livelihood and income. While they acknowledge the advance the sector has made since the turn of millennium, they also argue that the agricultural sector has yet to fully utilize its potential, despite the abundance of natural resources. Nevertheless, they note that the improved infrastructural conditions for access to large markets, especially with the completion of the Lao PDR–China railway, have opened up a window of opportunity for the development of agricultural production in Lao PDR. However, the productivity of agriculture remains low in the country, which among other reasons is due to traditional farming practices that heavily depend on natural conditions, climate change issues, degradation of soil quality with limited adaptive capacity, and the limited utilization of modern technology.

In chapter eleven, Naw Shareen notes that Myanmar's agricultural industry (which contributes to 21% of exports, employs 56% of the labour force, and generates 32% of the national GDP) has been negatively impacted by the COVID-19 pandemic and the 2021 military takeover, causing numerous challenges for its agricultural sector. The chapter evaluates the correlation between food security and climate-resilient agriculture, as well as sustainable agriculture in Myanmar. In doing so, the chapter addresses food security and the policy framework behind the implementation of Climate Smart Agriculture (CSA), which aims to address climate change issues in agriculture. The chapter uses an analytical approach to study Myanmar's agricultural productivity, food security and impacts from climate change. In doing so, the chapter examines the underlying causes of food insecurity in the country. It evaluates the effectiveness of these approaches and provides recommendations for the future implementation of sustainable agriculture initiatives involving relevant stakeholders, organizations, and research institutions.

Finally, in Chapter 12, Courvisanos et al. (2024) examine the development of agriculture in Timor-Leste. They argue that subsistence agriculture still dominates farming in the country. Also, two-thirds of households in Timor-Leste are engaged in farming, with more than a quarter of the working-age population dependent on subsistence agriculture, which creates a serious challenge task for the country to make a strong and successful transition to sustainable agriculture. Timor-Leste is unique in this book that it is the only country discussed that is not in the Association of Southeast Asian Nations (ASEAN), a political-economic union of some of the fastest growing economies in the world. The subsistence nature of its agriculture and ASEAN demands to diversify away from the dominant gas-export basis of its economy is the major backdrop to this chapter's analysis. This chapter focusses on the country's ability to diversify its economy from subsistence to sustainable agriculture. The chapter first examines the relative stagnation of the agriculture sector in the country, followed by charting past and present government policies on agricultural development. The chapter then examines the difficulties in addressing a shift to sustainable agriculture, including its initiatives to stimulate digitalization and climate resilience, self-sufficiency, and food security. Finally, Habermas' societal evolution framework is applied to address this transition into the future.

Rajah Rasiah presents a synthesis of the sustainable agricultural development experiences of the Southeast Asian countries in chapter thirteen. The chapter focuses on initiatives to raise agricultural productivity and end hunger, and to quicken the proliferation of digitalization and strengthening of climate resilience. Despite the wide range in GDP per capita and sustainability capabilities among these countries, the chapter argues that the prospects for establishing a climate smart sustainable agriculture system in Southeast Asia looks bright. Rasiah (2024) argues for effective implementation and constant appraisals to strengthen the planning process with strong coordination between central, provincial, and local policies, the constant identification and implementation of best climate smart agricultural practices to enable the member countries to meet the relevant SDGs, albeit the Philippines, Lao PDR, Myanmar, and Timor-Leste remain uncertain with policies. The achievement of the SDGs, however, depends very much on political stability, continued cooperation among the countries, and strong persuasion to seek diplomacy to achieve political, economic, and social consensus.

Taken together, the book brings a complex set of chapters representing large economies, medium-sized economies, and small economies on the one hand, and high-income, medium-income, and low-income economies on the other hand. Two small countries in Southeast Asia where SDG 2 is not of concern, i.e. Singapore and Brunei Darussalam, have long solved the problem of hunger. The remaining countries still face problems of hunger but the most serious are found in Timor-Leste, though Cambodia, Lao PDR, and Myanmar still have a long way to go. As middle-income countries with fairly good distribution, Malaysia, Thailand, Indonesia, and Vietnam face less serious hunger problems. The deleterious liberal policies described by Ofreneo (2024) obviously means that hunger has remained a critical problem in the Philippines.

It is also important to note that greening policies have become clearer in most of Southeast Asia, except for the Philippines and Timor-Leste where carbon net zero targets remain unclear. Timor-Leste is still faced with serious problems of poverty with the economy still very much immersed in early-stage primary processing. Singapore and Brunei lack natural non-fossil endowments, such as hydropower but have moved swiftly to raise the share of renewable energy in total energy consumption, suggesting that these countries are poised to meet SDG 7 with increasing use of clean energy, including in agriculture while intensifying traditional and organic farming. While hydropower resources have ensured that Thailand, Vietnam, Malaysia, Indonesia, and the Philippines enjoy relatively high share of renewable energy in overall energy consumption, the renewable shares show a trend fall owing to increasing use of fossil fuels and rising carbon emissions. Cambodia, Lao PDR, and Myanmar show high use of hydropower but these countries should quickly reverse the reliance on fossil fuels. Meanwhile, with exception of Timor-Leste and the Philippines, the remaining countries have introduced master plans to focus on sustainable agriculture through the smartification of agriculture and building climate resilience. It is interesting to note that Singapore, Brunei Darussalam, Vietnam, Malaysia, Indonesia, and Thailand have started to emphasize the smartification of agriculture to strengthen climate resilience. Also, Singapore, Brunei, and Malaysia have resorted to IR4.0 technologies to reverse food import dependence. While these developments augur well for the region, it is pertinent for these countries to strengthen their review process to ensure that climate change targets remain contemporaneously upgraded to meet the challenges that emerge.

Although the pressures facing these countries, including Timor-Leste (which only became independent in 2022), each of these countries has taken on efforts to pursue the sustainable development goals launched by the United Nations in 2016. While pace and extent of success achieved by the individual countries vary significant, all of them have taken initiatives to meet these goals. The chapter narratives provide collectively a rich analysis of how agriculture has evolved in Southeast Asia, the pressures each country in the region is facing, and what strategies can possibly be undertaken, both nationally and by the region together, to address the problems of food security, as well as to smartify agriculture to strengthen climate resilience.

References

Ahmed, H. O. & Rose, A. (2025). Modern Agriculture and Sustainability: Towards Economic Sustainability and Food Security in Brunei Darussalam. In R. Rasiah, A. Venus, N.I. Annizah (Eds.), *Sustainable Agriculture and Food Security in Southeast Asia: Towards Climate Resilience and Digitalization,* Singapore: Springer Nature.

ASEAN. (2015). Economic Community, Downloaded on April 19, 2022 from asean.org/our-communities/economic-community/

Afzanizam, M. A. R. (2023). Digitalizing food production through the fertigation system. In Rasiah, R., Low, W.Y. & Kamaruddin, N. (Eds.) *Digitalization and Development: Ecosystem for promoting Industry 4.0 Technologies.* Routledge.

Balisacan, A. M., & Hill, H. (Eds.). (2003). *The Philippine economy: Development, policies, and challenges.* Oxford University Press.

Baumol, W., & Oates, W. (1988). *Theory of environmental policy.* Cambridge University Press.

BBC. (2018). Khmer Rouge: Cambodia's years of brutality, November 16, Downloaded on April 19, 2023 from bbc.com/news/world-asia-pacific-10684399.

Foray, D. (2018). Smart specialization strategies as a mission-oriented policy—A case study on the emergence of policy practices. *Industrial and Corporate Change, 27*(5), 817–832.

George, S. (1977) *How the other half dies: The real reasons for world hunger.* Allanheld, Osmun & Co.

George, S. (1988). *A fate worse than debt.* Grove Press.

Hardin, G. (1968). The tragedy of the commons. *Science, 162*(3859), 1243–1248.

Hill, H. (2000). *The Indonesian economy.* University Press.

Lappe, F. M., Collins, J., & Fowler, C. (1977). *Food first: Beyond the myth of scarcity.* Houghton Mifflin.

Lappe, F. M., & Collins, J. (1986). *World Hunger: 12 Myths.* Grove Press.

Lewis, W. A. (1955). *Theory of economic growth.* Allen & Unwin.

Ljunggren, B., & Perkins, D. H. (Eds.). (2023). *Vietnam: Navigating a rapidly changing economy, society, and political order.* Harvard University Press.

Mackill, D. J., Khush, G. S. (2018). IR64: A high-quality and high-yielding mega variety. *Rice, 11*(18), downloaded on February 18, 2025 from https://doi.org/10.1186/s12284-018-0208-3
Malthus T. R. (1798). *An essay on the principle of population.* Chapter 5, pp. 39–45. Oxford World's Classics reprint
Marshall, A. (1890). *Principles of economics.* Macmillan.
Mazzucato, M. (2013). *The Entrepreneurial State: Debunking public vs. private sector myths.* Anthem Press.
Mazzucato, M. (2018). Mission-oriented innovation policies: Challenges and opportunities. *Industrial and Corporate Change, 27*(5), 805–815.
Meadows, D. H., Meadows, D. H., Randers, J., & Behrens III, W. W. (1972). *The limits to growth: a report to the club of Rome.* 2004 reprint, Chelsea Green Publishing Company.
Miliband, R. (1969). *The state in capitalist society: An Analysis of the western system of power.* Basic Books.
Nazia, N., & Rasiah, R. (forthcoming). *Industrial Upgrading in the Textiles and Clothing Industry: Pakistan and Late Industrialization.*
Nordhaus, W. (2008). *A question of balance—Weighing the options on global warming policies.* Yale University Press.
Nordhaus, W. (2018). Projections and uncertainties about climate change in an era of minimal climate policies. *American Economic Journal: Economic Policy, 10*(3), 333–360.
Pigou, A. C. (1932). *Economics of welfare.* Macmillan.
Pongpaichit, P., & Baker, C. (2002). *Thailand: Economy and politics.* Oxford University Press.
Rasiah, R., Salih, K., & Cheong, K. C. (Eds.). (2022a). *Malaysia's leap into the future: The building blocks for balanced development.* Springer.
Rasiah, R., Gopi Krishnan, V., & Azleen, O. R. (2022b). *Financing sustainability with east Asian experiences.* Universiti Malaya Press.
Rasiah, R., Low, W. Y., & Kamaruddin, N. (Eds.). (2023). *Digitalization and development: Ecosystem for promoting IR 4.0 Technologies.* Routledge.
Rowthorn, R., & Wells, J. (1987). *Deindustrialization and foreign trade.* Cambridge University Press.
Samuelson, P. (1954). The pure theory of public expenditure. *Review of Economics and Statistics, 36*(4), 387–389.
Stern, N. (Ed.). (2007). *The economics of climate change: The stern review.* Cambridge University Press.
Teng, P. (2024). *Food security issues in Asia.* World Scientific Publishing.
World Bank. (2022). *World Development Indicators by Country.* World Bank.

Agricultural Development and Food Security in Singapore

2

Elyssa Kaur Ludher and Paul Teng

Abstract

Singapore's history of agriculture and food has been tumultuous. Agricultural production and efficiency advanced quickly in the first decade after independence, subsequently declined over the next few decades, but resurged due to food security concerns thereafter. Food policy, conversely, has been growing in strength through the decades, particularly in relation to food safety and acceptance of novel foods. This chapter utilises secondary data and available documentation to provide an overview of Singapore's food and agriculture policies over the decades. It will also share the opportunities and challenges by the food and agriculture sector in Singapore, including the adoption of novel high-tech practices in food and agriculture. It will provide recommendations for improving the food and agriculture sector in Singapore.

E. K. Ludher (✉)
ISEAS-Yusof Ishak Institute, Singapore, Singapore
e-mail: elyssa_ludher@iseas.edu.sg

P. Teng
ISEAS-Yusof Ishak Institute, Singapore, Singapore
e-mail: paul.teng@nie.edu.sg

Keywords

Agriculture · Food security · Urban food policy · Climate resilience

Introduction: Singapore's Agriculture Policy

Singapore is a small island state of 728 square kilometres and 5.6 million people. Less than 1% of its land is dedicated to agriculture. It thus has features which differ from its much larger neighbours in Southeast Asia. More than 90% of its food is imported, especially staples (such as rice and wheat) and animal protein. As a result, Singapore is vulnerable to disruptions in the supply chains. Domestic production contributes less than 0.05% to Singapore's gross domestic product and accounts for approximately 3.7% of its import share (ASEAN Secretariat, 2021).

Singapore's current low agricultural production stems from its de-prioritisation of agriculture as the country industrialised after independence in 1965. Initially, Singapore was self-sufficient in some produce, namely vegetables, pork and chicken eggs. Over time, agricultural land was converted to other uses, and agriculture was limited to only small pockets. Singaporeans fed themselves through imports but nevertheless strengthened its food safety and

R. Rasiah et al. (eds.), *Sustainable Agriculture and Food Security in Southeast Asia*, Sustainable Development Goals Series, https://doi.org/10.1007/978-981-95-0477-0_2

quality programmes to ensure its public health. Policy makers, however, reviewed its food policies after the Global Food Price Crisis in 2007–2008, which eventually led to the establishment of an ambitious goal, announced in 2019, to produce 30% of the country's nutritional needs by 2030 under the "30by30" policy. This would equate to a 300% increase in food production in just 11 years. Singapore is planning to achieve this by introducing IR4.0 technologies and climate-resilient and adaptive agriculture methods, which includes the use of high technology urban vegetable, fish and factory-cultured protein "farms", and reducing food waste. Together these are expected to contribute to greater resilience, security and sustainability in Singapore's food system.

Singapore's strong focus on food resilience and sustainability over the years has contributed to it performing well in international comparisons such as the Economist Intelligence Unit's Global Food Security Index, coming in 28th out of 113 countries in 2023, outperforming all other Southeast Asian nations. Although the city state status of Singapore with little arable land left makes it uniquely different from the other countries in Southeast Asia (Rasiah, 2025), countries such as Brunei Darussalam are already showing signs of adopting its leveraging strategy to upgrade to higher value-added activities (see Ahmed & Rose, 2025). The larger Southeast Asian countries, such as Malaysia, Thailand and Indonesia, are also showing signs of introducing similar high technology farms, albeit on a small scale. Thus, Singapore offers a case study on food and agriculture policies and implementation plans that could be useful for others in the region, including for urban areas seeking to enhance food security.

This chapter draws its evidence from local and global secondary data sources, including statistics from the Singapore Department of Statistics, annual report by relevant government agencies, World Bank, the United Nations Food and Agriculture Organization and Statista. It also draws from past publications on the topic, of which only a few exist.

Changes in Singapore's Agriculture

Agriculture had always played only a minor role in Singapore's economy. Post-independence in 1965, Singapore focused on surviving as an independent nation with little arable land, and its policy imperative focused on higher-value creation economic activity that did not rely on the use of natural resources, which led to its focus on finance, trade, services, manufacturing and tourism sectors.

Then, agriculture was limited to a few items, and the rest was imported. Local production focused on rubber, market crops (fruits, food crops) and small livestock (chicken, ducks, pigs). Singapore also had a small but thriving fishing industry. Agriculture, fishing and quarries contributed 2.4% of Singapore's Gross Domestic Product (GDP) in 1960, which fell to less than 0.03% in 2020 (see Annex for more GDP information). At its peak, the sector contributed just under $450 million SGD in the mid-1980s, which then declined thereafter to approximately $150 million SGD in the last decade. The food and beverage sector, in contrast, contributes approximately 5 billion SGD to the GDP, contributed approximately 0.8% to the economy in 2020 (see Table 2.1).

Due to Singapore's focus on industrialisation and services, agricultural land was gradually converted for non-agricultural uses. In 1965, approximately 25% of Singapore's land was used for agriculture. By 2010, this had declined to less than 1%, or less than 700 hectares. Employment in the agriculture and fishery sectors also fell from approximately 170,000 and 4000, respectively, in 1966, to fewer than a thousand (combined) in both sectors in 2023 (PPD, 1966).

Due to Singapore's historical natural advantage as a seaport, Singapore had adopted free trade policies to enhance this. Singapore does not tax food and agricultural products, with the exception of intoxicating liquors and tobacco (Singapore Customs, n.d). Singapore's food trade balance continued to be heavy on imports and light on exports. Singapore's imports have generally been on an upward trend (see Fig. 2.1) aligned with

Table 2.1 Share of agriculture, and food and beverage services in GDP, Singapore, 1960–2020 (%)

	1960	1965	1970	1975	1980	1985	1990	1995	2000	2005	2010	2015	2020
Agriculture, Fishing and Quarries	2.37	1.89	1.50	1.14	0.84	0.62	0.24	0.13	0.09	0.05	0.04	0.03	0.03
Food & Beverage Services	na	na	na	na	Na	na	na	Na	na	1.25	1.16	1.22	0.81

Note: data for Food and Beverages was not available until 2004
Source: Singapore Department of Statistics

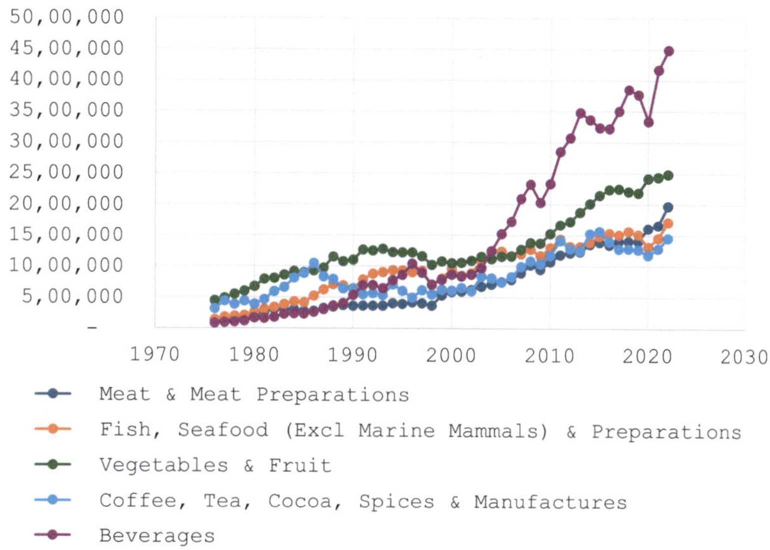

Fig. 2.1 Food imports, Singapore, 1976–2022 ('000 SGD). *Source*: Enterprise Singapore

population increase (from 2.2 million in 1976 to 5.6 million in 2022), rising affluence causing an increase in food consumption, as well as the rising price of food (see Fig. 2.2). Even though Singapore is not a net food producer, it still exports value-added food products, particularly in the category of coffee, tea, cocoa, spices and manufactures, which has grown in the last decades, as evidenced in Fig. 2.3.

Evolution of Policy Landscape

The policy landscape of Singapore's agricultural and food sector is presented here in three phases. The different phases reflect the changing focus of the government as the country faced new challenges from rapid economic growth and structural change.

Early Years to the Late 1970s: Productivity Optimisation Through R&D

Prior to independence, Singapore's agriculture mainly consisted of commodity crops, such as rubber and gambier (Henderson, 1949).The Japanese occupation and post-World War II saw diversification of crops with plantations replaced by food crops, such as tapioca, vegetables, coconuts, pineapples and fruits (Henderson, 1949). Even then, Singapore's high population and scarce availability of land resulted in most food items being imported, and the department of agriculture's main role was to increase productivity.

After 1965 a quarter of Singapore's land was used for crop production and livestock with (Chou, 2014; de Koninck, 2017) 9% of the population

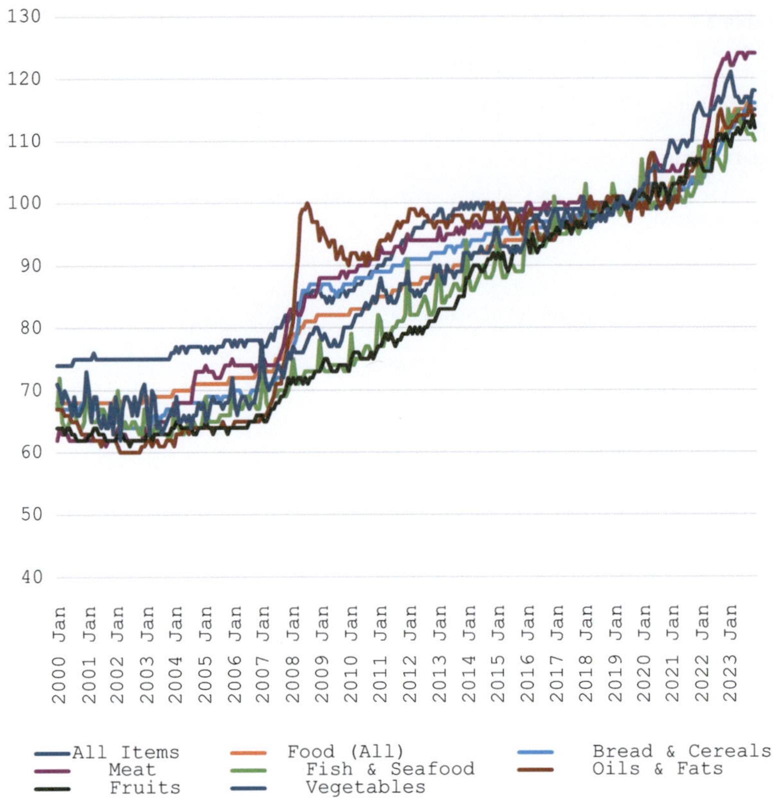

Fig. 2.2 CPI for food (excluding food servicing services), Singapore, 2000–2023. *Source*: Singapore Department of Statistics

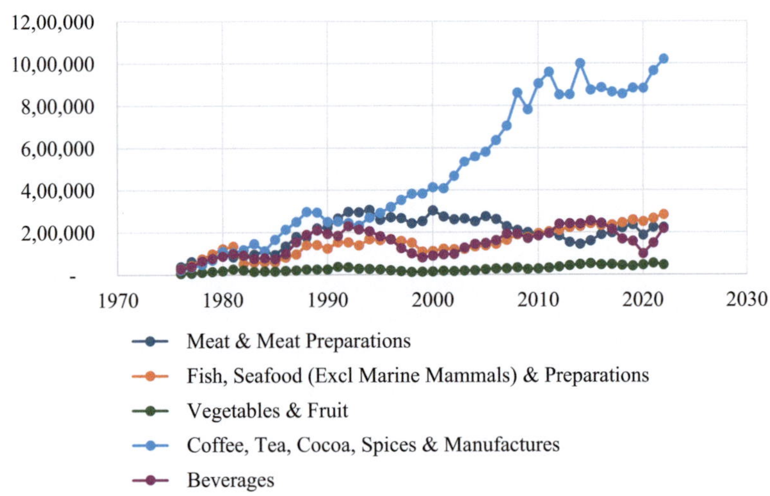

Fig. 2.3 Food exports, Singapore, 1976–2022 ('000SGD). *Source*: Enterprise Singapore

employed in agriculture or fisheries (Ludher & Paramasilvam, 2018). The administration of agriculture was by the Ministry of National Development's Primary Production Department (PPD). Its mandate was to increase productivity and employment. PPD sought international assistance to improve the sector, including from Japan and the United Nations Development Programme (UNDP). One of the latter's partnerships led to the establishment of the Marine Fisheries Research Department and the Pig and Poultry Research and Training Institute in 1971 for research and development (R&D) capacity (Ludher & Paramasilvam, 2018). Through these, PPD developed animal husbandry extension services, disease management protocols, vaccines, training and technology, which drove significant productivity leaps, and by 1977 so that Singapore's self-sufficiency in pork, hen eggs and chicken reached 104%, 100% and 80%, respectively (Chou, 2014; Ludher & Paramasilvam, 2018).

Late 1970s-2000s- Adapting to Imported Food Supply

As Singapore's population burgeoned and the city urbanised, agricultural lands and farm numbers dwindled. Supply was replaced by imports. Lucrative pig, broiler and duck farms were phased out by the end of 1980s due to environmental pollution, which conflicted with water security needs. By 2005, agriculture zoned land was only limited to six greenfield "Agrotechnology Parks" that featured regularised lots with access to utilities and incorporated sanitary and phytosanitary specifications (Ngiam, 2016).Domestic agriculture value addition fell from US$323 million in 1984 to less than US$90 million in 2003 (Teng & Montesclaros, 2023).

The rise in food imports led to questions of the safety of imported food, which spurred the development of programmes for food regulation, safety testing and monitoring. The *Sale of Food Act 1973* was subsequently enacted to regulate the food industry, setting out permissible limits of food additives, mineral and nutrient supplements, labelling requirements, and regulations for food manufacturing and retail. This was necessary as the industrialisation of food often led to the introduction of chemicals and preservatives above healthy levels.

To raise its scientific capabilities in food safety, Singapore introduced protocols for inspection, testing and control. The government established the Veterinary Public Health Centre (VPHC) in 1979 to house and develop its food testing laboratory. It expanded its food safety protocols to adopt the whole-of-food chain approach taking into account farms, middlemen, importers, retailers and food and beverage services. From the 1980s, this included field-testing capabilities for use on farms in Malaysia and Indonesia and outreach to train farmers on proper chemical use (Ludher & Paramasilvam, 2018).In 2000, a Veterinary Public Health Laboratory and the Animal and Plant Health Laboratory were established to address emerging health threats, and in 2019 the National Centre for Food Science was established with rapid testing capabilities to identify food borne pathogens within two hours of collecting swabs. These labs have functioned as regional reference centres for food safety, and animal and plant health.

To make import control and safety monitoring more efficient, fruit, vegetable and dried food imports were centralised at the Pasir Panjang Wholesale Centre in 1983. This had the added benefits of improving price competitiveness and monitoring. Slaughterhouses, which processed the imported live poultry and pigs, were also centralised and regulated for higher sanitation standards. The introduction of a national single window for trade declarations, "TradeNet", in 1989 further streamlined movement of food products and enforced health, safety and other regulatory requirements. This widened food safety function beyond the PPD to include Singapore Customs and other departments working on environmental health. These collective efforts achieved some success in improving food safety. For example, the proportion of imported vegetables failing to meet the minimum pesticide residue levels fell from 70% in 1980 to less than 5% in 1990 (Ludher & Paramasilvam, 2018). The complexity of agriculture and food governance eventually led to the centralisation of many departments dealing with food under the statutory board Agri-Food and Veterinary Authority (AVA) in 2000, which contributed to the autonomy and flexibility needed to respond to food issues.

2000s Onwards: Building Food Resilience as a Response to Global Supply Disruptions

The growth of agriculture productivity in Southeast Asia benefited Singapore. By the 2000s, more than 90% of Singapore's food was imported, largely from Malaysia and other countries in the region. Despite having little local production, there was a high amount of food available from overseas supply chains, especially as Singapore's strengthening dollar ensured that households had sufficient purchasing power (Teng & Montesclaros, 2023).

The PPD identified a case of Nipah virus from Malaysian pork, which subsequently led to the immediate ban of pork imports in 1999—a staple protein for Singaporeans. Pork was unavailable on shelves overnight. Singapore took time to identify alternate sources of accredited pork sources, which caused much public consternation.[1] The supply of food was again disrupted during the 2007–2008 global financial crisis. Rice, another major staple, saw its prices jump from US$350 to nearly $1000 per tonne (Reuters, 2011). Globally, staples such as wheat, maize and oil grains were also affected. Singapore's Consumer Price Index (CPI) for food spiked by 7.8 percentage points between 2007 and 2008 (Department of Statistics, 2009).

A post-mortem of the crisis (Mittal, 2009) pointed to structural challenges in the sector, including chronic underinvestment in agriculture, competing demands for biofuel, long-term unfavourable climate due to climate change and volatility of food trade (von Broaun, 2011), which led to a policy review in Singapore, in some ways a U-turn, towards restoring food supply resilience and security (Teng & Montesclaros, 2023).

The main outcome was the development of the *Singapore Food SecurityRoadmap* in 2012 (See Fig. 2.4). The *Roadmap* established "core", "secondary", "supporting" and "enabling" strategies to support Singapore's path to food security. Core strategies included diversification of import sources and food industry development. Secondary strategies included boosting stockpiling and increasing local food production. Supporting strategies included infrastructure improvements, R&D in agri and food, and expansion of the social welfare system. Lastly, enabling strategies related to market monitoring and emergency planning.

Singapore sought to diversify its food sources so that it did not have to depend on just one or two sources as in the event of a disease outbreak; importers have been unable to pivot to alternate sources. The strategy meant delving into the sources for its key food items—rice, vegetables, fruits, meats—and working with importers and stakeholders—through trade visits or regulatory tools—to expand sources. This ensured that there was no dependency on a single source or country of failure. The push to diversify also meant improvements in cold chain infrastructure, logistics and transportation. This strategy slowly saw change. For example, in the early 2010s, 77% of eggs were imported from Malaysia (AVA, 2009). In 2019, AVA required egg importers to diversify import sources, failing which they would lose their import licences (Wong, 2020). As a result, Singapore's eggs sources grew from 12 in 2019 to 17 by 2022 (SFA, 2023). Overall food sources also expanded to more than 180 countries and regions by 2022 (SFA 2023).

Policymakers prioritised items for local food production that were vulnerable due to short shelf lives, such as eggs and leafy vegetables, and those that could be farmed intensively, such as fish. Singapore's land, research and funding policies were updated to support these three food items. Grants of S$30 million were launched to support this in 2013, under the Food Fund. The grant was renamed the Agriculture Productivity Fund in 2014 with an additional earmark of S$63 million, and to the Agri-food Cluster Transformation (ACT) Fund in 2021 with another fund available to the tune of S$60 million. Co-funding quantum increased to S$2million in 2018 per farm, up from the previous S$700,000. These funds assisted in the set-up of capital-intensive fish, egg, leafy vegetable and other farms and encouraged the increase in productivity from adopting new knowledge and technologies.

Singapore also invested overseas in the Sino-Singapore Jilin Food Zone in Northeast China to secure pork and rice for Singapore. The zone

[1] Singapore eventually lifted the ban on pork from Malaysia only in 2015, 16 years after the event.

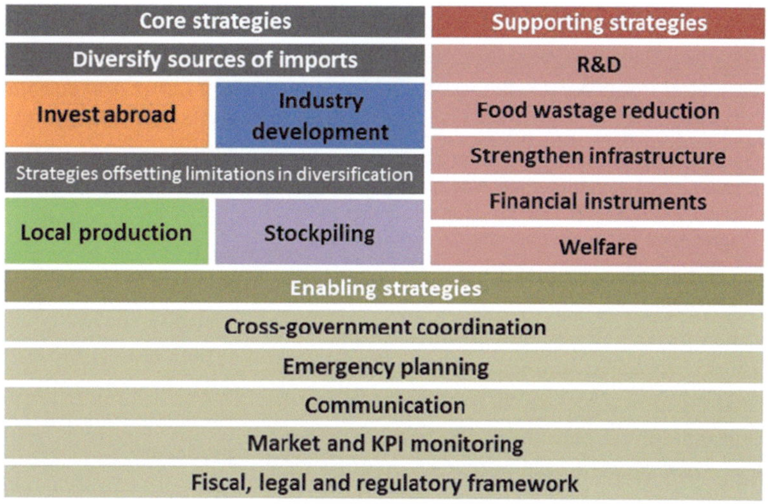

Fig. 2.4 Singapore's food security roadmap, 2012. *Source*: SFA (2013)

spanned 1450 square kilometres—double the size of Singapore. It was built to Singapore's Disease-Free Zone standards leveraging on the food safety experience in Singapore. The project was conceptualised in 2008, but over time faced considerable challenges. Today, the zone provides a limited amount of rice to Singapore as it has yet to achieve full capacity as envisioned. The Food Zone was possible partly due to the enabling strategy of whole-of-government coordination. Concurrent to the release of the food security roadmap, an "Inter-Ministry Committee on Food Security" was established consisting of 13 government ministries and agencies, including covering research, security, trade, health and environment. Five working groups were created to look at industry development, food defence, R&D, reducing food waste and overseas agri-investments (Ludher & Paramasilvam, 2018). These working groups contributed to the formulation of policies and their implementation to improve overall food security.

Agriculture improvement strategies were further developed over time. For example, Singapore introduced a "Farm Transformation Map" in 2016 to maximise space, develop viable innovation, build a future workforce and create an enabling ecosystem for the farming industry to thrive. Farms were matched with an "account manager" within AVA that could advise farms on extension services, financial resources availability and business development options (AVA, 2018). AVA also worked with higher education institutions to introduce modern farming skills training, for example for aquaculture or indoor farming. These strategies supported the growth of the agri and food sector to continue to improve its resilience.

While these efforts gained currency to improve food production in Singapore, overall, the sector still struggled due to limitations of land and support from government. For example, government authorities swapped agricultural land with the Ministry of Defence's Paya Lebar Airport which was converted for civilian use. This resulted in the loss of a number of farms in the early 2020s. Prioritisation was therefore needed to achieve the goals of food resilience and security in Singapore, which came later in the form of a national goal.

Post-2019: Ramping up Farming and Food Security

Singapore had previously deprioritised agriculture due to its pollutive runoff into potable water reservoirs, which was vital for its water security. But as agricultural pollution could be better managed with new regulations and technology,

Singapore could refocus on food production. In 2016, the Minister of State for National Development, Koh Poh Koon, stated (Koh, 2016):

> "We have a mission to achieve that is as crucial as water – and that is to strengthen our food security in an era of global food uncertainty. My Ministry and AVA will make the development of our local agriculture sector a priority for the coming years."
>
> - Koh Poh Koon, Minister of State for National Development.

For its new focus on raising domestic food security, AVA was restructured into the Singapore Food Agency (SFA) in 2019. SFA launched its food policy and plan in 2019, called the *Singapore Food Story*, which consisted of three food strategies (or food baskets)—to diversify import sources (including by avoiding dependency on single markets), to grow up to 30% of its nutritional needs locally by 2030 (30by30 policy) and to support local companies to expand overseas (Koh, 2016).

Increasing self-production by approximately 300% between 2019 and 2030 was particularly ambitious (Teng & Montesclaros, 2023). Although land was a limiting factor, existing lands were maximised for production. New land leases were extended to 20 years, though still short compared to global lease lengths, and were tied to production targets. Underperforming farms' leases were not renewed.[2] With only 700 hectares allocated for agriculture use, the space needed was allocated to farms using creative production technologies. Existing zoning regulations were revised to allow food for commercial food cultivation in alternatives spaces, such as in underutilised commercial, industrial and residential spaces. Eventually, rooftop spaces on industrial and residential estates were tendered out for commercial farming. SFA is also in the midst of establishing an 18-hectare Agri-Food Innovation Park (AFIP) with private sector partners to "catalyse innovation in the agri-tech ecosystem, by bringing together high-tech urban indoor farming and associated R&D activities" (MTI, 2019).

These efforts resulted in an overall increase in farm output and productivity. The number of vegetable farms grew from 83 to 109 between 2018 and 2022, but bean sprout, hen and quail egg farms, and coastal fish farms declined slightly (AVA 2018). By 2022, Singapore was home to 261 farms (see Table 2.2) (SFA, 2023).

Overall, food production has since 2018 been on an upward trend. However, external factors, such as cost of production, volatility of markets, climate change related damage and business viability, have impacted negatively on overall production in the recent past (see Figs. 2.5 and 2.6). Consequently, more concerted efforts are needed to improve the viability of farms and food producers.

To accelerate food production beyond conventional means, Singapore turned to IR4.0 instruments in food production and accelerate the adoption of novel foods. Funds were also allocated to support R&D through Singapore's National Research Foundation (NRF), which came to be known as the Singapore Food Story R&D programme (SFA, n.d.)—S$144 million (US$101 million) was earmarked in 2019, and another S$165 million (~US$121 million) in 2022 to support research projects in intensive farming for agriculture and aquaculture, future foods and food technology and innovation domains (SFA, 2022). Funds were mainly allocated to catalyse growth, through institutes of higher learning, incubators, accelerators and labs. Labs included high-tech facilities to develop novel foods, such as cell-cultured meats, plant-based meats, upcycled food products, precision fermentation and prolonged food storage.

Through these efforts, Singapore quickly grew to attract start-ups keen to develop novel foods. As of 2022, Singapore had 17 plant-based protein and dairy alternative companies, developed from pulses, soy, algae, wheat, peas and nuts—from nearly zero just ten years earlier. Plant-based proteins are widely available in restaurants and supermarkets. While it has not achieved price parity with conventional meats, the price gap has been narrowing. Singapore also made news as the first country to approve cell-cultured meats in January 2023, though cell-cultured products have yet to be commercially available. A selected list of edible insects have also been approved for sale

[2] Some agricultural leases were not renewed as the lands were diverted for other uses.

Table 2.2 Farm-level data, Singapore, 2018–2022

	2018		2020		2022	
	No	MT (exc eggs)	No	MT (exc eggs)	No	MT (exc eggs)
Vegetables	83	12,178 (13% of leafy veg)	95	11,137 (13% of leafy veg)	109	19,900 (3.9% of all vegetables)
Bean sprouts	6	na	6	na	5	Na
Hen and quail eggs	5	488 m pieces# (24%)	5	616 m pieces# (28%)	4	609.1 m pieces# (28.9%)
Coastal food fish	111	4,577 (9%)	110	3960 (8%)	109	4,400 (7.6%*)
Land based fish	12		22		24	
Other	10	na	9	Na	10	Na
TOTAL	227	-	247	-	261	-

Refers to hen eggs only
*Refers to local farm production as a % of seafood (life and chilled) for consumption only)
Source: Singapore Food Agency

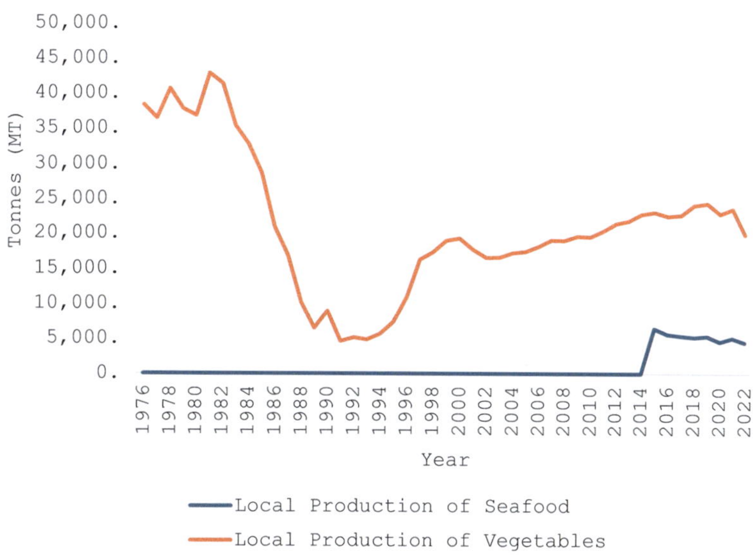

Fig. 2.5 Local production of seafood and vegetables, Singapore, 1976–2022. *Source*: Singapore Department of Statistics

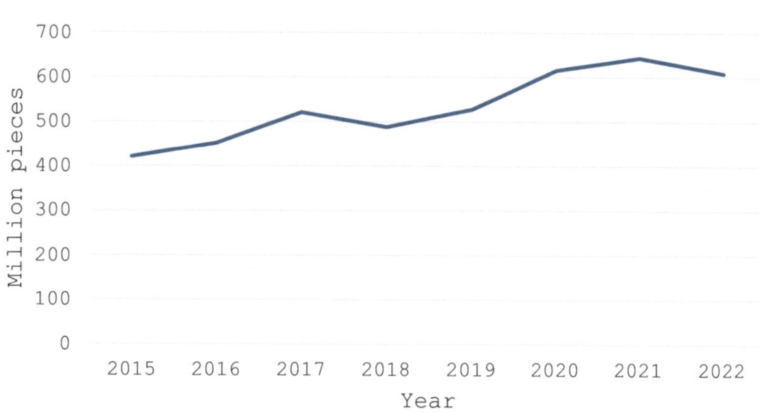

Fig. 2.6 Local production of hen shell eggs, Singapore, 2015–2022. Note: data unavailable prior to 2015. *Source*: Singapore Department of Statistics

for consumption, after the government completed its stakeholder engagement in 2022. Precision fermentation—which is to multiply proteins using a carbon source, such as sugars or gases and combining them with nitrogen gas—is another form of novel food that is gaining popularity, and in 2022, SFA approved the commercial sale of a protein grown from air by a Finnish company, though it is still not yet commercially available (Donna, 2022).

To address the safety of novel foods, SFA introduced a novel food regulatory framework and body, the Future Ready Food Safety Hub (FRESH), which facilitates companies to conduct safety assessments on foods for SFA's review to obtain licence for sale. "The government has also enacted the Food Safety and Security Act 2025" to assist it in strengthening the regulatory framework for novel foods.

Food Self-Sufficiency, Digitalisation and Climate Resilience Practices

Beyond food production, Singapore is looking to maximise self-sufficiency by reducing waste. Singapore's food waste is among the highest in the world, at approximately 149 kg per capita annually (NEA, 2023). It accounted for 12% of all waste in Singapore. Estimated food loss[3] per year is at S$2.54 billion, while food waste[4] costs households collectively S$342 million (SEC, 2019).

Recycling rates for food are low, stagnating at between 17 and 19% between 2018 and 2022 (NEA, 2023). The remainder is incinerated, and its ash dumped in Singapore's only landfill site on Semakau Island. As this landfill is filling up faster than anticipated, Singapore launched a *Zero Waste Masterplan* in 2019 to reduce incinerator bottom ash and non-incinerable waste by 30% by 2030.

Table 2.3 Food waste valorisation pathways currently explored in Singapore

Type of valorisation pathway	Examples currently in commercial development or research phase in Singapore (non-exhaustive)
Food to food	• Spent grains to beverages, noodles or functional food products • Non-fresh bread to alcoholic products • Okara waste to cricket protein
Food to feed	• Homogenous food waste (okara, soy bean waste, other) to black soldier fly feed protein
Food to fertiliser	• Food waste to biochar or compost
Food to fuel	• Food waste to biochar with hydrogen by-product • Food waste to biogas
Food to other products	• Coffee grounds to biomaterials • Fruit rinds to bio-packaging

Adapted from materials from NEA, 2023.

This was accompanied by a *Resource Sustainability Act* in the same year. Food waste was one of the three priority focus areas, along with electronic and packaging waste. The *Act* sought to prevent food waste at source, redistribute unsold or excess edible foods, promote recycling or recover energy. Food was envisioned to be recycled through five valorisation pathways—food to food, food to feed, food to fertiliser, food to fuel and lastly food to other products (see Table 2.3).

As a result of the plan, large food waste producers are required to segregate and treat food waste by 2024, which affects food from malls, hotels, industrial developments, caterers, food manufacturers and food storage warehouses (see Table 2.4).[5] It is calculated that this would not only reduce food waste to improve efficiency, but also expand the marketplace for recycling.

[3] This refers to food lost in the supply chain between the producer and the market.

[4] This refers to food discarded that is safe for human consumption.

[5] On site food waste treatment was required for new developments starting from 2021.

Table 2.4 Premises subjected to the mandatory segregation requirements, Singapore

Premises	Type of function and threshold	Exemptions
Shopping malls, hotels	F&B areas and function areas greater than 3000 sqm	–
Single user factory	Large food manufacturers with operation areas larger than 750 sqm	Manufacturers of dried foodstuff, additives, bottled water, spices, high pressure processing
Multi-user factory	Food tenants with gross floor area greater than 20,000 sqm or has more than 20 food tenants, or has at least one food manufacturer with operation area larger than 750 sqm	For large food manufacturer: Manufacturers of dried foodstuff, additives, bottled water, spices, high pressure processing

Adapted from materials from NEA, 2023.

One infrastructure that will support food waste recycling is the Tuas Nexus plant, expected to be completed in 2025. This facility is co-located as a waste management facility, which takes in organic waste, with a water reclamation plant. The discards are then co-digested to create a more potent biogas for use in energy generation. Tuas Nexus is expected to reduce CO2 emissions by over 200,000 tonnes a year (NEA, 2021).

Singapore has also encouraged its farms to adapt to climate change through the application of climate-resilient genetics, reduction of water use, increasing efficiency of nutrient inputs through sensors and expansion of e-commerce use to connect to markets. For example, Singapore's sovereign wealth fund, Temasek, partnered with Bayer on a new venture called Unfold in 2020 to develop vegetable varieties suited for indoor hydroponics and vertical farming (Ellis, 2020). Grants have been made available for co-funding of hydroponic and water recycling equipment, LED lighting, sensor technologies and automation machinery to reduce long-term operational costs, which are anticipated to increase production and yield efficiency, resulting in more climate-resilient crops production. To address the concern about sea level rise affecting the main farming area in Singapore—Lim Chu Kang—SFA commissioned a masterplan in 2023 to protect intensive farms with embankments and seawalls. The embankments would double to provide indoor food production and packaging facilities to subsidise the cost of construction.

Singapore's farms and food producers have also improved their resilience by selling direct to customers and having greater reach through online participation. This was accelerated during the COVID-19 period, as the Singapore government made available grants to cover up to 80% of costs to set up digital presences (Ang, 2021). While data is not available specific to food and agriculture, overall, Singapore's small and medium-sized enterprises (SME) technology adoption rate increased from 73.8% in 2018 to 94.3% in 2022 (IMDA & LKYSPP NUS, 2023). The COVID-19 pandemic also encouraged more consumers to switch to e-commerce for fresh foods and produce, creating larger markets and incentives for technology adoption.

Conclusion: Implications for Theory and Policy

Singapore continues to experience population increase and economic structural change, which has resulted in the continued need to import diverse foods and in higher quantum, particularly to satisfy Singapore's culture and identity as a food haven. The country has incredibly managed to achieve its policy goals primarily because of policy initiatives and the agile role of intermediary organisations to set targets and ensure that they are achieved.

Nevertheless, even if Singapore achieves its goal of local production to meet at least 30% of domestic demand of its nutritional needs by 2030, its food security will continue to be dependent on global supply chains and trade. Singapore's quest for higher economic value creation activities has resulted in its population having among the highest GDP per capita in the world. This has given it the purchasing power to rely on imports. However, in an emerging era of climate change and geopolitical conflicts and economic stress, protectionist policies among countries may result increasing volatility in global trade.

As such Singapore will need to continue to boost its food security policies. It will need to walk a fine balance between supply (imports, self-production and stockpiles) and demand (consumer diets, expectations and wastage), which requires that it develops the capacity to buffer demand against supply disruptions for longer periods. Some ways it can do so are to focus on the following areas:

a. Facilitate market access for local production

Singapore's food producers struggle to compete with cheaper imports, which limit its ability to scale up quickly. To improve local production, the Singapore government needs to focus on breaking down market barriers, such as access to supermarkets and markets, that currently limit producers. It could also consider contract bulk purchases of local produce for institutional facilities, such as army camps, prisons and other large facilities.

b. Encourage more "Controlled Environment Agriculture" such as indoor vegetable and fish farms, in anticipation of more variable weather extremes due to climate change, which should build on investments to date
c. Sharpen policies to improve food circulation solutions

Current policies on food waste recycling and treatment are not nuanced enough to promote food recycling towards higher valorisation pathways as shown in Table 2.3. Grants and policies could be tweaked to encourage higher adoption for valorisation.

d. Strengthen regional cooperation and partnerships

While Singapore has already developed experience in developing regional satellite farms, for example in Jilin in China and Bulan Island in Indonesia, perhaps more should be invested in agriculture and food companies in the region to ensure more resilient and sustainable food production that could benefit Singapore.

Due to its size and economic structure, the city state of Singapore is unique country in Southeast Asia. Some of its agriculture and food policies, such as food safety framework and standards, use of high technology for production and adoption of novel and circular food solutions, may have lessons for the Southeast Asia region as the region seeks to improve its food security. It may also be of particular importance to cities and towns seeking greater involvement in food security, especially as Southeast Asia continues to urbanise. For one, Brunei Darussalam's leveraging strategies on high value-added agriculture seems to follow the Singapore experience (Ahmed & Rose, 2025).

Finally, as a limited agricultural nation, Singapore has potential to contribute to the broader community through its high standards of scientific R&D capability, its financial and corporate governance systems, and its commercial base in Asia for many global equity and venture funds. Additionally, apart from using agriculture to give it a certain level of food self-sufficiency, even if it is low, Singapore has aspirations to be a regional hub for the development and export of new farming and food technologies (EDB, 2023).

Anneux

See Fig. 2.7.

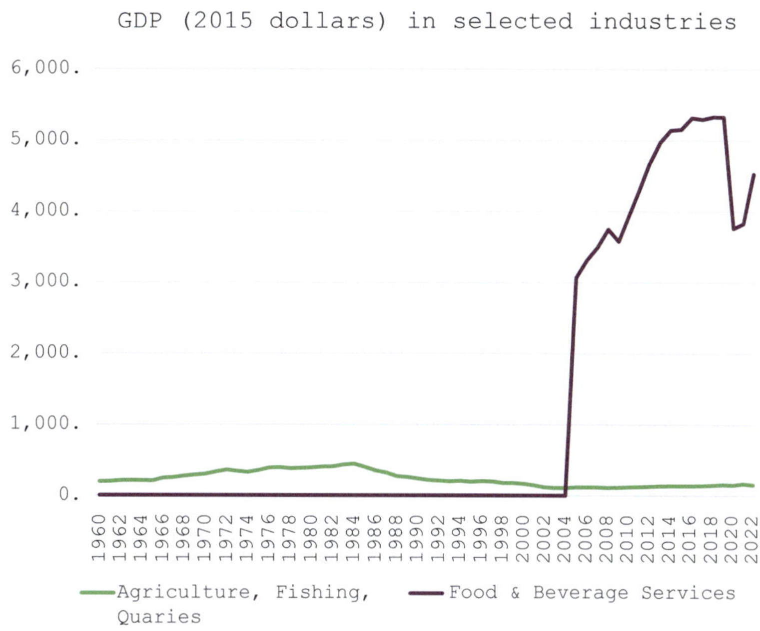

Fig. 2.7 Graphical representation of Table 1. *Data on agriculture aggregated with Fishing and Quarries. Information on F&B services was not available before 2005. *Source*: Singapore Department of Statistics

References

Ahmed, H. O. & Rose, A. (2025). Modern Agriculture and Sustainability: Towards Economic Sustainability and Food Security in Brunei Darussalam. In R. Rasiah, A. Venus, N.I. Annizah (Eds.), *Sustainable Agriculture and Food Security in Southeast Asia: Towards Climate Resilience and Digitalization,* Singapore: Springer Nature.

Ang, P. (2021, Sept 23). SMEs can tap funds for digital marketing with expansion of Productivity Solutions Grant. In *The Straits Times.* https://www.straitstimes.com/business/economy/smes-can-tap-funds-for-digital-marketing-with-expansion-of-productivity-solutions

ASEAN Secretariat. (2021). *ASEAN Statistical Yearbook 2021.* https://www.aseanstats.org/wp-content/uploads/2021/12/ASYB_2021_All_Final.pdf

AVA. (2009). *Annual Report 2008–2009.* Agri-Food and Veterinary Authority Singapore. p82.

AVA. (2018). *Annual Report 2017–2018.* Agri-food and Veterinary Authority.

Chou, C. (2014). Agriculture and the end of farming in Singapore. In T. P. Barnard (Ed.), *Nature contained: Environmental histories of Singapore.* NUS Press.

de Koninck, R. (2017). *Singapore's Permanent Territorial Revolution: Fifty Year in Maps* (p. 56). NUSPress.

Department of Statistics. (2009, Jan 21). *Singapore Consumer Price Index: December 2008 and Year 2008.* Press Release, National Archives of Singapore, 21 January 2009, http://www.nas.gov.sg/archivesonline/data/pdfdoc/20090128003/cpi+dec+08+vdd.pdf

Donna, L. (2022). All sizzle, no steak: how Singapore became the centre of the plant-based meat industry. *The Guardian.* https://www.theguardian.com/environment/2022/nov/06/all-sizzle-no-steak-how-singapore-became-the-centre-of-the-plant-based-meat-industry

Economic Development Board (EDB). (2023). *Advance, Accelerate, Access: Singapore, the innovation hub gateway to feeding Asia and the world.* Economic Development Board. https://www.edb.gov.sg/en/business-insights/insights/advance-accelerate-access-singapore-the-innovation-hub-gateway-to-feeding-asia-and-the-world.html

Ellis, J. (2020, Aug 12). Bayer & Temasek launch $30m seed genetics business targeting vertical farms. In *AgFunderNews.* https://agfundernews.com/bayer-and-temasek-launch-unfold-30m-seed-genetics-business-targeting-vertical-farms

Henderson, R. (1949). *Report on the Department of Agriculture, Singapore 1948.* Government Publications Bureau.

IMDA and LKYSPP NUS. (2023). *Singapore Digital Economy Report 2023.* Infocomm Media Development Authority and Lee Kuan Yew School of Public Policy National University of Singapore.

Koh, P. K. (2016 April 11). *COS 2016: Speech by MOS Dr Koh Poh Koon "Working Together for an Inclusive and Innovative City".* https://www.mnd.gov.sg/newsroom/parliament-matters/speeches/view/cos-2016-speech-by-mos-dr-koh-poh-koon-working-together-for-an-inclusive-and-innovative-city

Ludher, E., & Paramasilvam, T. (2018). *Food and the City: Overcoming Challenges for Food Security*. Centre for Liveable Cities.

Mittal, A. (2009). *The 2008 food price crisis: Rethinking food security policies* (pp. 1–40). UN.

Ministry of Trade and Industry Singapore (MTI). (2019). *Fact sheet: Agri-Food Innovation Park (AFIP)*. Ministry of Trade and Industry Singapore.

National Environment Agency. (2023). *Love your food, waste less*. Save more. National Environment Agency. https://www.nea.gov.sg/our-services/waste-management/3r-programmes-and-resources/food-waste-management

National Environment Agency. (2021, Jul 30). *NEA Commences Next Stage of Tuas Nexus Integrated Waste Management Facility to Develop Integrated Sludge Processing And Food Waste Treatment Facilities*. National Environment Agency. https://www.nea.gov.sg/media/news/news/index/nea-commences-next-stage-of-tuas-nexus-integrated-waste-management-facility-to-develop-integrated-sludge-processing-nd-food-waste-treatment-facilities

Ngiam, T. T. (2016). *Sustaining Singapore's Farming Heritage*. Centre for Liveable Cities Lecture Series. https://www.clc.gov.sg/events/lectures/view/clc-urban-pioneer-lecture-series-sustainingsingapore-s-farming-heritage

Teng, P., & Jose, Ma., Montesclaros, L. P. (2023). Agricultural Transformation for Small (Island and Developing) States. *Asian Journal of Agriculture and Development*, *20*(1), 13–30.

Primary Production Department (PPD). (1966). *Primary Production Department Annual Report 1966*. Primary Production Department.

Rasiah, R. (2025). Building Agricultural Sustainability and Food Security in Southeast Asia. In R. Rasiah, A. Venus, N.I. Annizah (Eds.), *Sustainable Agriculture and Food Security in Southeast Asia: Towards Climate Resilience and Digitalization* , Singapore: Springer Nature.

Reuters. (2011, June 10). Factbox: 2008 food price crisis—what caused it? *Reuters*. https://www.reuters.com/article/us-climate-crops-price-factbox-idUSTRE7590IJ20110610

Singapore Environmental Council (SEC). (2019). *Advancing a Circular Economy for Food: Key Drivers and Recommendations to Reduce Food Loss and Waste in Singapore*. Singapore Environmental Council.

Singapore Customs. (n.d). *Duties and Dutiable Goods*. https://www.customs.gov.sg/businesses/valuation-duties-taxes-and-fees/duties-and-dutiable-goods/

Singapore Food Agency (SFA). (2023). *Annual Report 2022*. Singapore Food Agency.

Singapore Food Agency (SFA). (n.d). *R&D programme*. https://www.sfa.gov.sg/food-farming/singapore-food-story/r-and-d-programme

Singapore Food Agency (SFA). (2013). *AVA's Food Security Roadmap for Singapore*. Singapore Food Agency. https://www.sfa.gov.sg/food-for-thought/article/detail/ava%27s-food-security-roadmap-for-singapore

Singapore Food Agency (SFA). (2022). *$165 Million Allocated to Singapore Food Story (SFS) R&D Programme 2.0*. https://www.sfa.gov.sg/docs/default-source/default-document-library/sfa-media-release-on-sfs-r-d-2-08543e0e9a81b43a699fd389f680dc71d.pdf

Tan, D. (2021). *Speech by Minister of State, Mr Desmond Tan—Strengthening Singapore's Resource Resilience for our Future Generations*. Ministry of Sustainability and the Environment. https://www.mse.gov.sg/resource-room/category/2021-speech-by-minister-of-state-desmond-tan-at-cos-2021

Teng, P., & Montesclaros, J. M. L. (2023). Agricultural Transformation for Small (Island and Developing) States. *Asian Journal of Agriculture and Development, 20*(1), 1.

von Broaun, J. (2011). *The Global Food Crises of 2008 and 2011- Causes and Remedies. A background paper for the Conference on—Global Cooperation for Sustainable Growth and Development—Views from G20 Countries*, New Delhi, ICRIER, September 13–14, 2011.

Wong, P. T. (2020, July 27). 4 firms contributing to 30% of supply blocked from importing eggs from Malaysia after failing licensing requirement. In *Today*. https://www.todayonline.com/singapore/egg-importers-contributing-30-supply-blocked-importing-malaysia-after-failing-meet

Smartifying Agriculture to Achieve Food Security and Sustainability in Brunei Darussalam

3

Osumanu Haruna Ahmed and Rose Abdullah

Abstract

Petroleum and natural gas sector is the dominant contributor to the economic growth of Brunei, accounting for over 50% of the country's GDP, which is why its economy fluctuates with the vicissitudes of oil prices. Although this dependence compels the country to import its manufactured goods and food, the country has since 2021 been on a significant economic diversification drive when the government rolled out a roadmap for achieving sustainable economic growth and development. The priority areas are the downstream oil and gas, food, tourism, information and communications technology, and services. It is worth noting that the ongoing economic diversification programme also prioritizes the agricultural and fisheries sectors. Although the contribution of the agricultural and fisheries sectors to the economic growth of the country is only approximately 1%, Brunei is poised to achieve self-sufficiency in broiler chicken, eggs production, day old chicks, fisheries, fertilized eggs and fruit production in the near future. The government's interventions also suggest that medium- to long-term self-sufficiency shall be achieved in cattle, goat, sheep, buffalo, vegetables, and paddy. The adoption of new farming strategies and technologies are expected to raise labour productivity by enhancing human capital, attracting investment, supporting resource optimization and value addition, reducing postharvest losses and through extensions into agrifood processing also suggests an expansion in the contribution to food to the country's GDP. This chapter discusses the economic sustainability of the agricultural and fisheries sectors in Brunei, especially since the introduction of leveraging strategies to develop agriculture and strengthen food security.

Keywords

Modern agriculture · Sustainable development · Food security · Climate resilience · Brunei

O. H. Ahmed (✉) · R. Abdullah
Faculty of Agriculture, Universiti Islam Sultan Sharif Ali, Bandar Seri Begawan, Brunei
e-mail: ahmed.haruna@unissa.edu.bn

R. Abdullah
e-mail: rose.abdullah@unissa.edu.bn

Introduction

Sustainable agriculture is a farming system or good agricultural practice which integrates new agricultural methods and technologies to

R. Rasiah et al. (eds.), *Sustainable Agriculture and Food Security in Southeast Asia*, Sustainable Development Goals Series, https://doi.org/10.1007/978-981-95-0477-0_3

boost agricultural, socio-economic and environmental efficiency to ensure that the present and future generations' needs are not compromised. Among the good agricultural practices for achieving food security and climate resilience are climate smart agriculture, regenerative farming, urban farming (including smart greenhouses and vertical farming), circular farming and indoor farming, which are facilitated by modern agricultural technology, such as vehicles, robotics, software, artificial intelligence, satellites, drones, mobile devices, computers and big data. Therefore, Brunei considers sustainable agriculture as consisting of farming systems to bolster sustainable food production in the country, which is supported by the following pillars: (i) economic sustainability through sustainable profit making from the agricultural value chains; (ii) social sustainability that places emphasis on the society rather than a privileged few; and (iii) ensuring a sustainable environment by safeguarding the quality of the environment for generations to come (FAO, 2014, 2018). These are part of the United Nations' Sustainable Development Goals that focus on transformations in agriculture and food systems to end hunger, achieve food security and improve nutrition by 2030, which is also in synchrony with Brunei's Vision 2035 that rallies its population to be globally recognized as educated, highly skilled and accomplished people with high quality of life supported by an economy that is dynamic (Wawasan Brunei 2035, 2019).

The major engine of growth in Brunei is the petroleum and natural gas sector, which accounts for over half of its GDP. To ensure sustainable development, the government launched a major economic diversification drive, which among other things resulted in the launching of an economic diversification programme that focuses on the agriculture, fisheries and livestock sectors. Consequently, the government rolled out a roadmap in 2021 for achieving sustainable economic growth and development through viable avenues with accompanying high-value employment opportunities. The blueprint priority areas are the downstream oil and gas, food, tourism, information and communications technology (ICT) and services (ASEAN Briefing, 2022). It is encouraging that the agricultural and fisheries sectors in particular, which in the 1920s dominated the economy of Brunei but subsequently declined following the discovery of the petroleum and natural gas, are being revived through policies as priority engines for the growth and development of the country. Also, the growing concerns over food security which peaked during the COVID-19 pandemic has compelled the government to raise food self-sufficiency through the adoption of new methods and state-of-the art technologies. Consequently, agricultural output has grown at an annual average rate of 6.9% per annum from 2013 till 2022. Output grew from B$314.8 million in 2013 to B$575.7 million in 2022. The expansion was particularly strong from 2018 (B$439.3 million) to 2022 (B$575.7 million). The livestock, crop and agrifood processing sectors contributed B$348.51, B$58.28 million and B$168.92 million, respectively, in 2022 (Department of Agriculture and Agrifood, 2022).

The total area of Brunei is 5,765-sq km and it is located on the north-west Borneo. Being a humid tropical country with annual precipitation of 2909 and temperature ranging between 24 °C and 32 °C, Brunei is suitable for agriculture and fishing. Prior to the discovery of oil and gas in the late 1920s, agriculture and forestry sectors were the foundation of its economy. Unlike petroleum became dominant around 50% of the population relied on the agriculture sector for their living, but this share has decreased significantly as most of the full-time farmers became part-time farmers who depended on low skilled migrant workers. Hence, the contribution of agriculture, fisheries and forestry to Brunei's GDP fell to 1.1% in 2021 (Department of Statistics, 2022).

Agriculture contributed 0.5% to Brunei's GDP in B$131.80 million in 2022 (Department of Agriculture and Agrifood, 2022), which is still negligible. Nevertheless, policy interventions show strong promise for the country to achieve 100% self-sufficiency levels (SSL) in a number of agricultural items. The SSL in 2023 for broiler chickens was 100%, 100% for eggs, 90.2% for day old chicks, fertilized eggs was 77.6%, and fruits was 68.6%. The country's fisheries sector

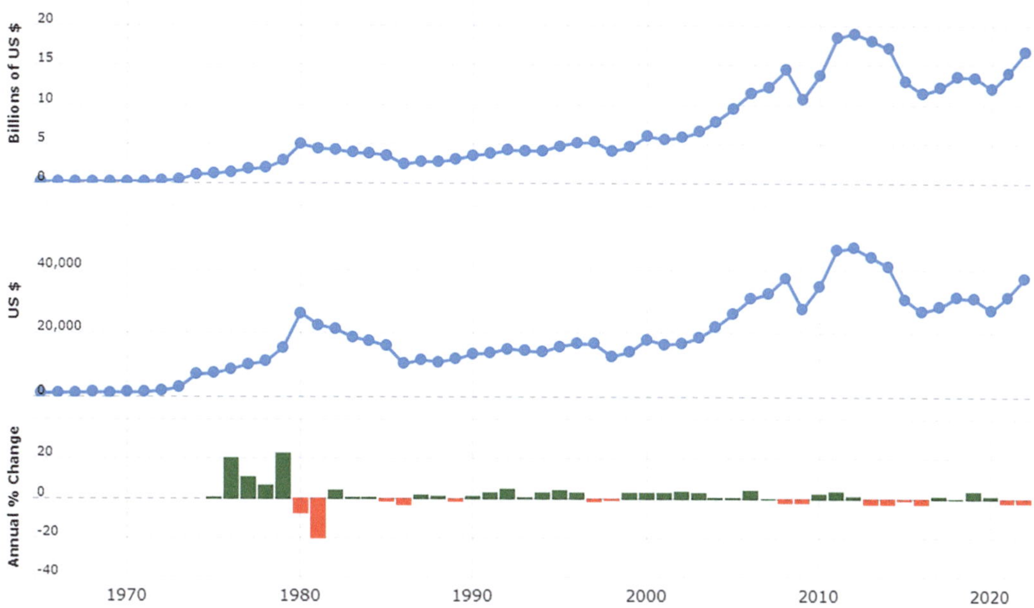

Fig. 3.1 GDP and economic growth, Brunei, 1965–2022. *Source*: https://www.macrotrends.net/countries/BRN/brunei/economic-growth-rate

(which focuses on capture fisheries, aquaculture and seafood processing) landed 24,285.9 mt of fish and seafood products, which amounted to B$191.0 million. The production of this sub-sector achieved 88.9% SSL and a GDP contribution of 0.5% in 2021. These targets were met in conjunction with meeting high conservation and climate resilience standards with the "Control and Prevention of Overfishing" regulation imposed in the country. Also the country's fisheries policy on permitting fishing exploitation limits by 20% lower than the permitted "maximum sustainable yield". These developments are expected to be boosted through the cooperation of Brunei, Indonesia, Malaysia, Philippines East ASEAN Growth Area (BIMP-EAGA) to accelerate the socio-economic development of less-developed, marginalized and geographically remote areas in these four countries. Following the problematization of sustainable development and food security undertaken by Rasiah (2025), this chapter examines the modern agricultural methods that have been promoted in Brunei since 2018 and their implications for agricultural expansion and food security in the country.

Economic Growth

As demonstrated in Fig. 3.1 and Table 3.1, the economic growth of Brunei has been inconsistent because it is characterized by fluctuations, slow growth and contraction, largely because of the instability of global oil and gas prices.

However, with the emergence of agriculture through scientific breakthroughs in both soil-less and soil-based farming, the contribution of the agricultural sector to the economic sustainability of Brunei has become important. This aspiration is being driven by: (i) increasing food premium by shifting to high-value agricultural production through the optimization of inputs; (ii) raising productivity using technologically advanced research and development; (iii) improving agronomic and economic efficiency that comes with the adoption of new agricultural methods and technologies, smart agriculture, including through foreign direct investment (FDI); (iv) reducing barriers on insurance and loans acquisition in agriculture in particular to make farming conducive by strengthening the enabling environment for boosting export of

Table 3.1 GDP, GDP per capita and economic growth, Brunei, 2010–2022

Year	GDP	Per Capita	Growth rate (%)
2022	$16.68B	$37,152	– 1.63
2021	$14.01B	$34,449	– 1.59
2020	$12.01B	$27,179	1.13
2019	$13.47B	$30,748	3.87
2018	$13.57B	$31,241	0.05
2017	$12.13B	$28,187	1.33
2016	$11.40B	$26,762	– 2.48
2015	$12.93B	$30,681	– 0.39
2014	$17.10B	$41,036	– 2.51
2013	$18.09B	$43,950	– 2.13
2012	$19.05B	$46,844	0.91
2011	$18.53B	$46,139	3.75
2010	$13.71B	$34,610	2.60

Source: https://www.macrotrends.net/countries/BRN/brunei/economic-growth-rate

processed or high value-added agricultural and fisheries outputs; and (v) shifting to sustainable agriculture and fishing through cutting-edge agricultural practices, including attracting private–public partnerships.

Brunei Vision 2035 Goal Three

Table 3.2 summarizes Brunei's vision 2035 goal three which aspires to make the economy dynamic and sustainable. In relation to this goal, agriculture plays a pivotal role *vis a vis* the set outcomes and key areas for the successful realization of the country's vision 2035 goal three. The 2021 Blueprint of Brunei identified food as one of its priority areas and to this effect, the country's Food Industry Roadmap was launched to increase safe and high-quality food production to strengthen the country's food system value chain.

Table 3.2 Brunei Vision 2035 goal three, national outcomes and key areas

Brunei Vision 2035 Goal Three	National outcomes	Key areas
Dynamic and sustainable economy	High sustainable growth	Output and productivity Investment
	Diversified economy	Sector diversification Export diversification Widening revenue base
	Low unemployment	Job creation Workforce mobility
	Macroeconomic stability	Price stability External balance Fiscal stability Financial stability

Source: Wawasan Brunei 2035 (2019). Sekretariat Tetap Wawasan Brunei 2035, Jabatan Perdana Menteri, Brunei Darussalam

Agriculture Development Models

This section looks at the different agricultural development models in terms of strengths and weaknesses from where the most relevant model for Brunei agriculture was considered for this chapter.

Resource Exploitation Model

The resource exploitation model, which stipulates that expanding cultivated agricultural land increases production or productivity on the assumption that excess agricultural land and capacity increases farm produce *vis a vis* stimulus new markets, regardless of the inferior technology used (Udemezue & Osegbue, 2018). This model does not apply well to Brunei sustainable and food system drive because of the following reasons: (a) it is unsustainable because it does not provide solutions for soil fertility loss in marginal lands or exhaustion of utilized excess land, (b) and land expansion is not possible under the country's zero forest cut policy, while several agricultural lands are facing soil fertility challenges from high acidity, iron, aluminium and sulphur contents.

Conservation Model

The conservation model recognizes that agricultural lands are scarce and that they are prone to degradation (Udemezue & Osegbue, 2018). This is important because the model assumes that because agricultural lands are scarce, farmers might be compelled to farm on infertile soils resulting in marginal labour productivity besides accelerating soil quality deterioration to the point of irreparable damage. Also, being a key pillar of climate change mitigation conservation discourages the pursuit of intense farming, though agricultural productivity per ha has significantly improved with little or no additional land in countries, such as Taiwan (Rasiah et al., 2018). The other weakness of this model is that it tends to understate the significance of high-yielding agricultural inputs, such as fertilizers, biopesticides and other technological inputs, which is unviable for achieving self-sufficiency in Brunei.

Location Model

Besides determining location for producing agricultural goods, the location model emphasizes the relationship between industrialization and urbanization, suggesting that agricultural activities near urban areas attract higher product prices, land values and rates of land use (Udemezue & Osegbue, 2018). Also, farms near industrial-urban areas are likely to have machinery investment/ha and man/ha yet having little effect on other inputs. This model is unlikely to suit Brunei's agricultural development agenda because the country is not fully industrialized—urbanized. Additionally, if the agricultural systems are insignificant or do not grow at a significant pace to absorb rural–urban migrant workers, the labour force will naturally go to the non-agricultural sectors, which is what the country is experiencing.

Diffusion Model

The diffusion model suggests that agriculture is developed by providing technical knowledge to farmers directly in addition to closing the inter-sectoral productivity gap within the country, and across regions among farmers/farms (Udemezue & Osegbue, 2018). Although this model has given significant intellectual foundation for research and extension efforts in farm management and production economics, the proponents of this model believe that farmers are well informed or are cleaver experimenters to face the realities involved in farming. The problem with this model is that it neither generates rapid modernization of traditional farms nor rapid growth in agricultural output. Hence, this model is unlikely to accelerate the needed agricultural transformation and job creation in Brunei.

High Pay-Off Input Model

The high pay-off input model calls for the transformation of traditional farming systems to a level where its productivity improvements shall spur economic growth, which shall attract high technology investment to raise farmers' competitiveness (Udemezue & Osegbue, 2018). This model identifies high productivity areas as: (a) centres of agricultural research; (b) industries which are capable of developing, producing and marketing new technical inputs; and (c) promotes effective use of agricultural factors of production by farmers. Apart from the successes this model has achieved in R&D, including knowledge transfer to farmers for boosting productivity leap, agriculture policies associated with this haves raised agricultural productivity growth. The merits of this model are consistent with Brunei agricultural development agenda *vis a vis* the country's economic diversification programme, and hence, why it is chosen as the model for this chapter.

Government Policies for Promoting Agriculture and Fisheries

Rice cultivation is facilitated via price support, improved infrastructure, irrigation and drainage facilities. To fill the gap being created by the decreasing participation of smallholding rice production, large-scale pilot rice mechanization projects to increase output by at least 30% of the country's need were initiated. In 1986, the end-product subsidy scheme for rice producers was established to: (i) revive paddy growing activities;

(ii) encourage farmers to return to farming activities; and (iii) socio-economically empower rural and low-income households. With this scheme, 1 kg of paddy is bought by the government at B$1.60 (Department of Agriculture and Agrifood, 2022). In 2019, the largest commercial paddy farm begun operations, targeting annual yield of 6,000 to 8,000 tonnes of paddy by 2025. The expectation is that 500 ha commercial paddy farm will increase the country's rice self-sufficiency level from 11 to 15% by 2025.

The poultry industry has also received a significant boost to facilitate the small and medium-scale enterprises growth and development in tandem with the processing and marketing centres. For example, the Mulaut Abattoir, Brunei serves as market outlets for the poultry producers. This approach encourages the private sector to take the leading role in growing the poultry industry whilst the government provides the basic infrastructure, amenities and production-support services. This has resulted in being self-sufficient in eggs and broiler production and almost self-sufficient in poultry and fish production (Department of Agriculture and Agrifood, 2022; Department of Fisheries, 2021). The livestock production is low, thus compelling live cattle importation from a cattle ranch in Australia own by Brunei government. The Fisheries Act (1982) delimits a maritime area of about 38,600 sq km extending to about 200 nautical miles (370 km) from the coast. Although it has more than 100 kms of coastline, and 85% of its population live on the seaboard, Brunei Darussalam imports 40% of its fish and related products. Thus, the government encourages small, modern and offshore fishing industry. The Eighth Plan allocates B$90 million for sustainable exploitation of fisheries resources. The country deliberately set the current permitted exploitation level at the estimated maximum economic yield and it is managed through the number of fishing licences issued for each fishing method. Besides providing infrastructure and amenities for the aquaculture sector, the country welcomes joint ventures with external investors for financing and expertise.

Land acquisition for commercial agricultural and fisheries activities being is on rental basis and it is granted by the Agricultural Development Area Unit (Table 3.3). The rent clustering based on the agricultural system type is to accelerate their sustainable productivity in relation to land use classification which emphasizes on precision farming or site-specific management farming for resource optimization which translates into sustainable agricultural, socio-economic and environmental efficiency. Moreover, it enables successful agricultural services including extension, training, advisory, knowledge transfer, monitoring and evaluation activities being provided by the different units of the Department of Agriculture and Agrifood, Brunei, universities and many more. This zoning approach is yielding good results as evident in the significant progress in the agricultural and fisheries sectors as discussed. The agricultural stations (including workshops), industrial agriculture parks and research centres based on districts and farming system type also add on to the acceleration of the agricultural and fisheries sectors in the country. Another important futuristic policy is designating areas called Rural Agricultural Development Area and Young Farmer Areas with good packages or incentives to attract the youth into agricultural, fisheries and related areas besides improving rural agriculture with modern farming systems and technologies. There are several successes adopting this approach; as an example, the highly skilled youth are able to participating in the global digital agriculture which embraces solar technology, smart farming, IoT, drones, sensors and data analytics hence the growth of urban farming in Brunei. Furthermore, there are research and development grants to facilitate collaboration between researchers and the agricultural and fisheries industries for knowledge transfer, commercialization, job creation, product development, value addition and so on. Excellent entrepreneurs and agricultural and fisheries enterprises are rewarded by the Department of Agriculture and Agrifood, Brunei. Yearly Agro-shows or Expositions are organized to draw attention and encourage the public to get involved in the agricultural and fisheries sectors.

Besides being recognized for being a clean environment, Brunei does not experience serious

Table 3.3 Agricultural development areas, rental rates, Brunei (2022)

Commercial agricultural systems	Rental in B$ per hectare per year
Goat	B$5.00
Deer	B$5.00
Buffalo and cattle	B$5.00
Paddy	B$12.50
Vegetable	B$25.00
Sugarcane plantation	B$25.00
Fruit	B$25.00
Nursery (Floriculture, grass, among others)	B$50.00
Tissue culture	B$50.00
Poultry	B$100.00
Duck	B$100.00
Hatchery	B$100.00
Abattoir and feedlot	B$125.00
Agri-Mart	B$500.00

Source: Department of Agriculture and Agrifood (2022)

natural disaster caused by severe climatic effect. The government policy areas identified for aquaculture are provided with infrastructure and amenities such as roads, water, electricity and telecommunication facilities. Annual licences to fish are granted to locally registered enterprises of farms and this also includes joint ventures. It must be noted investors are given fishing area by taken into account the available resource and estimated fishing capacity. The other condition is the catch of the licence holders must land in Brunei. The licence holders are not permitted to have total annual harvest exceeding the maximum economic yield (20,000 tonnes per year). In summary, pursuing the government's initiatives such as co-capital matching scheme, pilot projects, youth contract farming, recruitment of the Institute of Brunei Technical Education graduates, business expansion programmes, facilitation of domestic market, promotion and facilitation of the export market, and optimization of idle farm lands for the agricultural and fisheries sectors is essential.

Trade and Investment

The total international trade amounted to B$ 32,287 million (Department of Statistics, 2022). In 2021, the fisheries sector imported 7,948.71 tonnes at B$ 41.22 million whereas exportation of 5,549.59 tonnes amounted to B$ 46.67 million (Brunei Darussalam Fisheries Statistics in Brief, 2021). In 2021, the 7851.67 tonnes of agriculture and agrifood products exported raked in $6.66 million, the bulk of which was contributed by the livestock industry whereas in 2022, there was significant increase in the exportation of muskmelon (143 tonnes of fresh) Japanese Muskmelon at a market value of B$ 0.67 million. In this same year, in Brunei, 8,764.21 tonnes of agriculture and agrifood products were exported, earning the country $10.37 million (CIF value). Again, the livestock sector was the main contributor as 8.233.14 tonnes (94.13%) poultry feed, followed by the agrifood and crop sector. The general improvement in the output of the agricultural and agrifood products is related to the continued enabling environment through pilot projects, contract farming for the youth, business expansion programmes, domestic and local markets facilitation, diffusion of new agricultural methods and technologies particularly in the newly opened areas for paddy, other crops and livestock production. The dominance of the livestock and agribusiness sectors is partly related to the strong participation of the private sector unlike the crop sector which is trailing in this aspect as most of the farms are small scale and dwindling. Among the reasons being that ageing of, for example, paddy farmers and the low interest of the youth in production agriculture although modern farming systems such as smart greenhouses, fertigation, aquaponics, hydroponics and vertical farming being put in place could reverse this trend. The report of the Department of Statistics (2022) did not indicate major direct foreign investment in the agricultural and fisheries sectors.

Tariff Structures, Trade Balance and Agricultural Labour Productivity

There is no personal income tax and there is no export, sale, payroll or manufacturing taxes. Only companies are subject to tax under the Income Tax Act 1949. Corporate income tax rate has been gradually reduced from 30% (2008) to the current rate of 22%. Pioneer status companies are exempted from corporate income tax, import duties on machinery, import duties on raw materials and tax exemption for 11 years. In the case of expansion of established enterprises, they are exempted from corporate income tax and tax exemption for 20 years. The minimum capital expenditure is US$650,000. Foreign Loans for Productive Equipment are exempted from withholding tax for interest paid for a minimum of US$130,000. Besides signing bilateral investment treaties and double taxation agreement with several countries, Brunei is involved in Free Trade Areas through its membership in ASEAN in addition to plurilateral and bilateral agreements.

In 2022, the GDP value at current prices of Brunei for the industry was B$15,625 million (66.9%) followed by the services sector at B$7,480.70 million (32%), and the Agricultural, forestry, and fisheries sector contributed B$ 252.20 million (1.10%) (Department of Statistics, 2022). The GDP of the agricultural and fisheries sectors as the share of the national gross domestic product from 2020 to 2022 is presented in Table 3.4. The total GDP contribution of these sectors is approximately 1% with the fisheries and animal sectors contributing the major part compared with the plant sector whose contribution decreased in 2021 and 2022 because of the ageing and low interest of the youth (Department of Statistics, 2022).

The exports and imports (Tables 3.5 and 3.6) by trading partners suggest that from 2020 to 2022, the export values were greater than import values suggesting the positive trade balance. For example, the international merchandise trade amounted to B$ 32,287 million out of which exports amounted to B$19,623.60 million whereas imports amounted to B$12,664.30 million and the trade balance was B$6,959.30 million (Department of Statistics, 2022).

Table 3.4 Share of agriculture and fisheries in GDP, Brunei, 2020–2021

Sector	2020	2021	2022
Vegetables, fruits and other farming enterprises (%)	0.20	0.10	0.10
Livestock and poultry (%)	0.40	0.40	0.40
Fisheries	0.50	0.50	0.50
Total	1.10	1.00	1.00

Source: (Department of Statistics, 2022)

Table 3.5 Exports of agricultural and fisheries products by trading partners

Exports by trading partner	2020	2021	2022
ASEAN (B$ million)	3,665.80	4,497.50	7,706.6
European Union (B$ million)	2.50	6.60	14.80
Other trading partners (B$ million)	5,453.50	9,629.70	11,902.10
Total	9,121.80	14,133.80	19,623.60

Source: (Department of Statistics, 2022)

Table 3.6 Imports of agricultural and fisheries products by trading partners

Imports by trading partner	2020	2021	2022
ASEAN (B$ million)	2,801.80	3,454.30	4,567.40
European Union (B$ million)	373.20	257.10	231.00
Other trading partners (B$ million)	4,163.60	6010.13	7,865.90
Total	7,338.60	9,271.70	12,664.30

Source: Department of Statistics, 2022)

Labour Productivity

Labour productivity, which is enhanced by capital investment, technology and human capital, is measured as output per unit of labour input. Globalization, digital economy and human ageing also affect agricultural labour especially when these factors affect the economic conditions of a country or farming systems. Globalization has effect on productivity through, for example, trade liberalization, exposure to new technology, foreign direct investment, among others (ASEAN Secretariat, 2021). The inroads of robotics, big data, satellites, drones, computers, mobile devices, software, artificial intelligence, internet of things and many more as the drivers of the Industrial Revolution 4.0 are significantly shaping the world-wide economies of which the economy of agriculture is no exception. In the agricultural and fisheries sectors of Brunei, digital farming such as smart farming, climate smart agriculture (CSA), precision farming, cloud farming, smart greenhouses, integrated farming, vertical farming, rooftop farming, indoor farming and digital marketing are transforming the value chains right from production to consumption particularly running business and production methods, smart working, new business models, and new inventions or innovations replacing old management practices.

Among the ASEAN countries, the per-worker labour productivity of 142.01 thousands of USD and per-hour labour productivity of 62.46 thousands of USD for Brunei were second to those of Singapore which were 149.05 thousands of USD and 66.50 thousands of USD, respectively (Table 3.7). For Singapore, these productivities were due to the high growth in capital investments from foreign direct investments with time unlike Brunei whose own probably relate to the oil and gas sectors. The contribution of the agricultural sector of Brunei to the employment sector decreased from 1971 to 2018 although it slightly improved (Table 3.5) from 2011 to 2018 (ASEAN Secretariat, 2021). The former decrease was the shift from the agricultural sector to the oil and gas sector whereas the latter was the response from the policy interventions to revive the agricultural sector. This observation is also true for the agricultural labour productivity and labour productivity growth rate (%).

Changes in the Share of the Agriculture and Fisheries Sectors

The crop sector has the highest number of entrepreneurs because of the broadness of this sector. In addition, these entrepreneurs are involved in the two major categories of cropping, namely soilless and soil agriculture. Moreover, the support given to the agriculture by the government of Brunei is manifesting in the gradual growth in the number of the crop entrepreneurs since 2019. Unlike the crop sector, there has been significant decrease in the number of the livestock entrepreneurs because of consolidation and the dominance of the large-scale private enterprises or companies (Table 3.8). The opposite is true for the crop and fisheries sectors which have more small-scale entrepreneurs than large-scale entrepreneurs.

For the fisheries sector, most of the fishermen are part-timers (Table 3.9). In spite of this, the fish selfish sufficiency level of Brunei increased to 88.89%, suggesting the potential of achieving 100% self-sufficiency level and beyond should the number of the full-time fishermen increase significantly. The growth of the agrifood sector entrepreneurs is relatively stable although from 2013 to 2022, the revenue generated is on the increase from B$86.24.00 in 2013 to B$168.92 (Department of Agriculture and Agrifood, 2022). The gradual decrease in the total number of the livestock, crop and agrifood entrepreneurs is related the significant decrease in the number of the entrepreneurs in the livestock since 2019. Unlike the livestock and crop sectors which might have been affected by the COVID-19 pandemic, the agrifood sector labour force grew even pre-, during, and post-COVID-19 pandemic. In spite of the relatively stable number of the entrepreneurs in the agrifood sector, it employs the highest number of workers and this is almost the combined jobs provided by the livestock and crop sectors, hence explaining increase in the total number of the labour force of the three sectors. This is consistent with the countries vision 2035 which emphasizes

Table 3.7 Labour productivity growth, agriculture, Brunei, 1975–2018 (%)

Country	1975–1980	1985–1990	1995–2000	2005–2010	2015–2018
ASEAN	2.04	−0.77	1.39	2.6	4.66
Brunei Darussalam	6.84	8.97	5.38	−5.65	0.77
Cambodia	−5.45	7.38	−0.11	2.23	2.53
Indonesia	3.92	−1.52	−1.49	3.55	5.05
Lao PDR	1.98	−1.19	2.20	3.47	0.75
Malaysia	4.51	3.19	1.40	1.76	4.03
Myanmar	3.57	−4.72	4.07	1.41	0.75
The Philippines	1.78	1.22	4.53	1.29	2.41
Singapore	6.68	0.99	−8.39	−8.20	−9.01
Thailand	−2.72	−1.75	3.93	0.72	3.23
Viet Nam	2.49	−0.08	3.69	3.72	6.91

Source: Adapted from ASEAN Secretariat (2021)

Table 3.8 Entrepreneurs and labour, agricultural sector, Brunei, 2018–2022

Entrepreneurs	2018	2019	2020	2021	2022
Livestock	831	827	794	497	494
Crop	4217	3849	3756	4069	4119
Agrifood	385	381	377	397	381
Total	5433	5057	4957	4963	4994
Labour					
Livestock	2160	1971	2055	1690	2246
Crop	2086	1877	1835	2786	2441
Agrifood	4047	4073	4294	4638	4680
Total	8293	7921	8184	9114	9367
Grand total (entrepreneurs + labour)	13,726	12,978	13,141	14,077	14,361

Source: Department of Agriculture and Agrifood (2018, 2019, 2020, 2021, and 2022)

on job creation and value addition in the agricultural and fisheries sectors. It appears that the grand total agricultural employability has been on the increase during the post-COVID-19 pandemic. This is another indication of the positive results from the enabling environment being created by the government of Brunei for the agricultural and fisheries sectors to thrive. A further attestation is the continued growth of the income generated from processed farm produce and value-added products over a period of ten years.

The major and consistent contributor to processed farm food is the livestock sector whose share since 2013 to 2022 increased from B$44.20 to B$121.01 (Table 3.10) although its number of entrepreneurs is smaller compared with the crop sector. Despite this significant contribution domestically, there is more room for improvement for the export market. In the case of the crop sector, the production is not able to meet the demand of the domestic. The 2021 data indicate that 5,549.39 tonnes, the bulk of which was processed fish, were exported for an export earning of B$190.95.00 million (Department of Fisheries, 2021). Obtaining HACCP certification made it possible for the exportation of processed seafood products including frozen shrimps (Department of Fisheries, 2021).

Changes in Agricultural Land Use

Table 3.11 shows the changes in the areas occupied by livestock and crops. The total size of the livestock area decreased in 2021 and 2022

Table 3.9 Small-scale fishermen, Brunei, 2015–21 (number)

	2015	2016	2017	2018	2019	2020	2021
Full-time fishermen	986	921	905	935	704	737	828
Part-time	1359	1127	1020	1040	941	1106	1235
Total	2345	2048	1925	1975	1645	1843	2063

Source:Department of Fisheries (2021)

Table 3.10 Revenue from processed farm produce, Brunei, 2013–2022 (B$ million)

Sector	2013	2014	2015	2016	2017	**2018**	**2019**	**2020**	**2021**	**2022**
Processed livestock and product value addition	44.20	43.87	45.27	62.63	66.66	77.22	87.88	107.08	123.08	121.01
Processed crop and product value addition	4.10	5.34	5.53	5.59	5.19	5.42	5.09	4.63	5.40	6.43
Others (Food and drinks)	37.94	53.47	59.69	49.90	48.34	45.33	39.55	38.51	38.21	41.48
Total	86.24	102.68	110.48	118.11	120.19	**127.98**	**132.47**	**150.23**	**166.69**	**168.92**

Source: Department of Agriculture and Agrifood (2022)

Table 3.11 Area occupied by commercial livestock and crop farming with the corresponding revenues, Brunei, 2018–2022

	2018		**2019**		**2020**		**2021**		**2022**	
Sector	Area (Ha)	Revenue (million$)	Area (Ha)	Revenue (million$)	Area (Ha)	Revenue (million$)	Area (Ha)	Revenue (million$)	Area (Ha)	Revenue (million$)
Livestock	3135.74	244.63	3,050.91	237.39	4778.53	261.17	2125.94	358.77	2108.35	348.51
Crops	4552.46	63.69	4,760.64	58.41	5242.20	53.46	6193.45	61.10	6920.59	58.28
Total area	7,668.20	–	7811.55	–	10,020.73	–	8319.39	–	9,028.95	–
Agrifood	–	127.98	–	132.47	–	150.23	–	166.99	–	168.93
Total revenue	–	436.30	–	428.27	–	470.86	–	586.56	–	575.71

Source: Department of Agriculture and Agrifood (2022, 2021, 2020, 2019, 2018)

but the opposite is true for the revenue generated as it increased from 2018 onwards and this is consistent with the decrease in the number of the entrepreneurs in the livestock sector because of consolidation and the dominance of the large-scale private companies. The vegetable, fruit and paddy farms contributed the crop sector land increase because their land sizes increased with increasing year (Table 3.12). However, the total revenue generated rose with increasing land size. The agrifood sector also contributed to the increase in the total revenue with the increasing year that of the agrifood also increased with the increasing year. Furthermore, the decrease is related to the significant decrease in buffalo rearing (Table 3.13) because in 2021 and 2022, the land size reduced significantly. It is also possible that the companies involved in this sector are being efficient in space optimization or shift their interest in this sector. Unlike the livestock sector, the crop sector land size increased over the period of five year because of the increase in the entrepreneurs although the increase did not significantly increase the revenues generated.

Table 3.12 Agricultural land use, selected crops, Brunei (hectares)

	2018	2019	2020	2021	2022
Vegetables	720.15	779.31	816.85	1,259.78	1,053.99
Fruits	2,926.00	3,043.85	3,207.84	3,422.52	4,345.50
Paddy	782.95	772.79	852.97	927.14	890.79

Source Department of Agriculture and Agrifood (2022, 2021, 2020, 2019, 2018)

Table 3.13 Land occupied by commercial livestock, Brunei, 2018–2022 (hectares)

	2018	2019	2020	2021	2022
Broiler (Chicken Meat)	274.17	302.07	294.11	300.73	276.93
Layer (Egg)	59.78	90.52	94.72	92.72	92.72
Day old chicks	73.90	76.60	74.19	93.69	154.69
Fertilized eggs	40.00				112.69
Buffalo	1,925.74	1,883.14	2,809.14	781.04	738.92
Cattle	176.18	166.78	420.31	184.04	205.46
Goats	281.62	280.52	470.42	306.12	214.17
Deer	29.05	29.05	49.25	36.14	16.30
Sheep	56.46	60.66	348.85	148.32	116.02
Livestock feed	4.70	4.70	7.70	135.74	7.7

Source: Brunei, Department of Agriculture and Agrifood (2022, 2021, 2020, 2019, 2018)

Intensive Food Farming and Food Security

Brunei's development plans not only focuses on achieving food security, but also production for exports. The policy aims to: (i) increase rice, vegetables, poultry, livestock and fish production; (ii) develop the agrifood industry; (iii) produce high value-added products using new agricultural and fishing methods and technologies; (iv) conserve and protect the country's biological diversity; and (v) attract foreign direct investment for the agricultural, fisheries and food processing sectors. To achieve these goals, local farmers are provided with and are still being provided subsidized: (i) infrastructure facilities (roads, irrigation, telecommunication facilities and electricity); (ii) seeds; (iii) liming materials and fertilizers; (iv) vaccine for livestock; (v) farm equipment; and (vi) financial and technical assistance.

Broilers and eggs production has been on the increase such that by 2022, Brunei no longer imports broilers and eggs (100% self-sufficiency). For the broilers, the large-scale companies contributed 80.68% whereas the small and medium-scale companies contributed 19.3% whereas for the eggs the large-scale and small-medium companies contributed 98% and 2%, respectively, whereas the small and medium-scale companies contributed 2.0%. The large-scale companies, the day old chicks and fertilized eggs production self-sufficiency increased to 90.2% and 77.7%, respectively (Department of Agriculture and Agrifood 2022). These impressive achievements are a reflection of the adoption of advanced technologies such as Light Emission Diodes technology, good feed, automated feeding system, automated water system, ventilated fans, air-conditioned system with air-conditioned wall, efficient ventilation system and many more.

The low beef, goat and sheep, and livestock self-sufficiency of 39.9, 3.9 and 39.7%, respectively (Department of Agriculture and Agrifood 2022), is being fixed through introducing good breed, good feed and free range on some of the unutilized agricultural lands. The possibility of producing high-quality livestock feeds locally is being explored. Plants-based local livestock feeds could be produced on the unused agriculture lands in the country, for example corn, napier grass, silage, hay, among others. Wireless IoT sensors should be used to monitor the animals in terms of location, well-being and health condition to separate the sick one from the herd to prevent spreading of diseases. Furthermore, Brunei should intensify attracting more new entrepreneurs at the same time encouraging the existing farmers to increase production via approaches such as: (i) expanding farm capacities and production areas, (ii) enhancing the capacity of farmers through guidance and skills training, (iii) adopting new farming methods and technologies and (iv) encouraging efficient farm management and good agricultural practices.

Brunei's tropical vegetables self-sufficiency is 37.7% (Department of Agriculture and Agrifood 2022) and the practices being put in place to increase the sufficiency level are improving soil quality, seed quality, pests and diseases management, and fertilizer management through, for example, precision farming or smart farming including smart greenhouse for soilless agriculture. Through precision farming, the aforestated parts of agriculture could be precisely managed not only to increase yield but to also ensure economic and environmental efficiency. Furthermore, the approach enables practising high-density planting for optimum yield with less farm inputs and space maximization as required by Agriculture 4.0. The basis of Agriculture 4.0 is incorporating technology in agriculture to capture the potential of every farm (Greenhouse and non-greenhouse) so as to increase yield with reduced input volumes leading to making significant profit without degrading the quality of the environment or being able to significantly reduce the environmental footprinting of farming. Moreover, because most of the agricultural lands of Brunei are weathered soils which are noted for or are characterized by low organic matter, acidity, and high contents of aluminium and iron, it is imperative to adopt regenerative farming. Regenerative agriculture advocates for soil health restoration, conservation farming, integration of crops and trees with livestock, and re-carbonation of the terrestrial biosphere. Additionally the approach reduces soil disturbance, reduces use of chemical inputs, increases plant and animal biodiversity, ensures sustainable covering of soil surface with crops and adaption of farming systems to the local environment including microclimate. Precision irrigation, drip irrigation and rainwater harvesting should be among the farming strategies to maximize water consumption efficiency and minimizing environmental degradation. This must be in tandem with sustainable nutrient management strategies such as organic fertilizer use, precision fertilizing and crop rotation to increase soil fertility yet reducing nutrient runoff and the emissions of greenhouse gases from chemical fertilizers.

The local fruit production has improved marginally to self-sufficiency of 68.6% (Department of Agriculture and Agrifood 2022). For both annual and perennial crops, machine learning is being considered because apart from it being precise in prediction, it models routine farm activities such as fertilization, pesticides, weedicides and water management as part of CSA. For example, the frequency of fertilizing crops, applying pesticides and weedicides every growing could be modelled such that the same amount is not used but only varied amounts are used in subsequent planting when needed. This is another cost-effectiveness approach of boosting crop productive. In the era of global warming and climate change, it is essential that farmers practice CSA improved: (i) productivity and profit; (ii) efficient agricultural practices for climate adaption; and (iii) agricultural practice which do no degrade or redeems the environment from pollution. The near infrared technology is another important technology needed for boosting Brunei Darussalam agriculture. This technology should be deployed for the following benefits: (i) detection of soil organic matter; (ii) measuring nutritional status of crops; (iii) determination of harvest time; (iv) determination of grain composition; (v) fruit processing; (vi) fish quality control; (vii) drought monitoring; (viii) fruit quality control; (ix) breeding programmes; agricultural soil testing; and pests, diseases and toxin detection. Indoor and greenhouse agriculture in Brunei needs to be intensive for high value crops. Thus, urban farming should be accelerated in the urban areas of the country. This can include vertical farming, fertigation (soilless and soil), aquaponics, hydroponics and aeroponics to attract the youth participation to bring on board their innovations or creativity in areas such as fertilizer technology, waste management, smart greenhouse technologies, product lines, business models and so on.

The local paddy production has significantly improved in terms of the quantity of the paddy grains and self-sufficiency since 2020. In 2022, the total consumption of rice was 27,627.0 tonnes worth B$33.09 million. The local production was 3,562.6 tonnes with a market value of B$2.89 million against the importation of 25,199.0 tonnes at a retail value of B$30.2 million. The self-sufficiency rice is 8.4% (Department of Agriculture and Agrifood 2022). The low self-sufficiency is related to the year productivity of 4.4 tonnes/ha. The potential yield of the varieties is partly challenged because of the acidic nature of the paddy soils, lodging of paddy plants, pests, diseases, water problem, among others. The paddy soils' acidity, aluminium, iron, and sulphur are only temporary fixed through liming, hence the need to rehabilitate the soils using high-quality organic amendments. In introducing high-quality organic amendments in the paddy production in the country, being them from plant or animal sources should be thoroughly mixed with the paddy soils a day before seeding or transplanting because based on our experience, this approach effectively improves the soil productivity (Ali et al., 2023). Furthermore, it is essential to practise site-specific management in the paddy cultivation for farm inputs including water optimization. With this approach, drones with hyperspectral camera with high resolution could be used during land preparation to detect spatial variability and at the later stages of the paddy growth and development, they could be used to monitoring growth of the paddy plants, weeds, pests, diseases, nutrient deficiency, water balance, lodging and many more.

Of another paramount importance is building silica in the paddy soils because paddy plants absorb significant amount of silica from paddy soils such that continued production of paddy on them without silica replenishing depletes silica to cause several problems to paddy farmers and among the problems is lodging of the paddy plants with grains and also susceptibility of the paddy plants to pests and diseases (Dorairaj et al., 2017; Gong et al., 2021; Hu et al., 2022). Lodging causes approximately 20% grain yield loss, and thus it is recommended that silica application in the forms of, for example, zeolite and calcium silicate are introduced in the paddy production in Brunei. The added advantage of including zeolite in the paddy cultivation is that besides neutralizing soil acidity, aluminium and iron, it captures essential nutrients and releases them slowly from being lost from soils through, for example, leaching and runoff. Moreover, they are reputed for capturing greenhouse gases and ammonia. To increase the paddy and other productivity, locally produced high-yielding varieties should be commercially cultivated on the approved idle agricultural lands supported by integrated pest management system, improved irrigation systems in addition utilizing underground or harvested water (Rain) to irrigate areas with incomplete or inadequate irrigation system. Integrated pest management for pest control could be adopted.

Fisheries Sector

In 2021, Brunei's fisheries sector focuses on capture fisheries, aquaculture and seafood processing produced 24,285.93 mt fish and seafood products and this earned the country B$190.95 million (Department of Fisheries, 2021). With significant local research breakthroughs in, for example, quality feed production, nature-inspired solutions for disease prevention or control, breeding, among others, the country could exceed 100% self-sufficiency. This is essential because, in 2021, the fisheries GDP was B$82.4 million with a contribution of 0.5% to the country's GDP (Department of Fisheries, 2021). Under capture fisheries, 15,295.68 mt was produced with equivalent revenue of B$109.53 million followed by aquaculture which produced 4771.72 mt with income of B$44.90 million whereas the seafood processing produce was 4,218.54 mt with a generated income of B$36.51 million (Department of Fisheries, 2021). It is worth noting that in 2021, Brunei Darussalam fisheries importation of 7,948.71 mt cost the country

B$41.22 million compared with exportation of 5,549.59 mt earning B$46.67 million (Department of Fisheries, 2021).

Diffusion of Digitalization and Digitization into Agriculture

Brunei is resolute in using the latest agricultural and fisheries methods and technologies to boost agriculture and fishing sectors right from upstream to downstream. There is proliferation of smart farming systems such as fertigation, hydroponics, aquaponics, closed-house ventilated poultry, Light Emission Dioxide lighting, greenhouses or protected houses, water management, crop and animal monitoring, mechanized ploughing, transplanting, and harvesting in paddy cultivation and equipment for agrifood processing.

A free online platform or portal Agrome IQ which is compatible with android by a Bruneian serves as a marketplace to enable producers to sell their produce to buyers or consumers. Also, this smart farming platform enables farmers to efficiently manage their farms in a manner that translate into improved yield and quality. For example, the Agrome IQ serves as a farm management software which is capable of assigning daily farm tasks and also enables tracking the assigned tasks by farm managers and their farmhands. With this approach, this smart farm software empowers small-scale to medium-scale farms lacking administrative resources to document their operation task management efficiently. This is essential for achieving GAP status as well as for exportation of produce or products because adoption of GAP increases yield and quality. When the products are branded especially with GAP logo, they attract better prices at the local and international levels. The manual booklet of GAP which farmers are mandated to use keeps their records; record-keeping (Booklet) has been digitized and integrated into portal Agrome IQ and being used by some farmers. Farmers being able to document every planting cycle activities enable them to trace previous planting cycles' productivity to guarantee spatial and temporal variability management in a form of site-specific management not only for efficient resource utilization but to also maximize yield, profit and environmental quality. The target is to training as many farmers as possible in this direction for the benefit of the country because these modern approaches are contributing significantly to the food self-sufficiency of Brunei. The Authority for Info-communications Technology Industry (AITI) of Brunei through its Smart Farm Testbed is also playing a leading role in smart farming entailing IoT and precision farming capable of increasing crop productivity by at least 10%; reducing chemical fertilizers use by 20%; and reducing labour particularly foreign workers by 30% using IoT and Data Analytics.

Moreover, there is a collaboration between a technology-based university and Economic Research Institute for ASEAN and East Asia on "Digital transformation high-tech and high efficiency Agriculture Model Project in Brunei". There is also an ongoing diffusion of new methods farming such as regenerative farming by the Faculty of Agriculture, Universiti Islam Sultan Sharif Ali, Brunei, for improving the soil and paddy productivity with some of the private sector paddy farms. There are other formal avenues of diffusing digital farming through method demonstration, community outreach, trainings, extension service and exhibitions.

Climate Resilient Practices in Agriculture and Fisheries

Brunei is prone to flooding, forest fires, strong wind and landslides. Since 1970, its mean temperature increases 0.25 °C every ten years (Observed change). The climate change projection suggests that in 30 years, temperature of the country will increase at 0.4 °C every ten years. The rainfall data of Brunei Darussalam suggest the rainfall intensity in terms of the total amount of rainfall is increasing at 100 mm every ten years. The 2021–2051 rainfall projection suggests that there will be increase in rainfall at 5 mm every year (climate change projection). Moreover, the available data suggest that 30 to 50 years to come, the country may experience sea level rise.

One of the approaches of Brunei Darussalam's climate resilience practices for agriculture is that through the private sector initiative waste management plants are capable of producing organic amendments between one and three days for use in crop cultivation. One of the waste management plants in the country is able to produce approximately three tonnes of organic amendments in three days from agro-industrial wastes. The other organic amendments such as composts produced from other agro-wastes in the country are used in vegetables and paddy cultivation. At five tonnes per hectare of the organic amendment application, some paddy farmers have reported six tonnes paddy grain yield per hectare compared with the existing three tonnes to four tonnes grain yield per hectare using chemical fertilizers only. The added advantage of conditioning soils particularly acid soils with organic amendments improves soil and crop productivity and at the same time reduces the use of chemical fertilizers which have been implicated in soil and water pollution besides causing the greenhouse effect. Moreover, this climate smart type of farming is significantly improving soil pH, organic matter and other soil properties for climate change resilience or adaption. Furthermore, relatively high yielding and hardy paddy varieties such as Sembada188 came into being as a result of the research collaboration between the Department of Agriculture and Agrifood, Brunei Darussalam and Syarikat PT. Biogene Plantation, Indonesia, is used by several paddy farmers. Apart from its early maturity (110 days), Sembada188 yields six tonnes of paddy grains per hectare unlike Laila (Local variety) which yields three tonnes per hectare. With the state of the technologies in water management and pests and diseases control in the paddy growing areas, paddy yield has increased from 1,904 tonnes in 2013 to 3,563 tonnes in 2022 although the production of 4,115 tonnes occurred in 2021 (Department of Agriculture and Agrifood, 2022). There has also been advancement in smart greenhouse farms including soilless and pesticides and weedicides free farming such as aquaponics, hydroponics, fertigation and vertical farming whose wide range of products including vegetables are sold to the public directly or through supermarkets, local markets, restaurants, among others.

To ensure that during production, harvesting and postharvest handling of agricultural produce are of high standard, Brunei Darussalam ensures that farms in the country adhere to the national and regional good agricultural practices standards, an example being the "Good agricultural practices for production of fresh fruit and vegetables in the ASEAN region". This Good Agricultural Practice (GAP) is made up four categories, namely food (ASEAN Secretariat, 2006) safety, environmental management, worker health, safety and welfare, and produce quality. In addition, the Authority for Info-communications Technology Industry (AITI) of Brunei through its Smart Farm Testbed is currently championing the cause of smart farming including IoT and precision farming capable of increasing crop productivity by at least 10%; reducing chemical fertilizers use by 20%; and reducing labour particularly foreign workers by 30% using IoT and Data Analytics. Among the other climate resilience and adaption features is real-time information such as soil properties, crop physiological and morphological, water management, temperature, humidity, precipitation, carbon dioxide, among others to farmers. This approach is already in place at the Agricultural Development Area, Brunei Darussalam, to service the broad spectrum of agriculture and related area with good results.

Conclusion

Long dominated by the expansion in petroleum mining, Brunei had experienced little growth in other economic activities, including agriculture. The returns from agriculture had stagnated so much that foreign workers dominated the limited agriculture that took place in the country until 2017. However, volatile fluctuations in oil prices, chronic dependence on food imports and the government's determination to smartify agriculture to strengthen food security and reduce import dependence have led to the introduction of policies to build sustainable development of agriculture. Innovation and R&D-driven activities are increasingly promoted in the country.

Although the contribution of the agricultural and fisheries sectors to the economic growth of Brunei Darussalam is relatively low (1%), Brunei is poised to mitigate this through several policy interventions especially since 2018, which had made it possible to achieve self-sufficiency in broiler and eggs production with the possibility of achieving self-sufficiency in day old chicks, fisheries, ornamentals, fertilized eggs and fruit production in the near future. Medium- to long-term self-sufficiency in cattle, goat, sheep, buffalo, vegetables, paddy and so on are also possible. Therefore, Brunei Darussalam has potential agricultural and fisheries outputs, which if realized through the adoption of smart technologies shall raise labour productivity through human capital upgrading, investment, resource optimization and reduction in postharvest losses, and agrifood processing is expected to significantly contribute to sustainable economic growth and development of this country.

References

Ali, M., Ahmed, O. H., Jalloh, M. B., Walter, C. P., Musah, A. H., & Ng, J. F. (2023). Co-composted chicken litter biochar increases soil nutrients availability and yield of Oryza sativa L. *Land, 12*, 1–20.

ASEAN Briefing. (2022). https://www.aseanbriefing.com/news/moving-away-from-oil-how-can-brunei-diversify-its-economy/. Dezan Shira & Associates.

ASEAN Briefing. (2021). https://www.aseanbriefing.com/news/brunei-launches-economic-blueprint-salient-features/. Dezan Shira & Associates.

ASEAN Secretariat. (2021). Regional study on labour productivity in ASEAN (p. 151). The ASEAN Secretariat. ISBN 978-623-6945-35-3 (EPUB)

ASEAN Secretariat. (2006). *Good agricultural practices for production of fresh fruit and vegetables in the ASEAN region* (p. 26). The ASEAN Secretariat.

Department of Agriculture and Agrifood. (2022). *Agriculture & Agrifood Brunei Darussalam Statistics in Brief, 2022* (p. 89). Statistics Management and SIPA Section, Department of Agriculture and Agrifood, Ministry of Primary Resources and Tourism.

Department of Agriculture and Agrifood. (2021). *Agriculture & Agrifood Brunei Darussalam Statistics in Brief, 2021* (p. 89). Statistics Management and SIPA Section, Department of Agriculture and Agrifood, Ministry of Primary Resources and Tourism.

Department of Agriculture and Agrifood. (2020). *Agriculture & Agrifood Brunei Darussalam Statistics in Brief, 2022* (p. 89). Statistics Management and SIPA Section, Department of Agriculture and Agrifood, Ministry of Primary Resources and Tourism.

Department of Agriculture and Agrifood. (2019). *Agriculture & Agrifood Brunei Darussalam Statistics in Brief, 2022* (p. 89). Statistics Management and SIPA Section, Department of Agriculture and Agrifood, Ministry of Primary Resources and Tourism.

Department of Agriculture and Agrifood. (2018). *Agriculture & Agrifood Brunei Darussalam Statistics in Brief, 2022* (p. 92). Information Management and SIPA Section, Department of Agriculture and Agrifood, Ministry of Primary Resources and Tourism.

Department of Fisheries. (2021). *Brunei Darussalam Fisheries Statistics in Brief 2021* (p. 49). Department of Fisheries, Ministry of Primary Resources and Tourism.

Department of Statistics. (2023). *Brunei Darussalam Key Indicators 2022* (p. 49). Begawan, Department of Economic Planning and Statistics, Ministry of Finance and Economy. ISBN 978-99917-72-45-5.

Dorairaj, D., Ismail, M. R., Sinniah, U. R., & Kar Ban, T. (2017). Influence of silicon on growth, yield, and lodging resistance of MR219, a lowland rice of Malaysia. *Journal of Plant Nutrition, 40*(8), 1111–1124.

FAO. (2014). *Building a common vision for sustainable food and agriculture: Principles and approaches* (p. 56). Food and Agriculture Organization. ISBN 978-92-5-108471-7.

FAO. (2018). *Sustainable Food Systems: Concepts and frame* (p. 9). Food and Agriculture Organization.

Gong, D., Zhang, X., Yao, J., Dai, G., Yu, G., Zhu, Q., Gao, Q., & Zheng, W. (2021). Synergistic effects of bast fiber seedling film and nano-silicon fertilizer to increase the lodging resistance and yield of rice. *The Scientific Reports, 11*, 12788. https://doi.org/10.1038/s41598-021-92342-5

Hu, Y., Javed, H. H., Asghar, M. A., Peng, X., Brestic, M., Skalický, M., Ghafoor, A. Z., Cheema, H. N., Zhang, F. F., & Wu, Y. C. (2022). Enhancement of lodging resistance and lignin content by application of organic carbon and silicon fertilization in *Brassica napus* L. *Frontiers in Plant Science, 13*, 217.

Ludher, E., & Teng, P. (2024). Agricultural development and food security in Singapore. In R. Rasiah, A. Venus, N.I. Annizah, *Sustainable Agriculture amd Food Security in Southeast Asia: Towards Climate Resilience and Digitalization,* Singapore: Springer nature.

Microtrends Net (2023). https://www.macrotrends.net/countries/BRN/brunei/economic-growth-rate.

Ministry of Finance and Economy, Brunei Darussalam. (2020). *Brunei Darussalam Economic Blue Print, 2020*

(p. 28). Ministry of Finance and Economy. ISBN 978-99917-72-30-1

Udemezue, J. C., & Osegbue, E. G. (2018). Theories and models of agricultural development. *Annals of Reviews and Research, 1*(5), 00134–00136.

Rasiah, R., Kari, F., Sadoi, Y., & Mintz-Habib, N. (2018). Climate change and sustainable development issues: Arguments and policy initiatives. *Journal of the Asia Pacific Economy, 23*(2), 187–194.

Rasiah, R. (2025). Building Agricultural Sustainability and Food Security in Southeast Asia. In R. Rasiah, A. Venus, N.I. Annizah, (Eds.), *Sustainable Agriculture: Towards Climate Resilience and Digitalization and Food security,* Singapore: Springer Nature.

Wawasan Brunei 2035. (2019). *Sekretariat Tetap Wawasan Brunei 2035*. Jabatan Perdana Menteri.

4 Agriculture, Food Security and Sustainability in Malaysia

Nur Annizah Ishak and Rajah Rasiah

Abstract

Food security has become a major national and global issue in recent years. According to the United Nations, nearly 9% of the global population shall face hunger, and an estimated 840 million will experience food insecurity by 2030. The relationship between agriculture, food security and poverty are profound. For many communities, especially in less developed countries and rural communities, agriculture is not just an economic sector, not just a means of income but it is a way of life. Agriculture remains the main source of livelihood for the majority of the population, making it a key player in the ongoing fight against poverty. In fact, food security and poverty are interrelated issues that impact the well-being and quality of life of individuals and communities around the world. This chapter tries to explore issues related to agriculture, food security and poverty especially in Malaysia. Although Malaysia has made significant progress in poverty alleviation since 1971, it has since 1990 faced food insecurity problems. Despite the implementation of various poverty reduction programmes and social safety nets to support the vulnerable communities, including promoting structural change into manufacturing and service to create job opportunities, and a more stable environment to enhance peoples' well-being, it is still grappling with low wage and food insecurity problems. The momentum achieved in both absolute and relative poverty reduction in the 1970s needs revisiting to address the country's main social problems. Its focus on climate smart agriculture targeted at raising productivity growth and building climate resilience in sync with the United Nations sustainable development goals (SDGs) is impressive but requires effective appraisal to ensure its success.

Keywords

Agriculture · Food security · Poverty · Malaysia · Digitalization · Climate resilience

We wish to thank The Department of Statistics Department of Malaysia and its Chief Statistician, Dato' Sri Dr Uzir Mahidin in particular for supplying us with recent data.

N. A. Ishak (✉)
Faculty of Business and Economics, Universiti Malaya, Kuala Lumpur, Malaysia
e-mail: annizah@um.edu.my

R. Rasiah
Asia-Europe Institute, Universiti Malaya, Kuala Lumpur, Malaysia
e-mail: rajah@um.edu.my

R. Rasiah et al. (eds.), *Sustainable Agriculture and Food Security in Southeast Asia*, Sustainable Development Goals Series, https://doi.org/10.1007/978-981-95-0477-0_4

Introduction

Malaysia, situated in Southeast Asia, is a country with a multi-diverse culture and ethnicity. Its modern capital city is Kuala Lumpur, while its administrative capital is Putrajaya. With a land mass of 330,241 square kilometres and a population of 34.3 million, the country enjoyed the third highest Gross Domestic Product (GDP) per capita in 2023, after Singapore and Brunei. Figure 4.1 shows the states and federal territories that make up Malaysia. While Malays form the majority of Malaysia's population, Chinese, Indians and others constitute other ethnic groups. The indigenous people in Sabah and Sarawak represent the majority of indigenous peoples in those states, which include Kadazan-Dusun, Bajau and Murut in Sabah as well as Iban, Bidayuh and Melanau in Sarawak. The Government of Malaysia is led by a prime minister and a constitutional monarchy, which employs a parliamentary system. It has three branches of government—the Executive, the Legislature and the Judiciary (Ministry of Economy, 2022). The major economic activities are manufacturing, especially export-oriented palm oil cultivation and processing, and electronics. Whereas the agriculture sector has shown a trend contraction in its value-added share in GDP, services and manufacturing has become the largest sector with manufacturing's contribution falling in trend terms since 2000 (Rasiah, 2011). The agriculture sector plays a pivotal role in socio-economic development for Malaysia. Since early years, this sector contributes significantly in Gross Domestic Product (GDP), providing employment opportunity, also ensuring food security for its people. Figure 1 depicts Malaysia's map with its fourteen states.

The agriculture sector has not only played a crucial role in global food production and sustainability, but also has a vital impact on the livelihoods of people through activities, such as cultivating crops, raising livestock and fishing. The sector contributes to poverty alleviation by generating income, particularly for those living in poor rural areas. For many farmers, agriculture is essential for their sustenance. The profound relationship between agriculture, food security and poverty will be explored further in this chapter with an emphasis on Malaysia's specific challenges. Additionally, the importance of agricultural productivity in poverty reduction and food security cannot be underestimated. It is widely recognized that agricultural growth, specifically increasing productivity, plays a pivotal role in combating rural poverty and addressing food insecurity (Okello et al., 2019).

As explained in the introductory chapter of the book (see Rasiah, 2024), this chapter seeks to discuss agricultural policies and agricultural expansion in Malaysia since independence with a greater focus on the period after 2000, in addition the status of agriculture as the country faced structural change, and its implications for agricultural productivity and employment, as well as the country's experience with food security, and digitalization and climate resilience policies. Given the close association of agriculture, especially small-scale agriculture with poverty, emphasis is given to the role of parastatals that played a key role in raising agricultural labour productivity in the 1970s, and the alleviation of rural poverty. As the other sectors grew rapidly, the trend fall in agricultures' contribution to GDP is also documented. Also discussed is the relative weakening in food security. In addition to a review of the national agricultural policies (NAPs) since the 1980s, the chapter also addresses government focus on stimulating the strengthening of climate resilience since 2000, and the diffusion of digitalization and Industrial Revolution (IR) 4.0 technologies since 2010.

Overview of Agriculture Sector

Malaysia's rich agricultural history has significantly influenced its economic development and cultural heritage. However, rapid industrialization and urbanization have shifted the country's focus to the industrial and service sectors. Despite these shifts, the agricultural sector remains a significant contributor to Malaysia's economy. In 2022,

Fig. 4.1 Map of Malaysia. *Note*: In addition to the 13 states, the federal territories include Kuala Lumpur, Putrajaya and Labuan. *Source*: Volina, (n.d), https://www.vectorstock.com/royalty-free-vector/map-of-malaysia-vector-1607161

the sector provided employment for 1.86 million people, which fell slightly from 1.87 million in 2021. In 2021, agriculture sector contributed about 7.2%, which is equivalent to RM98.9 billion of national GDP, reduced from 101.6 billion in 2019 (Ministry of Economy, 2022). Although the manufacturing and services sectors have played a remarkable role in transforming Malaysia towards an industrial society, with 24% and 57% national GDP, respectively, in 2021. The contribution of the agricultural sector is still very significant particularly during COVID-19 pandemic for its food sustainability.

The pandemic had devastated the sector due to mobility restrictions imposed by lockdowns. The Movement Control Order (MCO) implemented during COVID-19 pandemic hampered agriculture business and food industries disrupting agriculture production and distribution (Laing, 2020). In Malaysia, the government announced its first MCO on 18 March 2020 until 12 May 2020 as the outbreak started spread faster. The drastic lockdown measure eventually disrupted the food production and supply chains. The movement restrictions lead to limited supply of raw materials and inputs, labour shortages, restricted market access, unpredictable market conditions with rising food prices. Although the primary food production industry, including agriculture and fisheries activities, was considered as an essential sector, and hence, was permitted to continue operations, the disruption of transportation owing to these restrictions hindered timely delivery of agricultural products leading to post-harvest losses and food waste (Tan et al., 2023).

In Malaysia, palm oil and food production are the major types of agriculture. There are about 5.7 million hectares of land cultivated for palm oil compared to 1 million for food crops. The country is doing well on palm oil plantation and rank as the second largest producer with 91,393,666 million tonnes of production in 2021, just behind Indonesia (Ministry of Economy, 2022). But in terms of food production crops like paddy, cocoa, pineapple and pepper, Malaysia is far left behind from its neighbouring countries. There are many reasons for it, among them are it is harder to maintain adequate supply food crops as they have shorter life span compared to 25 years for oil palm plantation. Also growing food crops required ample manpower and facing higher labour turnover rates, other than higher cost due to fertilizers and pesticides. It resulted in decreasing of some important commodities and cash crops between 2019 and 2021. Besides, Mohamed Arshad et al. (2011) mentioned that

Table 4.1 Planted area of main crops in Malaysia, 2018–2021

Crops ('000 hectares)	2018	2019	2020	2021
Oil palm	5,849.3	5,900.2	5,865.3	5,737.7
Cocoa	15.6	5.9	5.7	6.0
Paddy	700	672.1	644.9	647.9
Coconut	83.4	86.5	84.9	82.6
Mustard	9.4	11.2	10.9	11.4

Source: Department of Statistics Malaysia, (2021a)

to boost agricultural value-added and alleviate poverty among farmers, the government encouraged farmers to target higher value-added crops, especially for export. Palm oil, cocoa, pineapple and pepper are among the main export crops covering roughly 75% of cultivated land. This strategy threatened the national sustainable rice production system in maintaining the domestic rice self-sufficiency level at 65 to 70%. Table 4.1 shows the decline in commodity food acreage under production in Malaysia from 2018 to 2021.

Agricultural Policies

In the early years of post-independence (1957), the focus was on achieving food self-sufficiency and addressing rural development challenges. The 1970s saw the introduction of the Green Revolution, promoting high-yielding crop varieties and modern farming techniques. In the 1980s, the Look East Policy emphasized learning from successful agricultural models, particularly from countries like Japan. The 1990s witnessed a shift towards agro-industrialization, encouraging value-added processing and diversification. The new millennium brought about a focus on sustainable agriculture, with an emphasis on environmental conservation and organic farming.

Rural development was foremost in the New Economic Policy (NEP) that was launched in 1971, embracing modernization (Rasiah & Salih, 2019). While such an incisive and performance-based set of strategies were gradually abandoned in the 1980s, modernization and large-scale agro-business became important from 1984 (Table 4.2). However, the broad agricultural policies covered short-term planning (1–2 years), medium-term (3–5 years) and long-term policy initiatives (more than 5 years) (see Malaysia, 1971, 1990, 2000). Short-term policies provide swift responses to immediate challenges, while medium-term plans aim for sustained growth and long-term strategies envision enduring transformations. These policies were viewed as comprehensive to reflect Malaysia's commitment to enhancing competitiveness, ensuring food security and nurturing sustainable agro-based industries. The multi-duration framework addresses not only current concerns but the agricultural sector's long-term success.

The establishment of state agencies was to plan and implement growth in this sector. Significant investments and institutional reforms were implemented, particularly in oil palm cultivation. Subsequent policies like NAP1 (1984–1991), NAP2 (1992–1997) and NAP3 (1998–2010) focused on both commodity and food crops. The separation of food-based agriculture from commodity agriculture in 2004 led to more targeted policies. The National Agro-Food Policy 2.0 (2021–2030) replaced the Agro-Food Policy (2011–2020) with a focus on enhancing productive capacity for food security in the agro-food sector (Table 4.2).

As the world increasingly focused more on digitalization and digitization, and climate resilience since the turn of millennium, governments began to chart policies to promote Industrial Revolution (IR) 4.0 technologies and building climate resilience. Malaysia's emphasis on these started with the 10th Malaysia Plan (2011–2015), 11th Malaysian Plan (2016–2020) and 12th Malaysia Plan (2021–2025). The 10th and

Table 4.2 Agricultural policies in Malaysia, 1984–2030

Agriculture Policy	Policy Objectives	Strategies
First National Agricultural Policy (NAP1) (1984–1991)	Modernize and revitalize the agriculture sector Maximize income through optimal utilization of resources Increase food production for local market	Invest in institutional building New land developments for oil palm and cocoa In situ development to resolve uneconomic farm size and low productivity among small holders
Second National Agricultural Policy (NAP2) (1992–1997)	Strengthen agriculture sector to contribute substantially to economic growth Enhance productivity, efficiency and competitiveness Increase land areas for oil palm Develop agro-based industries Accelerate the transformation of the sector into a dynamic and commercial sector	Promote resource use optimization to diversify export crop cultivation Develop agro-based industries to generate off-farm demand for smallholders Enhance R&D to strengthen production, labour upgrading to remove other constraints Greater participation of the private sector in agriculture Deepen human resource development
Third National Agricultural Policy (NAP3) (1998–2010)	Enhance food security Increase productivity and competitiveness of the sector Deepen inter-sectoral linkages Create new sources of growth Adopt sustainable development, utilization and management of natural resources	Adoption cluster-based agro-industrial development Tackle resource constraints problems and promote sustainable agriculture Create a large production base for agriculture and forestry
National Agro-Food Policy (NAFP) (2011–2020) National Commodity Policy (2011–2020)	Provide food security and safety Make agro-food a competitive and sustainable industry to raise farmer incomes of agro-based entrepreneurs Increase the contribution of plantation industrial commodities to the nation's economy Modernize and transform the commodity industry raise competitiveness and sustainability	Upgrade agricultural infrastructure Improve quality and safety of food by enforcing standards compliance Expand the agricultural value chain Stimulate income increase of smallholders in the commodity industry Promote Malaysia as the centre of excellence in R&D, technology development and the downstream processing of industrial commodities
New National Agro-Food Policy (NAFP 2.0) (2021–2030)	Drive income growth to facilitate better quality of life for food producers Raise output and its quality by increasing productivity Establish agile and resilient value chains with high value-added activities Improve food safety and nutritional well-being of Malaysians Embrace greater economic, social and spatial inclusiveness Encourage greater adoption of sustainable consumption and production	Embrace modernization and climate smart agriculture Prioritize market, and demand-driven and export-oriented products Develop talent that is demand driven Promote sustainable agricultural practices and food systems Create robust business ecosystem

Source: Compiled by authors

11th Malaysia Plans also marked the implementation of the National Agro-Food Policy (2011–2020), which aimed to propel the agricultural industry into a modern, profitable and sustainable era. To achieve this vision, a multifaceted approach was adopted, incorporating the infusion of climate smart technologies (CSA) into farming practices and a shift towards market-driven strategies. The policy also prioritized support for smallholders and implemented initiatives for long-term sustainability, creating a comprehensive strategy geared towards enhancing productivity, income and overall sustainability in Malaysia's agriculture sector. Table 4.2 provides a comprehensive review of Malaysia's national agriculture policies from 1984 until 2030, outlining the specific objectives and strategies implemented. It is imperative to highlight that climate change will be an important factor in future agricultural activities. To optimize the outputs, local problem-solving technologies are being developed simultaneously. The priorities are to provide financial platforms for CSA activities to be extended to vulnerable people such as isolated farmers, non-granary rice farmers, small-scale agro-businesses and farmers' organizations. As a frontier R&D organization, in-house innovation-driven technologies and other latest the new technologies, MARDI has embarked on knowledge sharing and national consultation to improve institutional arrangements, especially government policies. There are also significant engagement of private entrepreneurs fabricating robots and drones, such as Aerodyne and DreamEdge to support CSA activities (see also Mohammad Hariz et al., 2023).

Contribution to Gross Domestic Product and Exports

The agricultural sector in Malaysia has seen significant transformation, driven by the importance of the sector for country's economic growth and development. Figure 4.2 shows that the agricultural sector's share of Gross Domestic Product (GDP), agricultural value-added growth and labour productivity from 1960, 1970 and 1991, respectively, till 2021 has declined in trend terms, though there is a slight rise in 2021. Value added contracted in absolute terms in 1975, 1983, 1990, 1993–1995, 1998, 2016 and 2020. Agro-food exports only constituted 3.2% of total Malaysian exports in 2010–2020. Meanwhile, agricultural labour productivity over the period 1991–2019 was extremely low not exceeding 0.2%, and despite the special focus attributed to agriculture since 1984 value-added and labour productivity growth in the sector has significantly lagged behind GDP growth.

While the share of agriculture in GDP typically falls before stabilizing with GDP growth, Malaysia's agricultural sector has begun to decline in significant before reaching maturity, which is obvious from the low labour productivity growth over the period 1991–2019 (Fig. 4.2). Since the government has prioritized the sector's growth with massive expenditure to drive IR4.0 proliferation and climate smart policies, the focus now should be on correcting policy failures and ensuring that government support is accompanied by economic performance.

While the palm oil sector has shown significant improvements, food exports have also shown a relative fall. Table 4.3 shows nominal exports of selected food items. The cumulative average nominal growth rate (CAGR) of approximately 7.3% over the period 2010–2020 is relatively low (see also Chamhuri et al., 2022). Cereal and cereal preparation, meat and meat preparation, and processed milk exports grew nominally by 12.2%, 12.0% and 11.5%, respectively, per annum in the period from 2010 till 2020 (Table 4.3).

The trend decline in GDP share of agriculture in the ASEAN region has been accompanied by a significant reduction in the agricultural share of employment over the years. The number of employed persons in 2022 amounted to 15,391.7 thousand persons. Out of total, 1,540.8 thousand persons (10.0%) were involved in the agriculture sector. Instead of pursuing massive mechanization and automation strategies, Malaysian agriculture has remained labour intensive with strong reliance on low-skilled foreign labour. Indeed, employed persons in the agriculture had 67.5% Malaysian labour and about 32.5% foreign labour. The

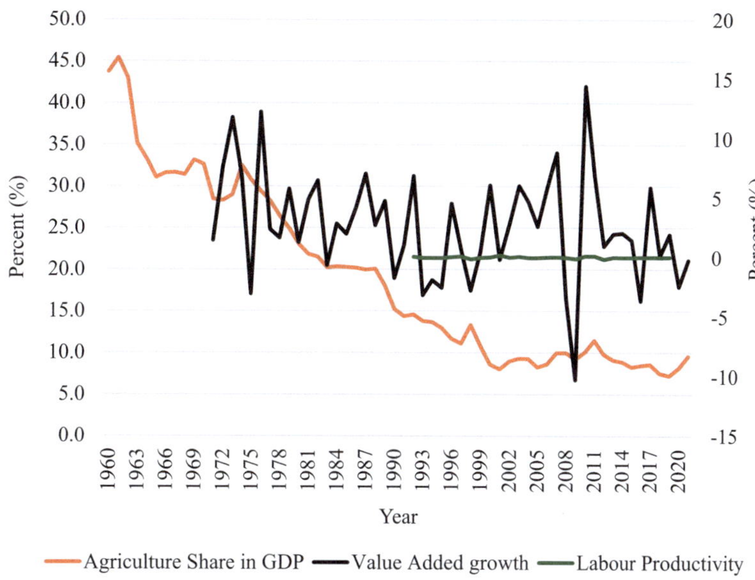

Fig. 4.2 Malaysian agriculture: Value added and labour productivity growth and share in GDP, 1960–2021. *Source*: Calculated from World Bank (2023) data

Table 4.3 Export of agricultural commodities, Malaysia, 2010–2020

Industry	RM Million			Cumulative Average Growth Rate (CAGR %)		Growth
	2010	2015	2020	2010–2015	2015–2020	2010–2020
Agro-food sector	18,096	27,310	36,479	8.58	5.96	7.26
Live animal (food)	551	705	863	5.05	4.13	4.59
Meat and meat preparation	263	569	813	16.69	7.40	11.95
Processed milk	550	1,419	1,631	20.87	2.82	11.48
Eggs	334	493	605	8.10	4.28	6.12
Fish, crustacean, Molluscs	2,586	2,614	3,070	0.22	3.27	1.73
Cereal and cereal preparation	1,405	2,982	4,459	16.24	8.38	12.24
Vegetables	682	1,097	1,676	9.97	8.85	9.41
Fruits	589	964	1,439	10.36	8.34	9.34
Sugar, sugar preparation and honey	873	934	970	1.36	0.76	1.06
Coffee, cocoa, tea and spices	5,323	7,403	9,497	6.82	5.11	5.96
Animal feed	974	1,448	2,086	8.25	7.57	7.91
Other food preparation	3,960	6,678	9,365	11.02	7.00	8.99
% of agro-food/ total export	2.83	3.51	3.87	4.40	1.97	3.18
Total of Malaysia Export	638,822	777,335	943,761	4.00	3.96	3.18

Source: Ministry of Agriculture and Food Industry, Malaysia, (2020)

high dependency on foreign labour for agriculture sector was evident, where employed foreign labour rose from 492.4 thousand in 2018 to 500.2 thousand people in 2022 (DOSM, 2023). Also, the COVID-19 pandemic had hampered this sector because a significant number of migrant workers in particularly agri-food industries returned to their home countries due to travel restrictions and health concerns, which led to a shortage of labour in the country.

The relationship between agriculture and poverty is profound, particularly in developing nations. Agriculture serves as a primary source of income and employment especially in rural areas. Also, it serves as effective tool for poverty reduction and economic development especially for Malaysia. This is evident as the majority of the poor in the country are associated with the agricultural sector. The median monthly salaries and wages received for skilled workers in agriculture, forestry, livestock and fisheries were RM1,764 with male workers at RM1,779 and female workers at RM1,531. Meanwhile, the mean monthly salaries and wages received were RM2,011 with male workers at RM2,033 and female workers at RM1,831 (DOSM, 2023). The agricultural sector known for its poor society and vital for poverty reduction assistance (Faisal & Rashid, 2023). The absolute poverty incidence in Malaysia increased from 5.6% in 2019 to 8.4% in 2020, due to the increasing rate of unemployment. Efforts to eradicate poverty in agriculture often involve policies aimed at improving productivity, ensuring resource access, providing financial support and promoting sustainable practices. By providing employment opportunities and income generation particularly in rural areas, the government's aim is to reduce poverty among the agricultural communities.

Besides, the poverty line income (PLI) was reviewed twice in 2005 and 2019. It aims to ensure meaningful measurement, taking account of changes in living standards and coverage of all inhabitants. Malaysia's incidence of absolute poverty experienced a downward trend from 49.3% (1970) to 0.4% 2016). In 2016, the incidence of absolute poverty was at the lowest, but a review of PLI saw the incidence of absolute poverty in the country updated to 7.6% in 2016 based on the PLI 2019 methodology (which changed the poverty line income from RM980 per household to RM2,028 per household) (Malaysia, 2021). Also, absolute poverty went down to 5.6% (in 2019) before increasing to 8.4% (2020). In 2022, the reopening and recovery of the economic sector has led to a decrease in incidence of absolute poverty to 6.2%. Figure 4.3 illustrates these details.

Role of Parastatals

The development of the agricultural industry in Malaysia, particularly through land resettlement schemes and support for key industries like palm oil, rubber and fisheries, has played a pivotal role in addressing poverty. Parastatals or government intermediary organizations have played a crucial role by providing essential infrastructure and institutional support to rural poor particularly in agriculture sector. The Federal Land Development Authority (FELDA) was the first parastatal established on 1 July 1956, aimed to reduce rural poverty by providing land and economic opportunities to landless families. It focused on developing agricultural land for crops like rubber and later expanded to include oil palm. FELDA's establishment marked a significant step in transforming Malaysia's agrarian landscape and improving socio-economic conditions (see Rasiah, 2018). The Sungai Siput project (1956–1961) was a pioneer effort to resettle low-income families on new agricultural land. Rapid expansion in the 1960s and 1970s saw FELDA's crucial role in cultivating cash crops, contributing significantly to Malaysia's economic development (www.felda.gov.my).

FELDA's land settlement programmes were not only focused on agricultural advancement but also had wider socio-economic goals. By relocating families to newly developed farmland, FELDA aimed to eliminate rural poverty and improve the overall welfare of its beneficiaries. The positive outcome of the Sungai Siput pilot

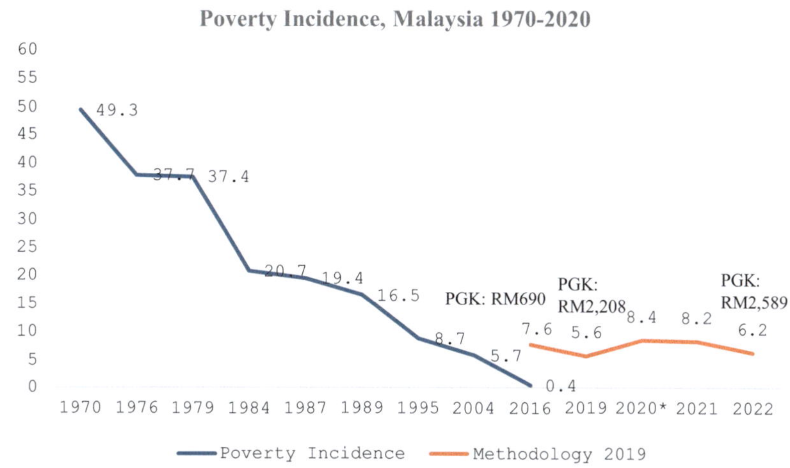

Fig. 4.3 Poverty incidence, 1970–2022. *Source* Department of Statistics Malaysia (2023a)

project set a precedent for similar schemes, establishing a model for sustainable rural development. Over time, FELDA expanded its role beyond agriculture, encompassing sectors like education, healthcare and infrastructure development within its settlements. This comprehensive approach by FELDA contributed to the establishment of thriving rural communities and played a pivotal role in Malaysia's holistic progress (O'Donnell et al., 2017). FELDA's achievements have positioned it as a significant institution in Malaysia's history, reflecting the government's dedication to rural development and poverty eradication. The organization continues to be an influential force in shaping the country's agricultural landscape and promoting socio-economic advancement in rural areas.

The Federal Land Consolidation and Rehabilitation Authority (FELCRA) is a wholly-owned corporate entity under the Ministry of Finance (Incorporated) and operates under the Ministry of Rural Development's supervision. With a focus on specific target groups, including community participants, FELCRA aims to enhance their quality of life through strategic initiatives (www.felcra.com.my). The organization is dedicated to adopting cutting-edge technologies for efficient management, incorporating advancements in agricultural practices to boost productivity. FELCRA's multifaceted functions demonstrate a commitment not only to economic development but also to improving the overall well-being of the communities it serves. This underscores a comprehensive approach to rural development and agricultural management in Malaysia, with the goal of reducing poverty and improving the agricultural sector.

The Padiberas Nasional Berhad (BERNAS) and Rubber Industry Smallholders Development Authority (RISDA) are vital components of Malaysia's agricultural sector. BERNAS, established in 1971, serves as the central agency for rice procurement, processing and distribution to ensure national food security through managing the rice supply chain. BERNAS plays a key role in stabilizing rice prices and making this staple food available nationwide. Meanwhile, founded in 1973, RISDA focuses on assisting smallholder rubber farmers by promoting and developing the rubber industry. It provides technical assistance, financial aid and market access to these smallholders. It contributing to both economic growth of the rubber industry and social well-being of communities engaged in rubber cultivation. BERNAS and RISDA's initiatives in the agricultural sector are crucial for poverty alleviation in Malaysia.

Political patronage particularly in agriculture sector is a major concern as the parastatals have become entrenched in partisan politics (Rasiah & Salih, 2019). They not only provide lucrative income sources and patronage resources to local officials but play a central role in drawing political support from the masses. These parastatals significantly influence resource distribution, economic benefits and government appointments. The leadership in these agencies often centre on political connections, shaping the policies and activities of these agencies. Thus, these parastatals further reinforce the economic disparities and perpetuate a culture of favouritism towards certain groups while marginalizing others. For instance, agricultural projects and subsidies are often allocated based on political connections rather than merit, leading to inefficiencies and unequal access to resources for farmers. As a result, small farmers and those without political connections are at a disadvantage, struggling to compete in the market and improve their livelihoods. Moreover, the prevalence of political patronage in Malaysia's agricultural sector has hindered overall development and modernization (Johanson, 2015).

Moreover, state-owned enterprises have often also exhibited negative effects, including inefficiency and bureaucracy due to the absence of competitive pressures (Gomez & Jomo, 1998). Political interference influences to ineffective decision-making and a lack of innovation. Parastatals become a financial burden on the state, requiring subsidies or bailouts, and their dominance can crowd out private investment. On the other hand, corruption and nepotism may thrive these parastatals, impacting resource allocation and fairness. The lack of accountability and transparency is problematic, contributing to market distortions and hindering social equality. Undeniably, parastatals played a crucial role in economic development until the 1980s (Rasiah & Salih, 2019), but they now require serious reforms to ensure that they become efficient again (see also Rasiah et al., 2024).

Transitioning into the 2020s, the policy landscape continued to evolve with the advent of the National Agro-Food Policy 2021–2030 (NAP 2.0). The policy outlines a strategic plan for the transformation and advancement of the agro-food sector over the next decade, with a focus on ensuring national food security. The policy aims to address key challenges and take advantage of opportunities in agriculture through measures such as establishing a centralized information centre to support investors. This updated framework placed a heightened emphasis on innovation, digitalization and climate-resilient practices, reflecting a commitment to staying abreast of contemporary challenges and opportunities. Despite these ambitions, the policy recognizes obstacles like limited youth participation in the industry, inadequate financial assistance for farmers and risks from natural disasters, diseases and unsustainable farming methods. NAP 2.0 is designed to confront these challenges and pave the way for a resilient, modernized agro-food industry that aligns with broader national development objectives (Makhtar et al., 2022). Table 4.2 above offers a detailed explanation of NAP 2.0 policy, which provides a comprehensive view of the policy's focal points, elucidating the key goals and targets set forth to guide and enhance agricultural initiatives.

Food Security

Malaysia has experienced a consistent decline in the agricultural sector's contribution to economic indicators, mirroring trends seen in many developing nations undergoing rapid industrialization. This shift is part of the country's economic transformation towards higher-income status. However, despite this decline, Malaysia faces challenges in achieving self-sufficiency for various food items, leading to concerns about food insecurity. Food security, which involves secure access to sufficient food for a healthy life, is a complex issue encompassing factors such as food quantity availability, resource access and proper food utilization. To address these challenges, Malaysia has implemented strategies, such as establishing a self-sufficiency level for rice production and building rice stocks domestically and internationally. Navigating the multifaceted dimensions of food security becomes crucial for ensuring the well-being of the population amid the evolving economic landscape.

The Self-Sufficiency Rate (SSR) is a measure that assesses a country's ability to produce enough of a particular commodity or group of commodities to meet its domestic demand without relying heavily on imports, which is an important indicator in assessing a nation's food security and resilience in the face of external market fluctuations or disruptions in the global supply chain. A higher SSR indicates a greater level of self-sufficiency, implying that the country is producing a significant portion of its needed food domestically. A lower SSR may suggest a higher dependence on imports to meet domestic demand for that particular commodity or group of commodities.

Based on the Supply and Utilization Account Selected Agriculture Commodities, 2018–2023, a total of 23 selected agricultural commodities recorded a SSR of 100% or more (DOSM, 2023c). The SSR for rice stood at 62.6%, while vegetables registered at 44.7%. Fruits demonstrated a robust SSR of 78.1%, whereas beef recorded a lower self-sufficiency rate at 14.7%. Chicken and duck meat showcased a high SSR of 94.4%, and chicken and duck eggs surpassed 100% with an impressive rate of 108.9%. Fresh milk achieved a SSR of 57.3%, and edible fish exhibited a substantial self-sufficiency rate of 90.2%. These figures provide insights into Malaysia's domestic capacity to meet its demand for essential food items and highlight variations in self-sufficiency across different commodities.

In 2022, Malaysia reached full SSR in most fruits production, except for mangoes. The country enjoyed positive net exports in papaya with 142.3% (2021: 146.9%), watermelon with 131.9% (2021: 139.5%), starfruit 120.2% (2021: 116.9%) and jackfruit 114.2% (2021: 110.8%). However, production of major commodity crops continued to decrease in 2022 as compared to 2021. Despite a stable land area for fruit production at approximately 192,000 hectares, Malaysia experienced a decline in production growth, decreasing by 0.31% from 2011 to 2020. The export value of fruits also decreased during the period from 1.62 million tonnes to 1.50 million tonnes between 2016 and 2020. To meet consumer demand, Malaysia heavily relies on imports, accounting for around 78% of its fruit consumption from global markets. The low actual yield is attributed to farmers replanting their crops, particularly the high-demand fruit Durian, incentivized by subsidies and grants provided by Department of Agriculture. This dynamics illustrates the complex interplay of factors influencing Malaysia's fruit production, including land use, export trends and government support initiatives.

Furthermore, Malaysia experienced a significant decline in paddy production from 2.74 million tonnes in 2016 to 2.34 million tonnes in 2.18 tonnes, marking a 7% decrease over the span of five years. This was due to low productivity, with the yield per hectare decreasing from 9.56 tonnes in 2016 to 8.32 tonnes in 2020 (DOSM, 2023). Despite domestic production challenges, paddy production still manages to supply approximately 70% of the local demand. However, Malaysians exhibit a preference for diverse rice varieties, particularly imported ones such as Basmati, Fragrance rice and Japonica. Reflecting this demand, Malaysia imported around 1.1 million tons of rice in 2020, sourcing primarily from countries such as Vietnam, Thailand, Myanmar, India and Pakistan to meet the nation's rice consumption needs.

Besides, there was a significant decline in the planted area for selected crops in Malaysia in 2022, amounting to 7,803.1 thousand hectares, which represents a 0.9% reduction compared to the planted area in 2021. Despite this reduction in planted area, the production of selected crops in Malaysia has shown a positive trend over the years. From 2018 to 2022, the production increased steadily, rising from 739,409 metric tonnes in 2018 to 770,339 metric tonnes in 2020 and further to 841,419 metric tonnes in 2022. For instance, in 2022, a total of seven selected vegetable types in Malaysia has demonstrated sufficient supply to meet domestic needs. Tomato is 118.0% (2021: 118.9%), Cucumber is 110.6% (2021: 111.7%), Spinach is 109.6% (2021: 112.0%), Lettuce is 107.3% (2021: 107.5%), Long Beans is 106.6% (2021: 109.3%), Brinjal is 106.6% (2021: 109.6%) and Lady's Finger is 103.4% (2021: 104.3%) (DOSM, 2023). This indicates improved efficiency or yield per hectare, contributing to the overall positive trajectory in

Table 4.4 Self-sufficiency rates, key food commodities, Malaysia, 2018–2022

	Percent (%)				
Fruits	2018	2019	2020	2021	2022
Coconut	66.6	68.2	66.6	69.6	71.6
Pineapple	105.5	105.0	104.7	104.3	102.5
Banana	102.3	98.7	100.1	102.6	99.3
Watermelon	168.0	161.3	139.4	139.5	131.9
Mango	23.4	32.1	20.2	16.2	32.0
Rambutan	101.7	101.2	98.1	99.4	96.6
Durian	104.8	105.2	105.2	108.7	108.9
Mangosteen	124.9	105.8	93.1	93.8	90.5
Jackfruit	110.6	110.5	109.9	110.8	114.2
Guava	94.0	100.0	96.9	100.7	98.0
Sweet corn	103.9	106.3	105.6	106.3	108.6
Papaya	149.7	153.1	156.0	146.9	142.3
Starfruit	146.1	132.8	117.4	116.9	120.2
Langsat	104.4	112.8	96.6	109.5	101.2
Vegetables					
Round cabbage	38.7	40.5	37.5	40.3	45.6
Tomato	126.5	131.2	123.7	118.9	118.0
Chilli	31.9	30.8	30.9	29.3	29.7
Cucumber	112.7	110.8	112.1	111.7	110.6
Mustard	97.6	98.9	97.6	97.0	95.9
Brinjal	115.4	119.3	112.3	109.6	106.6
Long bean	107.2	107.0	107.2	109.3	106.6
Lady's finger	105.8	104.4	102.2	104.3	103.4
Spinach	113.5	112.6	112.0	112.0	109.6
Lettuce	114.5	112.5	114.0	107.5	107.3
Other crops					
Cassava	92.9	100.1	98.4	100.2	100.5
Sweet potato	83.1	81.7	75.6	78.1	73.5
Sugarcane	182.1	207.2	157.7	149.4	144.6
Ginger	23.1	16.2	18.9	14.6	15.9
Lime	101.9	96.7	91.1	85.5	75.9
Livestock products					
Chicken/ duck egg	116.4	117.8	114.1	114.4	108.9
Pork	92.2	93.1	94.9	93.4	84.6
Mutton	10.9	11.8	9.4	10.7	8.7
Beef	23.7	23.2	21.3	18.9	14.7
Chicken meat	101.5	101.4	101.7	100.2	93.3
Duck meat	131.1	130.1	125.2	130.6	130.2
Fresh milk	61.3	59.3	64.2	56.7	57.3
Fisheries					
Shrimp	100.5	102.9	101.7	100.4	96.5
Tuna [1]	102.1	101.3	108.1	98.8	100.9

(continued)

Table 4.4 (continued)

	Percent (%)				
Fruits	2018	2019	2020	2021	2022
Mackerel [2]	86.5	84.5	88.2	100.7	78.2
Crab	86.9	94.0	94.3	104.4	97.8
Cuttlefish	104.4	109.3	96.6	80.0	73.8
Tilapia	99.4	98.9	99.9	99.0	97.3
Freshwater Catfish	102.8	101.0	98.9	108.3	102.2
Torpedo Scad	96.7	97.9	96.7	93.8	91.6
Sardine	86.3	85.1	80.7	74.9	72.1
Seabass	96.1	104.9	108.8	114.1	116.5
Threadfin bream	100.5	99.6	100.0	99.7	99.3
Stingray	76.2	84.9	80.6	81.7	79.7
River catfish	100.3	102.1	100.3	100.0	100.0

Source: Department of Statistics Malaysia, Supply & Utilization Account Selected Agricultural Commodities (2023c)

crop output, particularly in Cameron Highlands, Malaysia.

Malaysia faces a challenge in achieving a high Self-Sufficiency Rate (SSR) for livestock, indicating continued dependency on imports (IDR) to meet domestic demand for livestock-related products (see also Afzanizam, 2023). Livestock plays a crucial role as a protein source but production of meat products has decreased significantly. In 2022, few livestock items demonstrated a significant Import Dependency Ratio (IDR) exceeding 50.0%. For examples, Mutton at 91.5% (2018: 89.1%), Beef at 85.6% (2018: 76.7%) and Fresh milk at 63.6% (2018: 41.9%). Most production of livestock products recorded a decline from 2018 to 2022. Mutton and milk production contribute 9.39% and 7% to SSR, respectively. Beef production was declined from 46.9 million tonnes in 2018 to only 35.9 in 2022. Chicken meat production declined about 3% in 2022 from 1587.5 million tonnes in 2018 as shown in Table 4.4.

Meanwhile, poultry production has surpassed local consumption with an SSR of 108.9% in 2022. SSR for duck meat recorded the highest SSR livestock with 130.2% (2021: 130.6%). In 2021, the GDP from fisheries was 0.8% and the fish available for consumption was 34.08 kg/capita. The Department of Fisheries Malaysia under the Ministry of Agriculture and Food Industries is the leader in the transformation of a sustainable and competitive fishery in the country. In the fisheries category, SSR for seabass was 116.5% (compared to 114.1% in 2021), freshwater catfish recorded 102.2% (compared to 108.3% in 2021), tuna showed 100.9% (compared to 98.8% in 2021), and river catfish remained at 100.0% (consistent with 100.0% in 2021) (DOSM, 2023a).

In general, the low SSR reflects potential vulnerabilities in the country's livestock sector, highlighting the need for strategic interventions to boost domestic production. Addressing factors such as production efficiency, infrastructure development and supportive policies becomes crucial in improving Malaysia's SSR for livestock, fostering greater self-sufficiency and mitigating dependence on external sources.

The Malaysian government attempted to overcome a number of these obstacles through policy responses, including offering financial aid to farmers and agribusinesses, streamlining the transportation of agricultural goods and encouraging local food production for self-sustainability (Table 4.5). In addition, the government provided economic stimulus package to the agricultural sector. For instance, RM40 million grants were given to agriculture and food production for small and medium enterprises and RM1 billion was allocated for the Food Security Fund for farmers and fishermen to boost domestic production. Another RM100 million was allocated for the development of agro-storage and distribution infrastructure, as well as crop integration programmes. A Special Fund between RM100,000 and RM200,000 was allocated for each competitive Area Farmers Organization (PPK) and Area Fishermen's Organization (PNK) to develop short-term agriculture projects that can produce output within 3 to 6 months (Ministry of Finance Malaysia, 2023). Overall, the government increased its role in promoting sustainable agricultural practices and providing support for agricultural innovation to mitigate the impact of climate change on the agricultural sector in Malaysia. However, the challenges facing the government are somewhat daunting as it has not significantly reversed import-dependence on several commodities, including mutton, beef, chilli and rice.

Part of the problems affecting good cultivation comes from political patronage, which stands out as a key factor undermining livestock production in Malaysia, particularly in the beef, buffalo and mutton sub-sectors. The mismanagement of Approved Permits and the overpowering influence of conglomerates significantly contribute to challenges in Malaysia's livestock sector. This consolidated control hinders fair competition, limits entry for smaller players and negatively impacts local farmers. Urgent reforms are needed to establish transparent permit allocation, promote fair competition and foster a more inclusive and sustainable livestock industry in Malaysia. Key stakeholders in this sector include Ternakan Kamran Sdn Bhd (the largest goat meat

Table 4.5 National Agro-Food Policy 2021–2030 (NAP 2.0)

	The overall NAP 2.0 policy framework encompasses several elements including a policy statement, policy principles, policy objectives and policy thrusts as it aims to serve as an inclusive and extensive guide for the implementation period from 2021 to 2030				
	Policy Statement: *Asustainable,resilient and technology-driven agro-food sector that prioritizesfood security* *and nutrition while driving economic growth and enhancing the well-being of the rakyat*				
Principles	Economic		Social	Environment	
	Highly Competitive and Innovative Agro-food Sector		Well-being of Food Producers and Inclusivity in Sector Development	Paradigm Shift towards a Sustainable Food System, Adapted to Climate Change	
Policy Objectives	Drive income growth and facilitate better quality of life for food producers				
	Raise production output with quality harvest by increasing productivity				
	Establish more agile and resilient value chains with high value-added activities				
	Improve food safety and nutritional well-being of Malaysians				
	Embrace greater economic, social and spatial inclusiveness				
	Encourage greater adoption of sustainable consumption and production				
Policy Thrusts	Embrace Modernization and Smart Agriculture	Strengthen Domestic Market and Produce Demand Driven and Export-oriented Products	Build Talent that Meets Demand of the Industry	Advance towards Sustainable Agricultural Practices and Food Systems	Create Conducive Business Ecosystem and Robust Institutional Framework
	STRATEGIES	**STRATEGIES**	**STRATEGIES**	**STRATEGIES**	**STRATEGIES**
	STRATEGIES FOR PADDY & RICE, FRUITS & VEGETABLES, LIVESTOCK AND FISHERIES				

Source: Ministry of Agriculture and Food Industries, (2021)

importer in Malaysia), Federal Land Consolidation and Rehabilitation Authority (FELCRA), ESPEK Livestock Sdn Bhd (RISDA), Ihsan Permata Sdn Bhd (JCorp) and Sawit Kinabalu. In 2018, nearly 50% of import permits were distributed among five primary importers, including Lucky Frozen Sdn Bhd and Fatric Sdn Bhd. Notably, some existing import permit holders, like Ternakan Kamran Sdn Bhd, actively participate in local ruminant livestock businesses. Alongside conglomerates and large enterprises, the local ruminant livestock industry sees involvement from small-scale breeders, adding diversity and vitality to the Malaysian livestock market.

Table 4.6 Targeted self-sufficiency rates, Malaysia, 2025–2030 (%)

	2025	2030
Rice	75.0	80.0
Fruits	80.0	83.0
Vegetables	70.0	79.0
Seafood	95.0	98.0
Beef	50.0	50.0
Chicken/Duck Meat	120.0	140.0
Chicken/Duck Eggs	114.0	123.0
Fresh Milk	100.0	100.0

Source: National Agro-Food Policy 2021–2030 (NAFP 2.0), Ministry of Agriculture, Malaysia

Diffusion of Digitalization and Climate Resilience Practices

The smartification of agriculture through the application of the National Agro-Food Policy (NAFP) 2.0 relies on innovations through the proliferation of digitalization, which at the same time seeks to strengthen climate resilience in Malaysia's agricultural sector. It is through the NAFP 2.0 that government is promoting the adoption of modern technologies, such as precision agriculture, digital farming and biotechnology targeted at assisting farmers to raise yields, optimize resource use, reduce production costs and strengthen environmental, social and governance (ESG) in the country (Naguib et al., 2024). Smart agriculture focuses on precision farming that relies on sensors, drones and global positioning system (GPS) technology to collect and analyse data related to soil and weather conditions, crop growth and pest infestations, which is then targeted at optimizing fertilizer and pesticide use, reduce waste and improve yields. It is through the smartification of agriculture that the government seeks to raise self-sufficiency rates between 2025 and 2030 (Table 4.6). The Malaysian Productivity Corporation (MPC) launched its ESG Corporate Strategy on 27 January 2023, covering industries in the 11 productivity nexus in the pursuit of the UN's 17 sustainable development goals (SDGs) that the Malaysian government seeks to achieve (MPC, 2023).

Digital farming, on the other hand, leverages the power of digital technologies, such as the Internet of Things (IoT), big data analytics and artificial intelligence (AI), to enable precision farming and better decision-making. For instance, farmers can use IoT-enabled devices to monitor soil moisture levels, control irrigation systems and track crop growth in real time. Biotechnology, such as genetic engineering, can also play a role in enhancing crop productivity and resilience. Through genetic engineering, crops can be developed with desirable traits, such as disease resistance, drought tolerance and higher yields.

Overall, the adoption of these technologies can help farmers to increase the productivity and efficiency of food production, reduce the environmental impact of agriculture and improve the quality and safety of food, which should contribute to achieving the self-sufficiency targets set by the Malaysian government to enhance the country's food security situation.

The following examples that require strong investment into R&D, which the Malaysian government is promoting, are:

1. Development of high-yielding varieties that are also more resistant to pests and diseases. Vegetable breeders in the country have developed high-yielding varieties that meet these conditions, including better flavour and texture, and adapted to different growing

conditions. For example, the development of high-yielding tomato varieties such as Roma, Cherry and Beefsteak has significantly increased tomato productivity (Naguib et al., 2024).
2. Adoption of hybrid rice through cross-breeding. Hybrid rice has higher yields and is more resistant to biotic and abiotic stresses than traditional rice varieties, though traditional integrated farming methods with fisheries as part of the farming has been successful in some countries.
3. Precision agriculture technologies, including remote sensing, GPS and crop modelling are increasingly being used to improve the efficiency and productivity of rice farming, which enable farmers to optimize fertilizer and water use, reduce crop losses and increase yields (see also Carandang, 2019).
4. Hydroponic and greenhouse systems have been used in Malaysia since the 1980s but its application has been promoted more aggressively now as urbanization continues to squeeze agricultural land. Such farming methods allow year-round vegetable production with higher efficiency and productivity. They also offer a controlled environment for vegetable farming, optimize resource use and reduce the risk of pests and diseases.
5. Biotechnology tools, including genetic engineering, are promoted to develop rice varieties that are more resistant to pests, diseases, drought, as well as raise yield. Such biotechnology tools can raise soil fertility, increase nutrient uptake and enhance plant growth and health (Azman et al., 2018)).
6. Best practices, such as integrated pest management, soil health management and crop rotation, are increasingly promoted to improve rice productivity and sustainability, which go a long way to reduce crop losses, raise soil fertility and enhance the efficiency of water use.

To appropriate benefits from the above six practices, the government should consider the following improvements:

1. Efforts by the new Madani government should be extended to distribute precision farming technologies to particularly smallholders, with the requisite improvement in infrastructure, and knowledge sharing and training.
2. The government should introduce awareness programmes to convince farmers on modern technologies.
3. The government should extend further the technical and vocational education training programmes to equip small farmers with precision farming knowledge. Engineers and technicians should be easily accessible to the small farmers to ensure problem solving is achieved seamlessly.
4. The government should strengthen policies and regulations to encourage the adoption of modern technologies by easing bureaucratic procedures and allocating sufficient funds.
5. Technologies, such as precision agriculture and internet of things (IoT), generally raise cybersecurity threats, such as data breaches and hacking. The national cybersecurity council should include among its duties protection for digital farmers using IR4.0 technologies.

Interestingly, as Naguib et al. (2024) found the agro-firms that have ventured into smart farming started as technology-based firms rather than agricultural firms, they show the development of three interesting smart agro-firms—Singularity Aerotech (SAT) Asia, MERAQUE and GrasiCili.

Conclusions

Malaysia implemented various policies and strategies to drive its economic growth and development. These policies include the New Economic Policy, National Development Policy, National Vision Policy and the National Transformation Policy. These policies serve as guidelines and benchmarks for the country's medium- and long-term plans and initiatives. One of the key policy objectives in Malaysia is to ensure food security and promote self-sufficiency in food production. The government aims to achieve this by promoting agricultural development, enhancing productivity in the agriculture sector and reducing dependency on food imports. By implementing these policies and strategies, Malaysia seeks to strengthen its agricultural sector and ensure a sustainable and secure food supply for its population. The integration of social work practices with economic policies has been crucial for creating political unity, social stability and equality over time. Among the practices in history to reinvigorate food farming is to revisit the strategies deployed to support small-scale intensive farming that raised self-sufficiency rates in rice farming to 87.6% in 1976, and to raise yield in other food crop and livestock farming, and fish catch (see Rasiah & Salih, 2019).

Malaysia's journey towards achieving self-sufficiency by 2020, as outlined in the NAFP (2010–2020), underscores the pivotal role of its agriculture sector. To enhance this sector, a comprehensive strategy is imperative. The integration of advanced technologies like precision farming and IoT, coupled with a commitment to research and development for sustainable practices, can optimize yields and ensure long-term viability. Comprehensive education, training and financial support for farmers contribute to a knowledgeable and empowered agricultural community. Improving infrastructure, market access and implementing supportive government policies are integral to fostering a resilient and thriving agricultural sector. As Malaysia strives for food security, a holistic approach that encompasses technology adoption, farmer empowerment, infrastructure enhancement and supportive policies is paramount for sustained growth and resilience against evolving challenges.

While Malaysia has progressed considerably over the period 1957–2023, the country is still grappling with food security issues in critical items, such as vegetables, mutton and beef. For the country to break out from the middle-income trap to reach developed status, among other things agriculture should be targeted to achieve rapid value-added upgrading and significant efforts must be taken to quicken the diffusion of digitalization, digitization and IR4.0 technologies into agriculture. There is surely a need to strengthen CSA diffusion into Malaysian agriculture accompanied by a serious use of an appraisal mechanism that ensures stringent targets are met. Since food farming is significantly influenced by intensive small-scale farming, the government's emphasis on ecosystem focus should stimulate collective support that can synergize the small farmers. In addition, in sync with the United Nations' efforts on the achievement of the sustainable development goals (SDGs), government policies have also increasingly focused on promoting digitalization and IR4.0 technologies, and strengthening climate resilience in the country. The government seeks to achieve all the 17 UN's sustainable development goals as seen from the MPC's role on ESG. However, to achieve zero poverty (SDG (1) and zero hunger (SDG (2), climate friendly development emphasis also requires Malaysia to meet SDG 7 (provision of quality jobs with decent incomes), SDG 8 (building climate resilience) and SDG 9 (infrastructure and innovation)(UNESCO, 2021).

References

Abdullah, Md., Firuza, B. B. M., & Chowdhury, Z. Z. (2020). *COVID-19 Pandemic Impact on Agriculture, Food Security in Malaysia and Southeast Asian Nations.*

Afzanizam, M. A. R. (2023). Improving food production through digitalizing the fertigation system. In R. Rasiah, W. Y. Low, & K. Nurliana (Eds.), *Digitalization and Development: Ecosystem for Promoting I.R 4.0 Technologies.*Abingdon: Routledge.

Akhtar, R., & Masud, M. M. (2022). Dynamic linkages between climatic variables and agriculture production in Malaysia: A generalized method of moments approach. *Environmental Science and Pollution Research, 29*(27), 41557–41566.

Azman, M. A., Zain, C. R. C. M., & Osman, A. (2018). Plant biotechnology research and development in Malaysia. *Malaysian Journal of Analytical Sciences, 22*(1), 1–14.

Azra, M. N., Kasan, N. A., Othman, R., Noor, G. A. G. R., Mazelan, S., Jamari, Z. B., Sarà, G., & Ikhwanuddin, M. (2021). Impact of COVID-19 on aquaculture sector in Malaysia: findings from the first national survey. *Aquaculture Reports, 19*, 100568. https://doi.org/10.1016/j.aqrep.2020.100568

Carandang, J. (2019). *Precision agriculture: A beginner's guide to the latest farming tech trend, Forbes, downloaded on August 25, 2024 from* https://www.forbes.com/sites/juleecarandang/2019/02/19/precision-agriculture-a-beginners-guide-to-thelatest-farming-tech-trend

Chamhuri, S., Rospidah, G., & Ibrahim, M. (2022). Agriculture and food security. In R. Rasiah, K. Salih, & K. C. Cheong (Eds.), *Malaysia's Leap into the Future: The Building Blocks for Balanced Development.* Singapore: Springer.

Dardak, R. A. (2016). *New strategies for greater growth and development of the agriculture sector in Malaysia.* FFTC Agricultural Policy Platform. http://ap.fftc.agnet.org/ap_db.php?id=642.

Dardak, R. A. (2022). *Overview of the agriculture sector during the 11th Malaysian development plan (2016–2020) 2022.* FFTC Agricultural Policy Platform. https://ap.fftc.org.tw/article/3010

Dardak, R. A. (2015). *Transformation of agricultural sector in Malaysia through agricultural policy* (p. 7). Malaysian Agricultural Research and Development Institute (MARDI).

Department of Statistics Malaysia (DOSM). (2021a). *Selected agricultural indicators, Malaysia, 2021.* Retrieved from https://www.myagricensus.gov.my/

Department of Statistics Malaysia (DOSM). (2021b). *Supply and utilization accounts selected agricultural commodities, Malaysia 2016–2020.* Retrieved from https://www.dosm.gov.my/portal-main/landingv2

Department of Statistics Malaysia (DOSM). (2023b). *Household income estimates and incidence of poverty report, Malaysia.* Retrieved from https://www.dosm.gov.my/portal-main/landingv2

Department of Statistics Malaysia (DOSM). (2023c). Selected Agricultural Indicators 2023. Retrieved from https://www.dosm.gov.my/portal-main/landingv2

Department of Statistics Malaysia (DOSM). (2023a). *Household expenditure and consumer price index.* Retrieved from https://www.dosm.gov.my/portal-main/landingv2

Entezari, A F., Wong, K K S., & Ali, F. (2021, July 7). *Malaysia's Agricultural Production Dropped and the Impact of Climate Change: Applying and Extending the Theory of Cobb Douglas Production.* https://doi.org/10.18196/agraris.v7i2.11274

Federal Land Consolidation and Rehabilitation Authority. (2023). https://felcra.com.my/

Federal Land Development Authority (*FELDA*). (2023). https://www.felda.gov.my/en/public/felda/about-felda

Food and Agriculture Organization of the United Nations (FAO). (2023). *Malaysia.* http://www.fao.org/countryprofiles/index/en/?iso3=MYS

Govind, S. (2023, September 29). *Reviving agricultural sector solution for Malaysian economy.* NST Online. https://www.nst.com.my/opinion/columnists/2023/09/961136/reviving-agricultural-sector-solution-malaysian-economy

International Trade Administration (ITA). (2021). *International Trade Administration (ITA) Malaysia—country commercial guide: agriculture sector. 2021* (September 30, 2022). Retrieved from https://www.trade.gov/country-commercial-guides/malaysia-agricultural-sector

Ishak, A. (2021). *Cameron Highlands farmers fear "wilting" industry amid COVID-19* (7 April 2022). https://www.malaysiakini.com/news/579130

Johansson, A. C. (2015, October 21). *On the challenge to competitive authoritarianism and political patronage in Malaysia.* https://doi.org/10.1111/apel.12114

Jomo, K. S. (2006, September 30). *Growth with Equity in East Asia?.* https://doi.org/10.18356/a3119d0f-en

Laing, T. (2020). The economic impact of the Coronavirus 2019 (COVID-2019): Implications for the mining industry. *The Extractive Industries and Society., 7*(2), 580–582. https://doi.org/10.1016/j.exis.2020.04.003

Amir et al. (2020). *Agriculture food supply chain scenario during the COVID-19 pandemic in Malaysia* (pp. 1–7). https://ap.fftc.org.tw/article/2679

Lembaga Pertubuhan Peladang (LPP). (2017). *Laporan Penghasilan Sayur-Sayuran Mengikut Negeri Di Bawah Lembaga Pertubuhan Peladang Tahun 2017.* Retrieved from http://www.data.gov.my/data/ms_MY/dataset/penghasilan-sayuran-di-bawahlembaga-pertubuhan-peladang

Makhtar, S., Abidin, I. S. Z., & Islam, M. R. (2022, December 31). *Reviewing food security on paddy production: A conceptual paper.* Retrieved from https://doi.org/10.15282/ijim.16.1.2022.9012

Malaysia Government. (2021). *Malaysia Government Budget 2021* (September 16, 2022). Retrieved from https://www.malaysia.gov.my/portal/category/350

Malaysian Palm Oil Board (MPOB). (2022a). *Malaysian Palm Oil Board (MPOB) FFB received by Mill 2020.* https://bepi.mpob.gov.my/index.php/sectoral-status/303-sectoral-status-2023/1142-ffb-received-by-mill-2023

Malaysia - Agricultural sector. (2023). *International Trade Administration | Trade.gov.* Retrieved from https://www.trade.gov/country-commercial-guides/malaysia-agricultural-sector

Maxime Rushemuka. (2020, December 10). *A Background Review of Agriculture and Food Systems in Malaysia, in preparation.* Policycommons.net; The Economics of Ecosystems and Biodiversity. https://policycommons.net/artifacts/2041196/a-background-review-of-agriculture-and-food-systems-in-malaysia-in-preparation/2793639/

Ministry of Economy. (2010). *Tenth Malaysia Plan 2011–2015.* Retrieved from https://www.ekonomi.gov.my/en/economic-developments/development-plans/rmk/tenth-malaysia-plan-10th-mp

Ministry of Economy. (2015). *Eleventh Malaysia Plan 2016–2020. Anchoring Growth on People.* Retrieved from https://www.ekonomi.gov.my/sites/default/files/2021-05/Chapter%201.pdf

Ministry of Economy. (2021). *Twelfth Malaysia Plan, 2021–2025. A Prosperous, Inclusive, Sustainable Malaysia.* https://rmke12.ekonomi.gov.my/en

Ministry of Finance of Malaysia. (2020). *11th Laksana report. Implementation of the prihatin rakyat economic stimulus package (prihatin) and national economic recovery plan (penjana)* (June 30, 2020). https://www.treasury.gov.my/pdf/Laporan-Pelaksanaan-Pakej-Prihatin-Rakyat-12.pdf

Ministry of Agriculture and Food Industries. (2021). *Executive Summary: National Agrofood Policy 2021–2030 (NAP 2.0). Agrofood modernisation: Safeguarding the future of national food security.* https://faolex.fao.org/docs/pdf/mal211654.pdf

Ministry of Agriculture And Food Security. (2022). *Malaysia Global Food Security Index (GFSI) 2022.* https://www.kpkm.gov.my/en_US/gfsi-2022

Ministry of Economy. (2022, December). *The Malaysian economy in figures 2022.* https://www.ekonomi.gov.my/sites/default/files/2023-01/MEIF_2022.pdf

Mohamed, A., et al. (2011). Food security: self-sufficiency of rice in Malaysia. *International Journal of Management Studies., 18,* 83–100. https://doi.org/10.32890/ijms.18.2.2011.10349

Mohammad Hariz, A. R., Mohd Fairuz, M. S., Mohd Aziz, R., Saiful Zaimi, J., Nurul Atilia Shafienaz, H., Rozimah, M. R., Fauzi, J., Shamsul Amri, S., Rahiniza. K., Site Noorzuraini, A. R., Mohd Shahril Firdaus, A. R., & Zul Helmey, M. S. (2023). Research Initiatives towards Implementation of Climate Smart Agriculture (CSA) Practices for Rice and Other Crops in Malaysia. *The FFTC Journal of Agricultural Policy* 5: 26–33

MPC. (2023). *ESG for Productivity.* Downloaded on August 24 from https://www.mpc.gov.my/esg

Naguib, M. N., Rasiah, R., & Kamarul, A. M. (2024). Smartifying agriculture to strengthen food security. In R. Rasiah, K. Salih, B. Hafriza, , M.I. Tawfik, (Eds.), *Tun Abdul Razak: Imagining Malaysia's Future Through His Lenses.* Universiti Malaya Press.

O'Donnell, M., Mansor, N. B., Yogeesvaran, K., & Rashid, A. (2017, June 1). *Organisational change and success in a government enterprise: Malaysia's Federal Land Development Agency.* https://doi.org/10.1177/1035304617706847

Okello, D. M., Bonabana-Wabbi, J., & Mugonola, B. (2019). Farm level allocative efficiency of rice production in Gulu and Amuru districts, Northern Uganda. *Agricultural and Food Economics, 7*(1), 1–19.

Olaniyi, A. O., Abdullah, A. M., Ramli, M. F., & Sood, A. M. (2013). Agricultural land use in Malaysia: An historical overview and implications for food security. *Bulgarian Journal of Agricultural Science, 19*(1), 60–69.

Organisation for Economic Co-operation and Development (OECD). (2020). *COVID-19 and the food and agriculture sector: Issues and policy responses.* https://www.oecd.org/coronavirus/policy-responses/covid-19-and-the-food-and-agriculture-sector-issues-and-policy-responses-a23f764b/

Padiberas Nasional Berhad. (2023). *Nourishing the nation.* BERNAS. https://www.bernas.com.my/index.php

Prime Minister Office of Malaysia. (2020). *Prime Minister Office of Malaysia: official statements of prime minister of Malaysia* (September 26, 2022). https://www.pmo.gov.my/2020/03/perutusan-khas-covid-19-18-mac-2020/

Rasiah, R. (2018). *Developmental States: Land Schemes, Parastatals and Poverty Alleviation in Malaysia.* Bangi: Universiti Kebangsaan Malaysia Press.

Rasiah, R., & Salih, K. (2019). Poverty reduction. In R. Rasiah, K. Salih (Eds.), *Driving development: Revisiting Razak's role in Malaysia's economic progress* (pp. 75–94). Kuala Lumpur: Universiti Malaya Press.

Rasiah, R. (2022). Poverty and Inequality. In R. Rasiah, K. Salih, K.C. Cheong, (Eds.), *Malaysia's Leap into the Future: The Building Blocks for Balanced Development.* Singapore: Springer.

Shafiai, S., Abidin, N. Z., Rashid, I. M. A., Faisal, S. I., Esquivias, M. A., Elsiefy, E., & Roni, M. (2023, June). Normality test on agriculture growth and food security to improve poverty in high income country. In *AIP Conference Proceedings* (Vol. 2608, No. 1). AIP Publishing.

Shaharudin, AA. (2020). Protecting the agriculture sector during the COVID-19 crisis. Khazanah Research Institute. https://www.krinstitute.org/assets/contentMS/img/template/editor/200330%20Views_Agri%20COVID-19_EN.pdf

Tan, S. P., Ng, L. C., Lyndon, N., Aman, Z., Kannan, P., Hashim, K., & Ibrahim, M. S. C. (2023). A review on post-COVID-19 impacts and opportunities of agri-food supply chain in Malaysia. *PeerJ, 11,* Article e15228.

The Rubber Industry Smallholders Development Authority. (2023). https://www.risda.gov.my/en/

Volina. (n.d). https://www.vectorstock.com/royalty-free-vector/map-of-malaysia-vector-1607161

World Bank. (2022). *The World Bank Employment in agriculture (% of total employment) (Modelled oil estimate) - Malaysia | data.* https://data.worldbank.org/indicator/SL.AGR.EMPL.ZS?locations=MY

World Bank. (2023). *Agriculture, forestry, and fishing, value added (% of GDP)—Malaysia.* Retrieved from https://data.worldbank.org/indicator/NV.AGR.TOTL.ZS?locations=MY

Yap, G. B. (2019). Food supply Chain in Malaysia: Review of agricultural policies, public institutional set-up and food regulations. *Khazanah Research Institute. License: Creative Commons Attribution CC BY, 3.*

Zakaria, M. B., Hasan, J., & Abdullah, W. I. W. (2018). Sumber Perundangan Pembangunan Tanah Pertanian Menurut Perspektif Islam: Dasar Pertanian Negara (DPN). *Jurnal Islam Dan Masyarakat Kontemporari, 19,* 35.

Agriculture, Food Security and Sustainability in Indonesia

5

Dianwicaksih Arieftiara, Mohamad Fadhil Hasan, and Fachru Nofrian

Abstract

As with the countries in this book, Indonesia has experienced a gradual fall in the contribution of agriculture in GDP as the other sectors increasingly became more prominent. Nevertheless, agricultural value added has continued to grow steadily. Following the end of colonial rule, the early governments in Indonesia began to focus on food in general and rice cultivation in particular. Although cash crops, such as tea and rubber, were important earlier, oil palm gained prominence especially in Islands outside Java from the 1990s. The mechanization, the deployment of high yielding seeds, fertilizer and pesticides in agriculture increased in significance from time it was introduced in the 1960s. A major expansion that has made Indonesia prominent is Indonesia's rise to become the world's largest lander of tuna fish since 2006. Although Indonesia still showed net imports of rice in 2023, its self-sufficiency rate exceeding 90% over the period 2018–2023 is still better than some other countries in Southeast Asia, such as Malaysia. However, as climate change issues and agricultural acreage fell as urbanization and industrialization became important, government promotion of climate resilience from the late 1990s and digitalization from 2014 began to expand strongly. The government focus on environmental, social and governance practices has especially risen strongly since the introduction of the United Nations sustainable development goals.

Keywords

Agriculture · Food security · Poverty · Sustainable development · Indonesia

D. Arieftiara (✉) · M. F. Hasan · F. Nofrian
Universitas Pembangunan Nasional Veteran Jakarta, Depok, Indonesia
e-mail: dianwicaksih@upnvj.ac.id

M. F. Hasan
e-mail: mfhasa189@yahoo.com

F. Nofrian
e-mail: fachru.nofrian@upnvj.ac.id

Introduction: Agriculture and Sustainability

While typically the share of agriculture in GDP tends to fall as industry and services grow, it remains a critical sector for ensuring food security and generating exports (see Rasiah, 2025). However, while agricultural productivity has shown a trend rise in countries, such as Taiwan and China, it has shown a trend fall in several developing economies. Nevertheless, the sector remains important in most developing

R. Rasiah et al. (eds.), *Sustainable Agriculture and Food Security in Southeast Asia*, Sustainable Development Goals Series, https://doi.org/10.1007/978-981-95-0477-0_5

economies, especially to support the rural poor. Consequently, agriculture has remained important in Indonesia despite the rapid expansion of industry and services.

The early decline in the significance of agriculture took place when mining, especially oil and gas mining, became important from 1945, though its commercial extraction started in 1885 (Indonesian Petroleum Association, 2024). Subsequently, industrialization and urbanization resulted in agriculture contracting despite major policies targeted at opening up arable land for large-scale cultivation of crops, especially oil palm.

This chapter seeks to examine agricultural development in Indonesia with a focus on value added, employment, and trade. Despite the launching of several policies, and overall growth in value added, small farmers are still gripped with low productivity, labour-intensive operations and poverty. This chapter seeks to examine changes in the contribution of agriculture to the Indonesian economy, and to discuss government policy efforts to tackle problems and challenges faced by the sector, and to examine initiatives to promote innovation, digitalization and climate resilient practices in the country. In doing so, the chapter will address both intensive and extensive farming practices, including their relationship with sustainability, alternative sources of food, integrative management, modernization and artificial intelligence. Finally, the chapter will elaborate in greater detail palm oil production, product diversification and firms engaged in its value chain, such as start-ups and their links to the political economy of Indonesia.

Structural Change

While agriculture's contribution to Indonesia's GDP has shown a trend decline, which as noted by Rasiah (2025), is to be expected when countries' develop, this section discusses how Indonesia agricultural sector's contribution towards the national economy has changed over time. This chapter analyses agriculture's contribution to the national economy over the years with a focus on value added, employment and exports.

Value Added

Agriculture contributed around 24.1%of Indonesia's GDP in 1983–88 (Fig. 5.1), before falling in trend terms to 16.1% in 1997 and subsequently to 13.0% in 2006 before the sector's contribution stabilized at around 13.3% in 2023. There was a brief rise in 1997–98 but that is only because the other sectors of manufacturing, mining, construction and services fell sharply then following the 1997–98 Asian financial crisis.

Employment

As with value added, the share of employment in total employment in Indonesia has also shown a trend fall, falling from 54.6% in 1991 to 29.9% in 2019 (Fig. 5.2), which is largely a consequence of expansion in manufacturing and urbanization. The shares among male and female shares of the national labour force have closely followed the overall employment shares in the country.

Nevertheless, agriculture still accounted for 38% of overall employment in 2023, up by 10% since the share was 30% in 2019—following renewed efforts by the government to support agriculture (Fig. 5.3). The story of employment can also be seen in the data of total labour force as indicated below. Whereas manufacturing accounted for 18% of overall employment, industry as a whole enjoyed an employment share of only 27% once construction is added.

Part of the explanation for the trend fall in agricultural employment in overall employment can be attributed to wage differentials. As shown in Fig. 5.3, while the average mean wage of the labour force in Indonesia over the period 2017–2022 was significantly higher than the mean agricultural wage, agricultural workers work slightly lower hours (35–36 h) monthly than the overall workforce (40–46 h). Wages assume a central issue not only in the agriculture sector but also in the economy as a whole. As observed, the weekly average wages data demonstrates that average wages in agriculture, forestry and fishing were lower than in the overall economy during the period of 2017–2022.

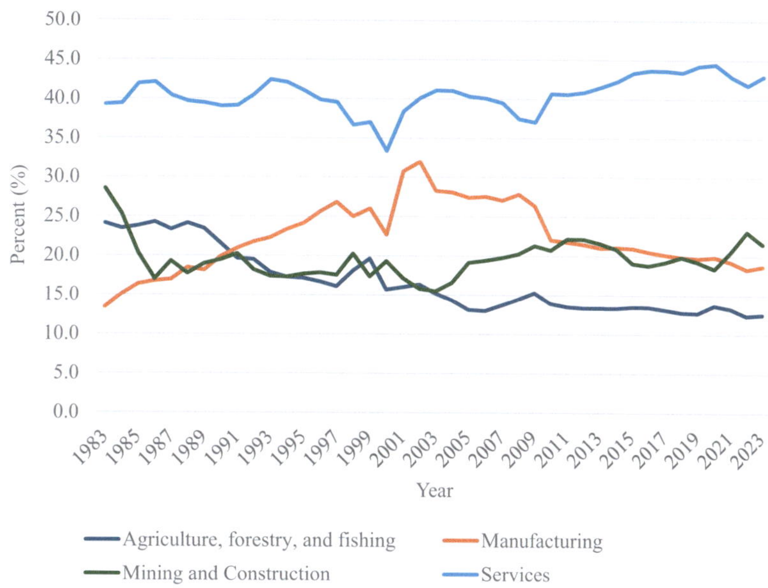

Fig. 5.1 GDP share by sectors, Indonesia, 1983–2023 (%). *Source*: World Bank (2024)

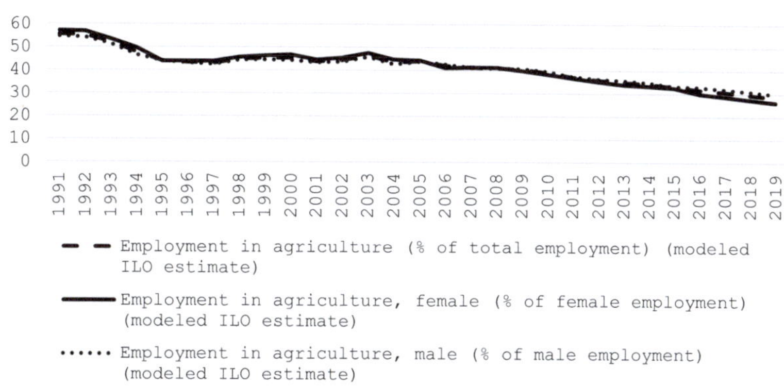

Fig. 5.2 Agricultural employment share in total employment, Indonesia, 1991–2019. *Source*: World Bank 2023

Trade

Figure 5.4 shows exports, imports and trade balance of Indonesian agriculture. While both exports and imports in current US dollars have grown in trend terms, except for a sharp rise in 2009–2010 and fall in exports from 2011 till 2016, the trade balance shows a trend decline between 1967 and 2001 with negative figures in 1998 till 2001 before a trend rise thereafter until 2021. The negative trade balance followed a fall in the agricultural trade from 1996 through the Asian financial crisis of 1997–1998 (Rasiah, 2020). Palm oil exports have been the prime reason for the trend growth in agricultural trade surplus since 2001. The surge in agricultural exports and the shift towards rising agricultural trade balance after 2001 shall be discussed in the section of agricultural policies.

Agricultural Policies

Agricultural policies in Indonesia can be structured into period starting with colonialism followed by the period under the periods 1945–1967, 1968–1998, 1999–2014 and 2014–2024. Whereas the emphasis in the first and second periods was on food production (especially rice cultivation), the government in addition also promoted aggressively cash crops, such as oil palm in the third and fourth periods.

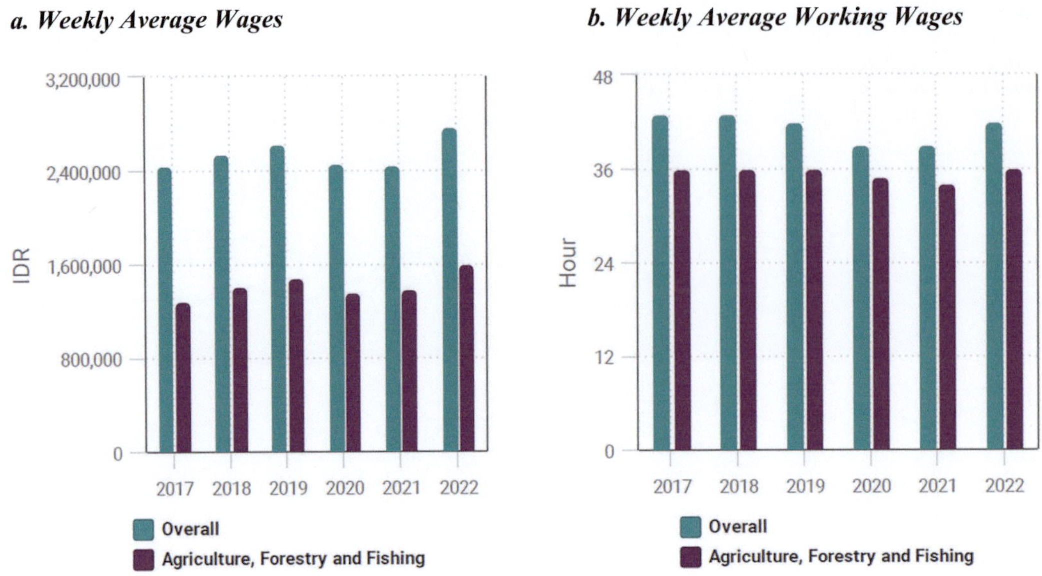

Fig. 5.3 Mean weekly wages and working hours, Indonesia, 2017–2022. *Source*: CEIC

Fig. 5.4 Exports, imports and trade balance, agriculture, Indonesia, 1967–2021. *Source*: Indonesia, National Statistics Agency, unpublished data, various issues

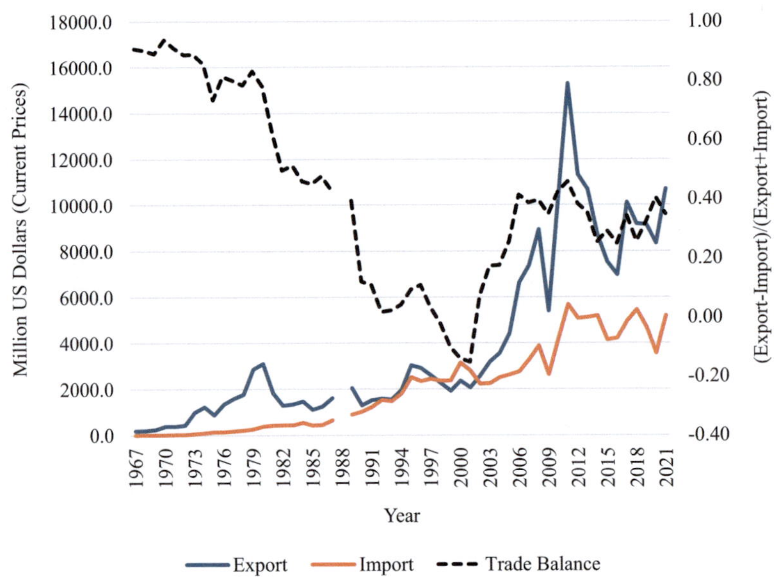

Colonial Period

Indonesia's agricultural policies can be traced to the role of *Vereenigde Oostindische Compagnie* (VOC) when General Governor Johannes van den Bosch adopted the *cultuurstelsel* in 1830 that resulted in 20% of its land to be cultivated for exports. Spices and tropical commodities were the major drivers of exports during Dutch colonial rule (Booth, 1998). While the Netherlands appropriated the profits from such a policy, the Indonesians toiled the lands for bare wages. This policy was stopped following the enactment of the *Agrariche Wet* and *Suiker Wet* in 1870. During Japanese occupation from 1941 until 1945, the focus shifted from plantation agriculture to food production.

The 1945–1967 Period

Following the declaration of independence, the agricultural focus of the Indonesian government turned to end hunger. The share of agriculture in GDP began to rise to exceed 50%, while its employment contribution in overall employment reached 71% (Glassbuner, 2007). In addition to the opening of government-owned parastatals, the government also promoted price stabilization, food availability and accessibility, through the provision of guarantees, such as the Kasimo Program in 1953–1956 and Agrarian Act No. 5/1960. These parastatals largely ventured into plantation cash crop agriculture, such as rubber, sugarcane, tobacco and coffee. The latter three were also promoted earlier by the Dutch colonialists.

However, the prime emphasis in the first period of 1945–1967 was on rice production to meet the needs of the Indonesian population, including the establishment of rice mills. The Badan Pendidikan Masyarakat Desa (BPMD), Komando Otoritas Ekonomi (KOTOE), Komando Operasi Gerakan Makmur (KOGM) and Pamong Tani Desa (PTD) were formed to increase self-sufficiency in particularly rice production. Meanwhile, the Badan Perusahaan Bahan Makanan dan Pembuka Tanah (BMPT) began to collaborate with Koperasi Produksi Pertanian (KOPERTA) and Organisasi Pelaksana Swasembada Beras (OPSSB) to specialize on extensive farming, especially on sugarcane, tobacco and coffee (Pembangunan Nasional Semesta Berencana (PNSB) (various issues). DEMAS and BIMAS focused on intensive farming. The rallying call by the Sukarno then was to target agriculture to meet the needs of the masses (*golongan marhaen*).

Although the respective parastatals were founded during the period 1945–67 in fishing and forestry, not much actually evolved in the 1960s until the period 1967–1998. Hence, much of the fishing did not translate into major exports, while forest products were mainly exported.

The 1967–1998 Period

Under the New Order government Badan Urusan Logistik (BULOG) was started in 1967 with the responsibility to purchase rice for the armed forces, the civil service and state corporation employees (University of Michigan, 1998). BULOG enjoyed monopoly control over rice prices and production, which was later extended to other basic commodities, such as wheat, rice flour, poultry, peanuts, mung beans, garlic, soybeans, sugarcane, onions, shallots and leeks. Price stabilization through the role of BULOG did significantly strengthen rice self-sufficiency by the 1980s.

Assistance from the International Rice Research Institute (IRRI) in Manila brought in high yielding varieties that are disease resistant. BULOG assisted with fertilizer, and disinfectants, as well as building irrigation canals. Meanwhile, the Bimbingan Massal (BIMAS) stimulated intensive farming. Farmer workers were also drawn to these areas through transmigration, which was promoted by the government directly at villages through *Capir* Group Forums. Livestock farming and farming for profits were supported by financing programmes through village and rural credit by Koperasi Unit Desa, and agrobusiness through the Perkebunan Besar Swasta Nasional (PBSN). These programmes helped raise farm output, even though the expansion of private plantations raised the income gap between the large and small farms.

Hiring educated personnel enabled BULOG to improve agricultural production. Simultaneously, its engagement in sugar and wheat flour processing reflected attempts to address farmers' concerns about low commodity prices relative to the high retail prices of processed sugar (World Bank, 1987; Bresnan, 1993). Consequently, the World Bank (1987) criticized BULOG for its handling of such essential commodities. The government also extended subsidies for rice farming and cattle rearing under the first, second, third, fourth, fifth, and sixth five-year plans (Repelita I, II, III, IV, V and VI) (Gie, 1999, 2019).

Fishing and forestry faced considerable policy shift from the late 1970s as export taxes were raised from 10 to 20% on timber in 1978 (Takeuchi, 1983: 3) and the government began to emphasize the development of domestic fisheries, including in tuna fishing (McElroy, 1989). Indonesia also became a signatory to the main ocean and sea treaties to ensure conservation and orderly governance of fishing alongside the Pacific Islands and the major fishing nations in the area, such as Japan, Taiwan, Korea and the United States (Rahmadi & Huang, 2014). Although Robison (1986) discussed timber export leakage through corruption from Indonesia to Japan, the tax had a major impact as it resulted in a contraction of timber exports and a rise in downstream processing.

Overall, the government combined price interventions and economic incentives to encourage agricultural production, especially of the staple crops. These policies included efforts from 1985 until 1998 when along with the early members of the Association of Southeast Asian Nations (ASEAN) launched a series of trade that were targeted at entry into the World Trade Organization in 1995 and to make ASEAN as single trading area, which was further boosted as the financial crisis of 1997–98 drove the government to accept a structural adjustment package as part of the debt bailout (Thomas & Orden, 2004).

The 1999–2014 Period

The early years following the fall of Suharto was characterized by intervention by the International Monetary Fund (IMF), which was focused on debt restructuring. External rice aid was complemented by the rationalization of agricultural subsidies. The fall in the Rupiah exchange rate in 1997–98 had already made food imports expensive.

The Asian financial crisis of 1997–98 undermined Indonesia's economy, as its short-term debt and balance-of-payments deficit exceeded international reserves (Rasiah, Chong & Doner, 2014). To mitigate the crisis, two measures were pursued: IMF aid to purchase rice, and government programs promoting extended paddy farming in Kalimantan and corn cultivation in Sulawesi. These programs, supported through subsidies on fertilizer, high-yielding seeds, infrastructure, facilities, and training, were designed to assist the poor. As a result, the national budget expanded to Rp20 trillion, which was distributed through various banking and financial institutions. In addition, the government under Abdurrahman Wahid published Tap MPR in 2001 to support farmers to access land for farming, including large plantations in 2003. In 2009, Water Act was issued, as well as Capital for Farmer Act and HGU up to 90 years. During the same period, the Sustainable Land Protection Act was enacted accompanied by Farmer Protection Act No. 19 and Food Act in 2012. It was also during Susilo Bambang's Presidency that emphasis began to rise to new levels on tuna fishing.

As with most Southeast Asian nations, tariff deregulation followed strongly after the ASEAN Free Trade Area (AFTA) was launched in 1992 (ASEAN, 2002). Indonesia is no different with tariffs particularly falling after 1998 when fiscal rationalization became a major policy instrument imposed by the International Monetary Fund as Indonesia defaulted its debt service during the 1997–1998 Asian financial crisis. Despite the use of subsidies, tariffs plummeted to below 5% since 2009 since Susilo Bambang became President (Fig. 5.5).

The 2014–2024 Period

Dramatic progress has been achieved under the Jokowi Widodo Presidency. Jokowi Widodo accelerated development as Indonesia experienced massive infrastructure development with the construction of metro rail and rapid train service between Jakarta and Bandung. Since 2015, irrigation was extended to 2.6 million hectares, while agriculture was modernized through the provision of tractors, pumping machines, combined harvesters, rice transplanters, power thrashers, corn sellers and RMU to farmers**.** The activity of seed subsidy is continued altogether with food estate, land certificate distribution for

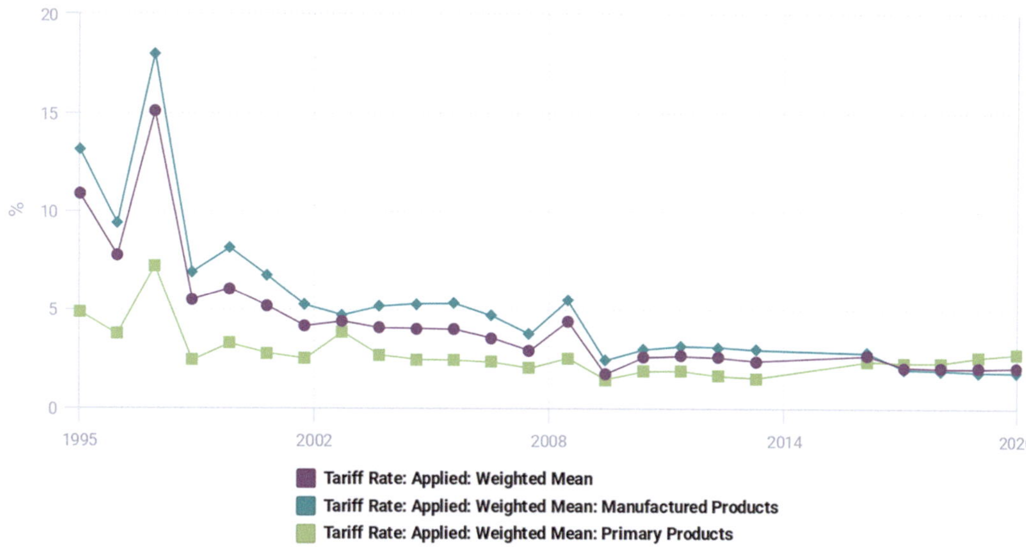

Fig. 5.5 Tariff rate, Indonesia, 1995–2020 (weighted mean). *Source*: CEIC

extensification, oil palm moratorium, development of super seed variety, rocketing illegal ship is discontinued and digitalization. Besides BULOG, the government set up National Food Agency or *Badan Pangan Nasional* (BAPANAS) to control price and food availability.

The government also increased focus on palm oil and coconut cultivation, and raised emphasis on tuna fishing. Oil palm cultivation increased particularly in North Sumatra, South Sumatra and East Java, whereas coconut cultivation expanded in East Java and North Sulawesi. Tuna fishing increased in North Sulawesi, North Maluku, Maluku, Bali, East Nusa Tenggara, East Java, South Sulawesi, West Java and Banten. Crab (*rajungan*) fishing was promoted in East Java, Central Java and South-East Sulawesi.

Of special significance experienced by the agricultural sector during the 2014–2024 period are the promotion of vertical integration, expansion of plantations, expansion in digitalization, expansion of irrigation, mechanization and emphasis on development of the outer Islands. Under Jokowi, the government also sped up the provision of land certificates and expanded the regulatory powers of BAPANAS (Rencana Pembangunan Jangka Menegah dan Panjang, various issues). Indeed, climate resilience and digitalization took on a new dimension during Jokowi's Presidency as the government began to emphasize the use of technology to both raise yields and strengthen climate resilience in conjunction with the initiatives of the United Nations sustainable development goals that were launched in 2016 (UNESCO, 2016).

Performance of Agriculture

Following the arguments by Rasiah (2025), it is understandable that the share of agriculture, forestry and fisheries in GDP, shall fall over time as countries develop. Figure 5.6 shows value-added growth in constant Rupiahs. Indonesia was gripped by volatile fluctuations in agricultural value-added growth in the 1960s until 1970, which is to be expected as the New Order government of Suharto took leadership under difficult circumstances. Except during the first oil crisis of 1973–1975, value added did not dip too low from 1976 until the 1997–1998 Asian financial crisis. Agricultural value-added growth has since stayed positive until 2023 (Fig. 5.6), though there was a slowdown in 2022–2023. The rise in value added since 2001 is largely a consequence of expansion in oil palm cultivation that drove a massive expansion

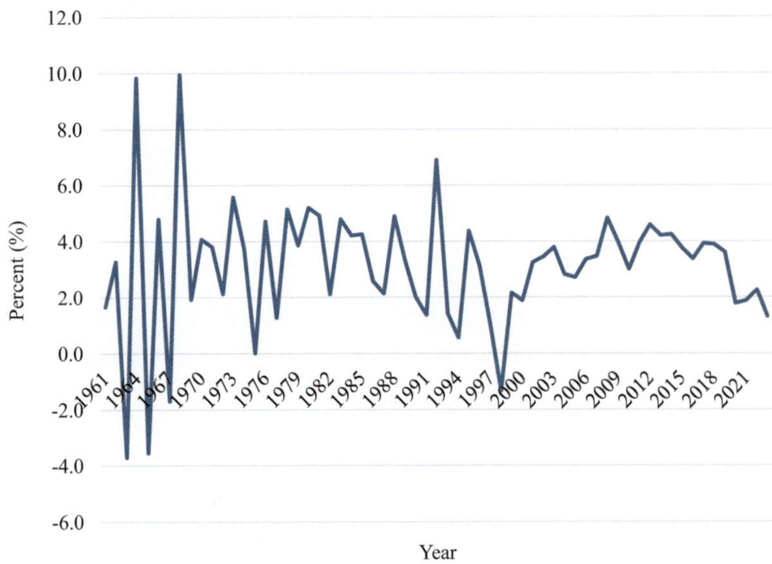

Fig. 5.6 Average annual growth in value added, agriculture, Indonesia, 1961–2023. Includes forestry and fisheries. *Source* Plotted from World Bank (2023)

in the export of crude palm oil (Rasiah & Azmi, 2005).

While productivity growth in paddy farming has largely stagnated (JICA, 2013), labour productivity in the agricultural sector has shown trend improvements in the period 1991–2019 (Fig. 5.7), which is largely driven by expansion in the palm oil industry.

Among the sub-sectors, agriculture still enjoyed 75% share of the sector's value added, followed by forestry at 20% and fisheries at 5%, respectively, in 2023 (Fig. 5.8). Within agriculture, plantations contributed 41% of value added, followed by food crops at 20%, horticultural crops at 13%, livestock at 13% and hunting at 13% (Fig. 5.9).

Broken down to just food and edible oils, rice and related items constituted 57% of value added within the food sector, followed by oil and fat at 12%, animal food at 12%, vegetables and fruits at 6%, sugar at 4%, tubers at 3%, nuts by 3%, others at 2% and oily fruit/seeds at 1% in 2023 (Fig. 5.10). This data shows that rice remains the dominant food item produced in Indonesia.

A major development that has helped Indonesia to become the leading exporter of fish in the world is the focus on tuna fishing. From 140 to 150 thousand metric tons in 1988 (McElroy, 1989: 285),

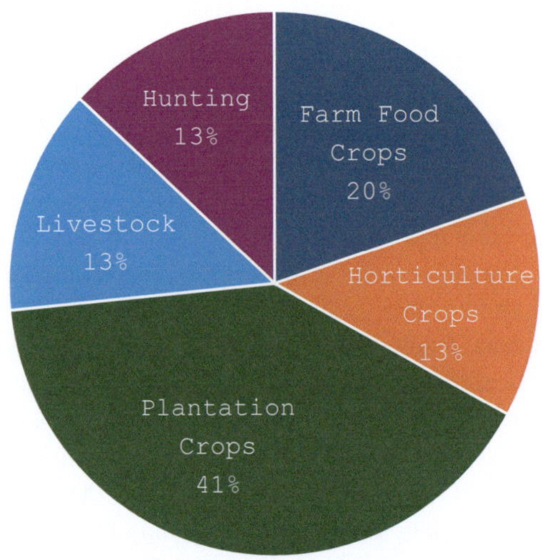

Fig. 5.9 Composition of agriculture, Indonesia, 2023. *Source*: CEIC, Statistics Indonesia

Indonesia's tuna catch rose gradually (Rahmadi & Huang, 2014: 174). It has since risen exponentially to reach 8.3 million metric tons by 2024 as Indonesia accounted. The figures from Rahmadi and Huang (2014), which is high, presumably include tuna landed by foreign fishermen on Indonesian waters but exported directly. The Indonesian fisheries ministry and the statistics department reported a smaller landing of tuna fish in Indonesian waters by Indonesian fishermen

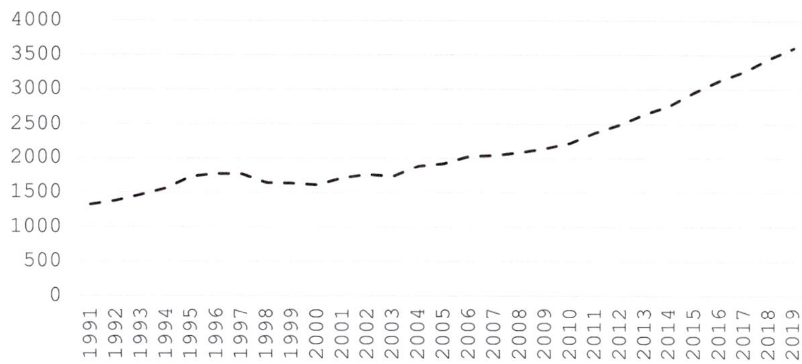

Fig. 5.7 Labour productivity, agriculture, Indonesia, 1991–2019. *Note*: In constant prices (US dollars). *Source*: World Bank (2020)

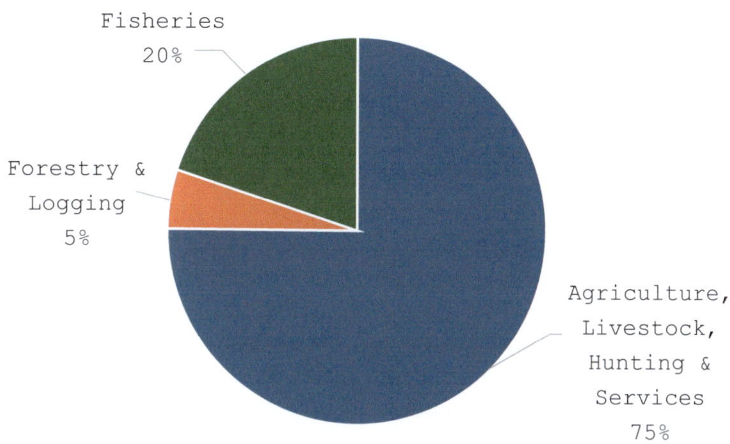

Fig. 5.8 Sub-sector breakdown, agriculture, forestry and fisheries, Indonesia, 2023. *Source*: CEIC, Statistics Indonesia

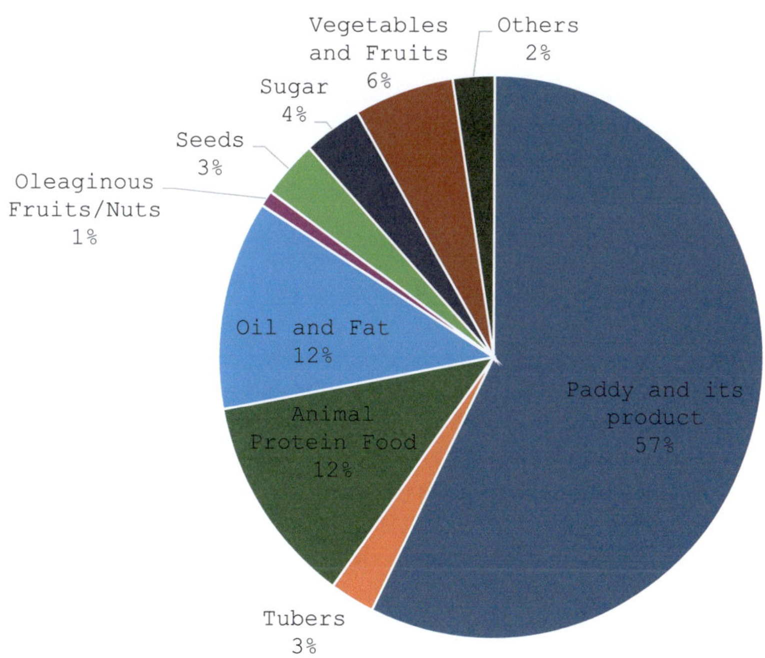

Fig. 5.10 Composition of selected agricultural output, Indonesia, 2023. *Source*: BAPANAS (National Food Agency)

(Fig. 5.11). While tuna landed rose from 241.4 metric tons in 2011 to 359.1 metric tons in 2021, the peak of 409.0 metric tons was achieved in 2018.

Agricultural Land Use

The data from the 1983–1993 Agricultural Census shows that there has been a change in the structure of land ownership among households in agriculture, namely the increasing number of smallholder farmers with extensive control which is getting smaller, and on the other hand there is an accumulation of land tenure on a small proportion of households own large areas of land (Sumaryanto and Rusastra, 2000). In 1997–2010, average land ownership in Java decreased from 1.13 ha to 1.05 ha, which has occurred in all types of land ownership, including rice fields and gardens. However, land ownership increased outside Java from 1.74 ha in 1997 to 1.79 ha in 2000. Meanwhile, arable land in total land area in Indonesia was stagnant at around 10% over the period 1961 until 1983; it has since shown a trend increase to around 14% in 2021 (Fig. 5.12).

Agricultural land area in arable land stayed stagnant at around 21% in 1961 until 1984 before showing a trend rise to 34% in 2021 (Fig. 5.13). Much of this expansion has taken place in the outer Islands, including Sumatra, Kalimantan, Sulawesi and Irian Jaya. Oil palm cultivation has become a major focus of this expansion.

Based on their significance to the Indonesian economy, this sub-section is devoted to examining paddy, rubber and oil palm cultivation in Indonesia. Whereas paddy is the main staple food of Indonesia, palm oil, through the production of crude palm oil, has become the chief agricultural export of the country.

Rice production is important in all regions in Indonesia, from the smallest value indicated by red areas to largest ones indicated by green areas (Fig. 5.14). The 20 top provinces producing rice in 2006–2023 were East Java, Central Java, West Java, South Sulawesi, Lampung, South and North Sumatera, Banten, West Sumatera, Aceh, West Nusa Tenggara, West Kalimantan, South Kalimantan, East Nusa Tenggara, Central Sulawesi, Southeast Sulawesi, Bali, DI Yogyakarta, Central Kalimantan and West Sulawesi. The largest value over the period 2006–2023 was achieved in 2017, while the whole country has generally faced low values from then on until 2023.

In total, paddy cultivation is expected to shrink between 2012 and 2023. The contraction is expected to be steepest from 15 million hectares in 2017 to 10 million hectares in 2023. Interestingly, despite the dry seasons that have plagued paddy cultivation, the fall in rice production has not seriously undermined self-sufficiency, which remained over 95% in 2018–2023.

Rubber Cultivation

Figure 5.15 shows agricultural land under rubber cultivation, which has risen from around 3.5 million hectares in 2011 to around 3.8 million hectares in 2021. Not only that smallholders dominate rubber acreage, the share of smallholders in rubber land ownership has also increased the most in 2011–2021. Whereas ownership among smallholders rose from 2.7 million hectares in 2011 to 3.3 million hectares in 2021, the commensurate ownership figure for government plantations fell from 0.2 million hectares in 2011 to 0.1 million hectares in 2021.

Oil Palm Cultivation

Meanwhile, oil palm cultivation has expanded sharply from around 9 million hectares in 2011 to 15 million hectares in 2021 (Fig. 5.16). The expansion is particularly most dramatic among private plantations, including through investment from Malaysia. By 2021 over 60% of the palm oil land was owned by private plantations. Around 35% of the oil palm land came under government plantations followed by around 5% owned by smallholders.

Figure 5.17 shows harvested area under paddy and tea cultivation. Except for 2018 when harvested area under paddy fell sharply,

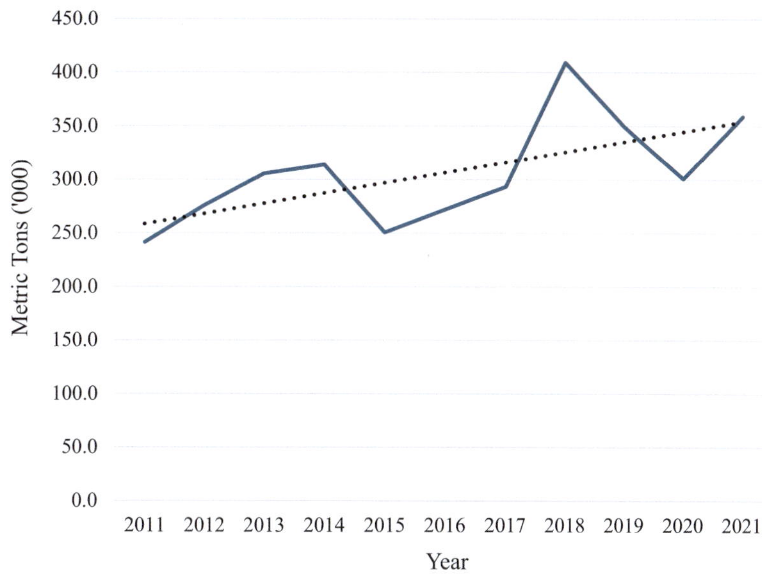

Fig. 5.11 Tuna landed in Indonesia, 2011–2021. *Source*: Indonesia, Ministry of Fisheries, 2022; Indonesia, Statistics Department, 2022

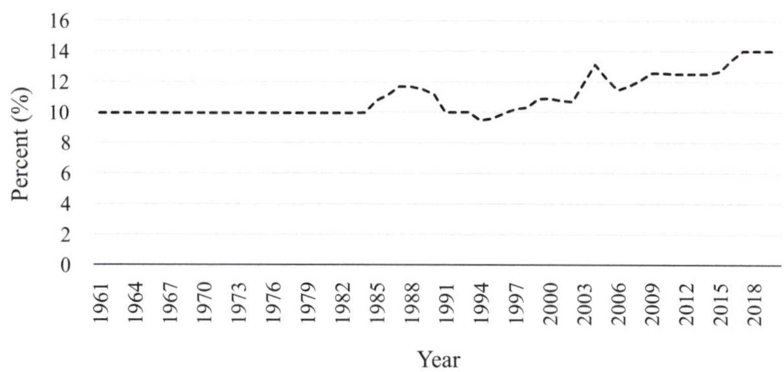

Fig. 5.12 Arable land area in total land area, Indonesia, 1961–2021. *Source*: World Bank (2023)

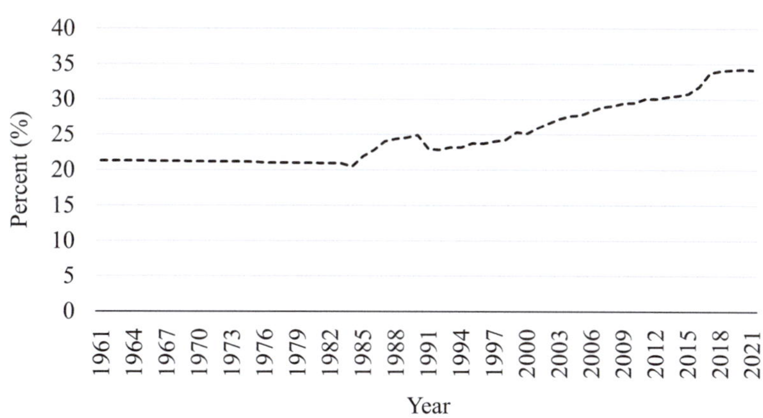

Fig. 5.13 Agricultural land area in arable land, Indonesia, 1961–2021. *Source* World Bank 2023

it has remained steady until 2023, though it has gradually shrunk owing to urbanization and industrialization. Tea cultivation area has also been shrinking with 2015 experiencing its biggest contraction.

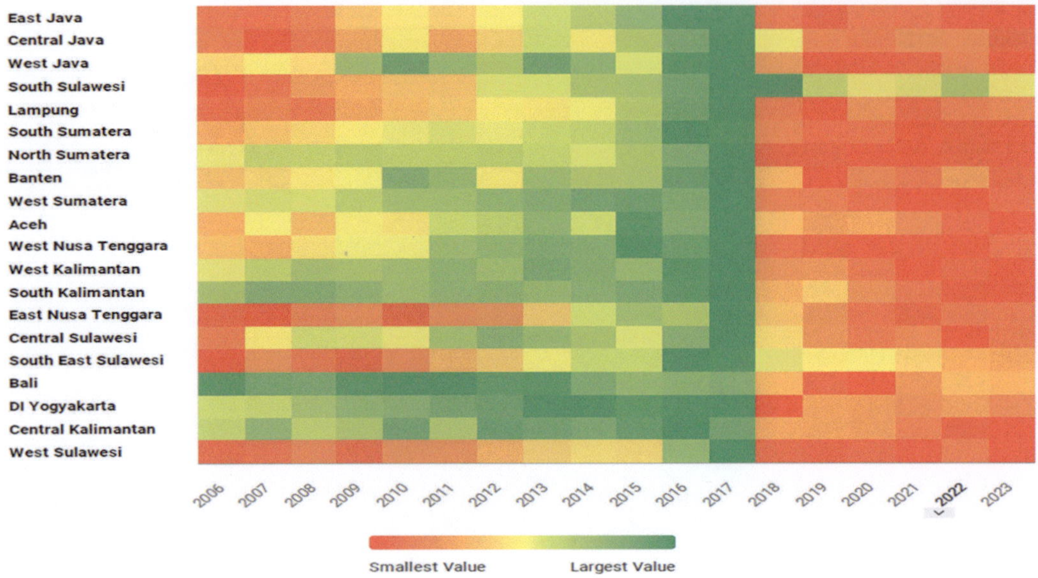

Fig. 5.14 Harvested area, paddy, Indonesia, top 20 provinces, 2006–2023. *Source*: CEIC

Food Security

There is contestation over whether Indonesia has evolved the developments essential to attain self-sufficiency in food and whether productivity in the food sub-sector has risen over time. JICA (2023) and ADB (2019) argue that the provision of subsidies and price controls over the years have undermined food self-sufficiency in Indonesia but much of the discussion is focused on how distortions have affected the sub-sector rather than connecting the accounts to changes in production and self-sufficiency levels. This section draws on evidence from the World Bank (2024) to show that not only has agricultural value added shown a trend expansion since 2001, but also the food production index has shown a strong expansion from 1961 until 2023 (Fig. 5.18). The growth in the food production index in particular has been impressive as over the years over the period 1961–2023 food production has risen by 1.7 times per year. The explanatory power of the model is high, as indicated by an ($R^2 = 0.99$). Meanwhile, except for 1967, 1978 and 1998, food imports in overall imports remained below 10%.

Although food security should account for those who are unable to purchase food at market prices (FAO, 2008), the simple use of market prices shows that Indonesia has enjoyed self-sufficiency rates over 95% over the period 2018–2023 (Fig. 5.19), which is much higher than Malaysia's rice self-sufficiency at 69% in 2023 (Annizah & Rasiah, 2025). Overall, the evidence reflects well on the government's efforts to support the food sector in general and rice cultivation in particular over the 1945–1967, 1967–1998, 1999–2014 and 2014–2024 periods but the emphasis on oil palm cultivation, especially in locations outside Java, was added from 1999.

Since paddy (rice) is the staple of Indonesians, we examine consumer and producer prices of this commodity over time. The mean consumer price for a kilogram of rice rose from IDR7,500 in 2000 to IDR12,000 in 2022 whereas the equivalent producer price rose from IDR3,950 in 2010 to IDR5,000 in 2022 (Fig. 5.20) demonstrating that consumer prices have risen far more than producer prices.

Fig. 5.15 Agriculture land under rubber, Indonesia, 2011–2021. *Source*: World Bank 2023

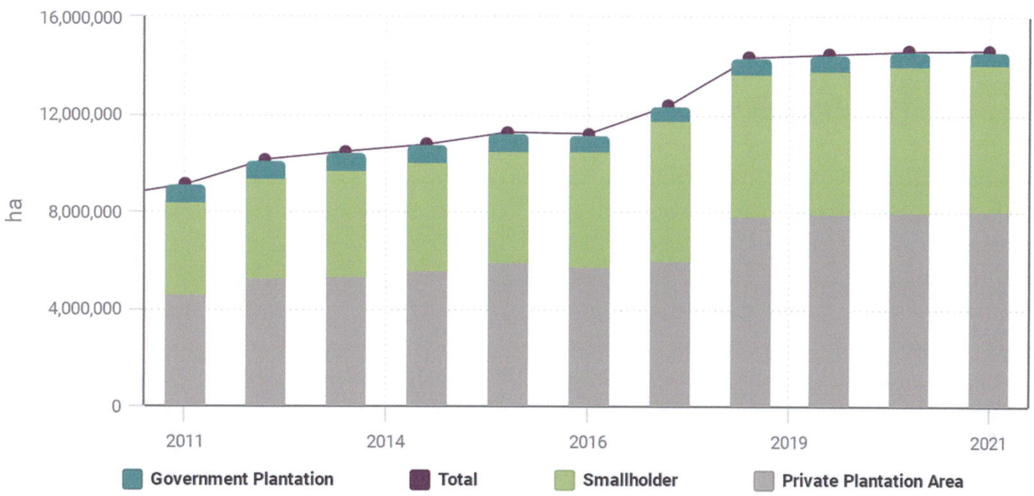

Fig. 5.16 Agricultural land under oil palm, Indonesia, 2011–2021. *Source*: World Bank 2023

Sustainability and Smart Agriculture Practices

The government, since the 1945–1967 period when mechanization, irrigation, fertilizer and insecticide use were begun, started to modernize agricultural farming. Table 5.1 shows government emphasis on mechanization since 1945. Machinery was used in rice cultivation over the period 1945–67 in Sekon, Pulau Timor, NTT, Kumber, Kruik and Merauke. Meanwhile, giant moldboard plow and crawler tractors were used in sugar cultivation in Jatiroto and East Java. Caterpillar type D4 and D6 with dumper, road grader, scraper, rome plow, Zeto tractors and hand tractors were used in Lampung. Rice milling unit, small tractor, rice huller, rice polisher and water pump were used in Java, Sumatra and Sulawesi. Rice milling units, mini tractors, diesel ploughers, rice huskers and polishers, and water pumps were used in West Sumatra, South Sulawesi, West Java, South Tangerang, Java, Sumatra and Sulawesi in 1967–1998. The explanatory power of the model is high, as indicated by an ($R^2 = 0.99$). The country also enjoyed further upgrades in tractors, threshers, irrigation pumping, transplanters and standardization in farming output in that period.

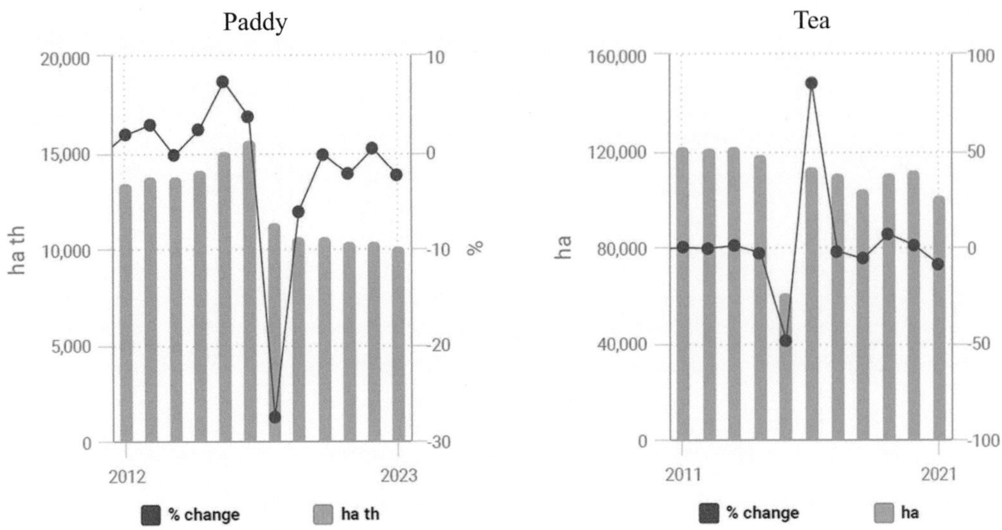

Fig. 5.17 Harvested area, paddy and tea, Indonesia. *Source*: CEIC

Modernizationof Agriculture

Rice cultivation in Java received significant focus in the drive to smartify agriculture. First, open horticulture production was introduced in Central Java. Second, rice cultivation was expanded in Central Java. Third, green house and smart greenhouse technology was deployed in Yogyakarta. Fourth, plant factories technology was introduced in Yogyakarta. The factories produce organic fertilizer, which is fertilizer composed of the remains of living things and decaying matter from plants and animals. Organic fertilizer production and its introduction in Java became important as C-Organic fertilizers in the soil had fallen sharply to make nutrient absorption ineffective absorbing nutrients and water holding capacity the soils had becomes poor to reduce plant vitality. One to two tons of organic fertilizer per hectare has been added for rice in Central Java and Yogyakarta constituting of manure, green manure, compost, artificial organics and humus.

Intensive-farming was expanded from 2010 to double cassava output and to focus on superior variety cassava to yield 100 tonnes per hectare in Darul Hidayah, Manggu and Gajah. The government identified 7.2 million hectares of dormant land according to PP no. 10/2010 for this development to assist traditional farmers to adopt modern agricultural technology. MSI formulated a vertically integrated cassava-based industrial development model, i.e. Integrated Cassava Agro-industrial Cluster to produce chips, cassava, krosok (tapioca 80 MES), tapioca starch, cassava flour, mocaf, bioethanol and degradable plastic in Lampung, Sumatera Utara, Jawa Barat, Banten, Yogyakarta, Jawa Tengah, Jawa Timur, Nusa Tenggara Timor (NTT), Nusa Tenggara Barat (NTB), Kalimantan Barat, Kalimantan Timur, Kalimantan Tengah, Sulawesi Selatan, Sulawesi Tengah, Gorontalo, Maluku Utara, Papua and Papua Barta. The use of new superior hybrid seed varieties with strong adaptation science research institutes has raised production, early maturity that is adaptive to the environment. Further improvements in cultivation have also enhanced productivity. In addition, farmland development supported by greater use of organic fertilizers and improved irrigation methods has significantly increased the yields of rice, sugarcane, corn, soybeans, peanuts, green beans, cassava, and sweet potatoes (Jarwo; Hazton; SRI). Figure 5.21 shows combined harvesters harvesting sugarcane in Singkong. Meanwhile, Fig. 5.22 shows cassava plants in Manurong.

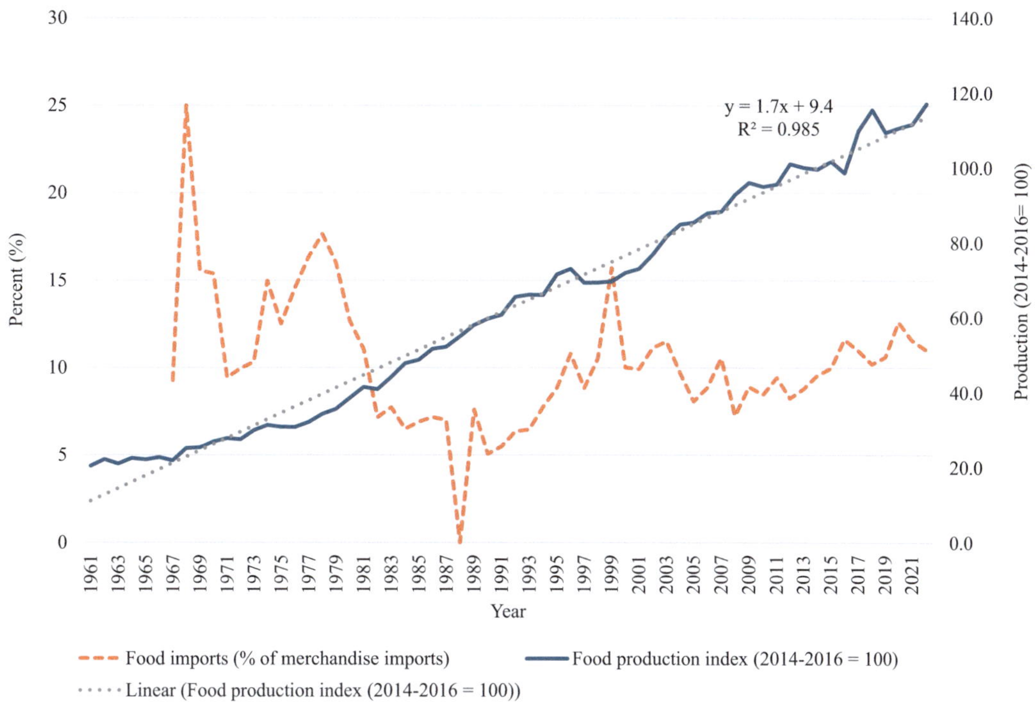

Fig. 5.18 Food imports and food production index, Indonesia, 1961–2023. *Source*: Plotted from data gathered from World Bank (2024)

Digitalizationand Climate Resistance

Digitalization and related start-ups have sprung up to solve aggrotech problems. The following developments have increased in Indonesia since 2014.

a. Problems and Aggrotech Solutions

Indonesia has historically struggled with issues related to land scarcity, capital availability and unstable market pricing in its agricultural industry, which has reduced the number of farmers in the country. Aggrotech businesses, such as TaniHub, SayurBox and Tanijoy, have emerged to solve these issues and are providing creative solutions to breathe new life into agro-industries by eliminating middlemen and bringing farmers and customers directly together.

b. Market Diversity and Financing Solutions

Indonesian aggrotech companies work in a number of markets. While some, such as TaniHub and SayurBox, concentrate on business-to-consumer (B2C) models, others engage in business-to-business (B2B) models, like Panen ID and Chilibeli, which serve hotels and restaurants. Furthermore, peer-to-peer lending is being provided by companies, such as TaniFund to help farmers access cash, which is more flexible that compliments the agricultural production cycle.

c. Technological Innovation in Agriculture

The Internet of Things (IoT) has given farmers a big boost to automate and standardize agricultural processes. For example, Habibi Garden's HabibiGrow system enables remote management of greenhouse temperatures and plant watering. Other inventions include smart fish feeding

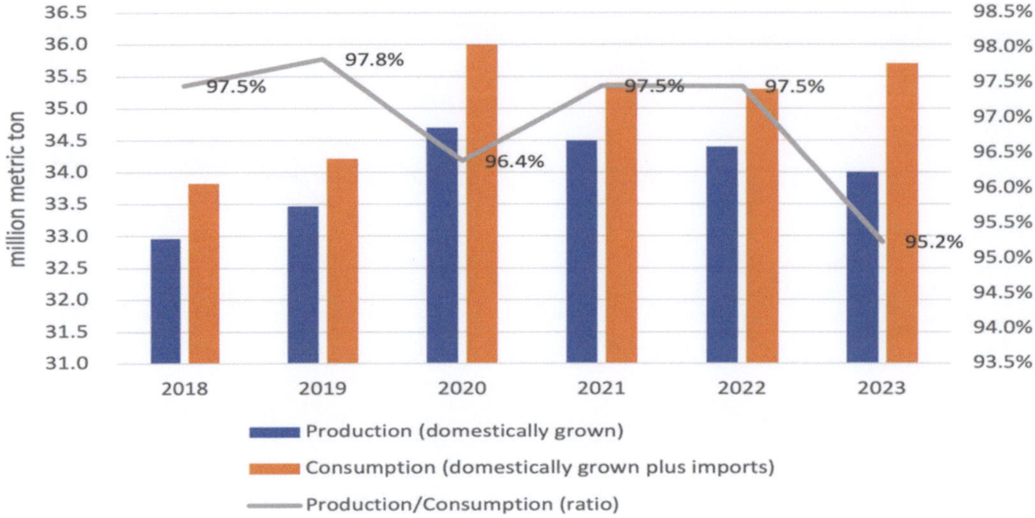

Fig. 5.19 Rice self-sufficiency ratio, Indonesia, 2018–2023. *Source*: Official data compiled by the US Department of Agriculture; Reproduced from Manggi and Siwage (2024)

systems in fisheries and drone technology for surveillance and soil/weather sensors.

d. Increase in Market Expansion

To increase market penetration, AgriAku, a B2B marketplace for farmers, obtained $6 million in pre-Series A investment to offer business software solutions and a more transparent pricing by bringing together retailers, wholesalers and manufacturers. With a notable rise in both its user base and gross sales value, AgriAku has begun to expand rapidly.

e. IR 4.0 Technologies

Especially since 2014 Indonesian agriculture has experienced the proliferation of new business models, blockchain, cloud computing, AI, drones, robots and machine learning technologies into Indonesian agriculture to raise yields, efficiency, revenues and profitability. The developments suggest the establishment of a digital agriculture ecosystem that encompasses supply chain management, digital infrastructure, consulting services, market connections and financial access. To facilitate the seamless emergence of this system, emphasis has increased on data, innovation and collaboration between the public and commercial sectors. Following the advent of the COVID-19 pandemic, there is increasing move to introduce climate-smart agriculture to promote sustainable farming practices, cross-sector growth connecting AggroTech and FinTech, and climate resilient methods.

The modernization of agriculture in Indonesia has also attracted firms to list in the IDX (full first please!), which is reflected in the rise in the number of listed companies in agriculture from 5 before 2000 to 6 in 2000–2010 to 13 after 2010 (Fig. 5.23).

ESG Reporting

The Indonesian Financial Services Authority (OJK) introduced rule POJK 51/2017, which requires sustainability reporting, which is a significant step towards responsible and open corporate governance. All Indonesian listed firms are required by this rule, which came into force in 2019, to produce non-financial reports that cover a wide variety of environmental, social and governance (ESG) features, including agricultural

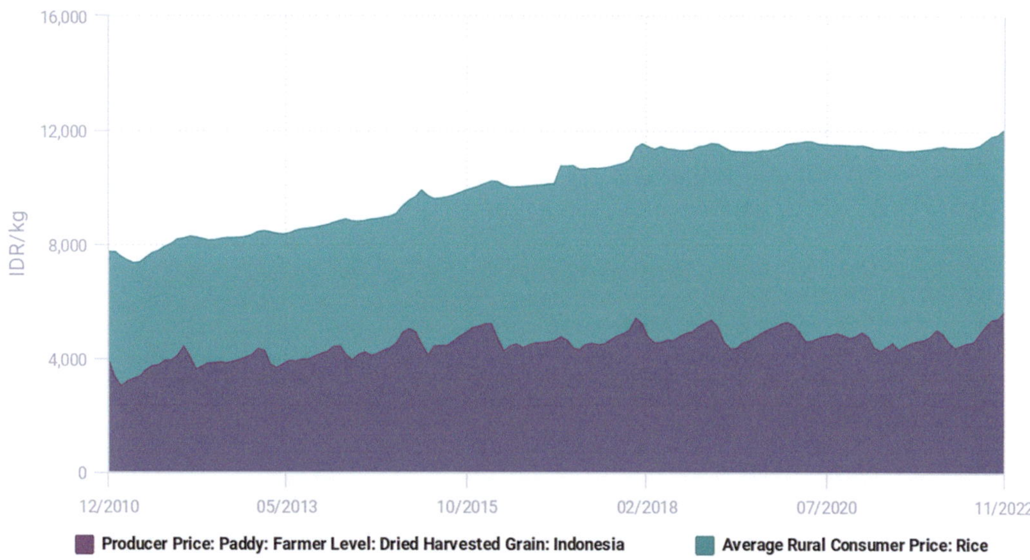

Fig. 5.20 Producer and consumer prices, rice, Indonesia, 2010–2022. *Source*: CEIC

Table 5.1 Mechanization in agriculture, Indonesia, 1950–2024

No	Period	Mechanization	Region
1	Guided Democracy (1945–1967)	The use of machinery in rice estate	Sekon, Pulau Timor, NTT and Kumbe, Kruik, Merauke
		The use of giant moldboard plow and crawler tractor	Sugarcane in Jatiroto, East Java
		Caterpillar type D4 and D6 with dumper, road grader, scraper and rome plow	Lampung
		Zetor tractor, hand tractor	Lampung
		Rice milling unit, small tractor, rice huller, rice polisher, water pump	Java, Sumatra, Sulawesi,
2	New Order (1967–1998)	Mini Tractor, rice milling, huller, rubber roll, diesel machine, rice husker, rice polisher, hand tractor, rome plow, power thresher, sprayer, mostly import	West Sumatra, South Sulawesi, West Java, South Tangerang,
3	Reformation Period (1998–2014)	Geographic information system, tractor, thresher, irrigation pumping, transplanter, standardization, certification	Almost all regions, Indonesia
4	Post-Reformation Period (2014–2024)	Up to 2015, 65.431 units of machinery and equipment were distributed in 8,220 locations	Almost all regions, Indonesia

Source: Sulaiman et al (2018)

sustainability. The adoption of this law is a demonstration of Indonesia's support for the sustainable development goals (SDGs) of the United Nations. The core feature of POJK 51/2017 is not just compliance but also cultivating a sustainable growth-oriented company culture, acknowledging the unbreakable relationship between financial success and non-financial social elements, such as

Fig. 5.21 Sugarcane cultivation, Indonesia, 2023. *Source*: Masyarakat Singkong, Indonesia

Fig. 5.22 Cassava plants, Manurung Plant Factories, 2023. *Source*: Manurung, ITB, workshop KEIN

social responsibility, environmental stewardship and ethical governance.

From the perspective of agriculture, the enactment of POJK implies that businesses are required to report on how they use agrochemicals, manage water resources, conserve biodiversity and use land. In addition, social issues including labour practices, community involvement and the influence on nearby farming communities must be reported. Corporate governance must exhibit ethical behaviour, openness and legal compliance. The government reported that the changes will be progressively implemented but by 2024, to largely allow businesses (especially small businesses) time to adjust to the new standards. Additionally, it makes it possible to develop internal ESG competences, which is especially important in agriculture, as many businesses will be experimenting with sustainability reporting for the first time. Nevertheless, the consequences of non-compliance are severe as these reports are now an essential resource for stakeholders and investors to evaluate the risk profile and long-term viability of businesses. In the agriculture industry, sustainability reporting is especially important because of its intrinsic connections to natural resources and community well-being. A growing number of investors are allocating capital to businesses that exhibit a dedication to sustainable practices, as they understand that such businesses are arguably more robust and better positioned for long-term success than others. The integration of sustainable practices into corporate strategy is becoming more and more popular. Examples include the introduction of organic agricultural techniques to fulfil customer demand for sustainable products and minimize environmental effect, as well as the use of precision farming to boost efficiency,

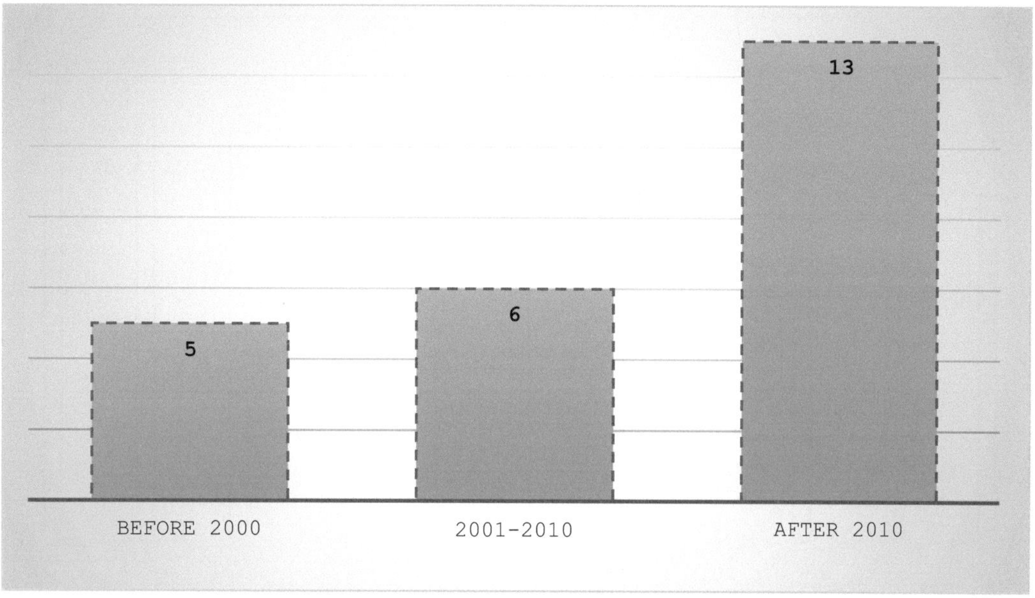

Fig. 5.23 Number of IDX-listed companies in the agriculture sector, Indonesia. *Source*: Extracted from IDX

reduce waste and greater transparency. Moreover, given that the agricultural sector directly affects rural people and natural resources, it is extremely important for businesses operating in this area to have strong corporate governance to ensure meeting the OJK regulation POJK 33/2014.

Furthermore, the social facets of agricultural activities, such as worker rights, community well-being and land rights, are within the jurisdiction of the POJK to ensure that business operations promote fair labour practices and community development to strengthen the social fabric of the communities in which they operate. To guarantee that all inputs, including seeds and fertilizers, are supplied ethically, corporate governance in agriculture also includes supply chain supervision. The implementation of sustainable agricultural methods, which can increase output while reducing the negative environmental effects, such as integrated pest management, organic farming and precision agriculture, will also be monitored by the governance structure of POJK.

Agriculture firms are obligated to periodically reveal their sustainability performance, including the effects of their activities on society and the environment, since transparency is of the utmost importance, which is not only to meet legal standards but also fosters confidence among stakeholders, including investors and customers who are calling for more responsibly produced food. To ensure the agriculture sector's resilience and sustainability throughout time, effective governance necessitates a comprehensive strategy that strikes a balance between production, environmental preservation and social equality development. Figure 5.24 shows mean ESG scores for Indonesia demonstrating that after a slight fall in 2012–2015, it has improved steadily since until 2022.

Given the usual complaints on the palm oil industry, some discussion is made of the industry in Indonesia. In the long term, production in this sector has increased significantly. For example, Indonesian Palm Oil Association (IPOA) data states that in 2018, palm oil production reached 41,983 million tons and continued until 2021 at 51,583 million tons. By exports, palm oil rose from 24.0 million metric tons in 2014 to 30.0 million metric tons in 2019 (Statista, 2024). Following the COVID-19 pandemic affected years of 2020–2022 when exports fell gradually to 26.2 million metric tons in 2022, exports rose again to 27.5 million metric tons in 2023.

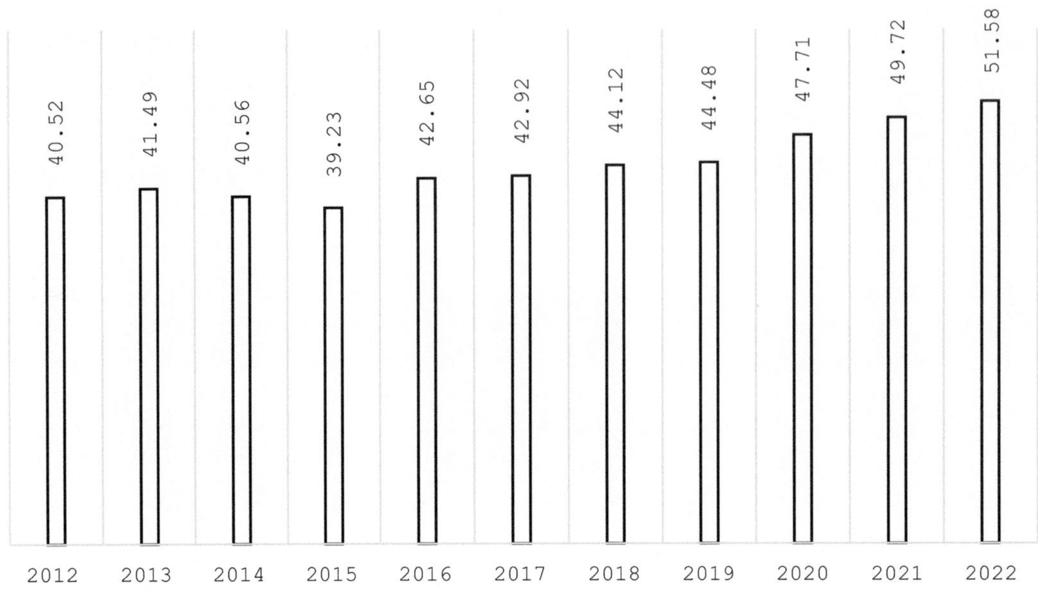

Fig. 5.24 Average ESG score per year. *Source*: Refinitiv Eikon, Thomson Reuters

Among the programmes the government implemented to strengthen sustainability in the oil palm industry, which includes partnership programmes (i.e. cooperation between large palm oil companies and farmers and/or people's companies). Although the majority of oil palm production is undertaken by large companies, a significant portion is still owned by smallholder farmers. The cooperation programmes focus on reducing inequality between the large plantations and the smallholders, which is an integral feature of ESG.

Deforestation has become a perennial complaint levelled against oil palm cultivation. Others include emission of greenhouse gases, and forest and land fires that cause haze. Consequently, the palm oil industry in Indonesia has been subjected to stringent sustainable certification, i.e. Indonesia Sustainable Palm Oil (ISPO) and Roundtable Sustainable Palm Oil (RSPO)). The European Union (EU) implemented the EU Deforestation Regulation (EUDR), which was passed on 31 May 2023 that came into effect on 29 June 2023 with a deadline of 30 December 2024. Companies have since been required to prove that commodities exported comply with the regulations to avert deforestation. Figure 5.25 shows ISO certification in Indonesia, which shows a trend rise from 6 in 2012 to 1005 in 2023.

Following the government's acceptance of ESG practices and in recognition of the EU deforestation regulation in 2023, the Indonesian palm oil industry has become more committed to adopting sustainability practices in the management of oil palm cultivation. Collaborations with non-governmental organizations (NGOs), international organizations and government have increased to enforce these ESG practices, especially at smallholder farmers. For example, the Indonesian Palm Oil Association (IPOA) collaborates with non-governmental organization (KALEKA) to support sustainable palm oil programmes in Kab. Ketapang: firstly, accelerating data collection and certification of independent farmers; second, HCV protection and preventing forest and land fires; third, reducing and handling social conflicts; and fourth, tracing the supply chain involved. The IPOA publishes a guidebook on the protection of women who work in oil palm plantations and is committed to protecting female workers. Furthermore, the IPOA prevents the potential for forest and land fires breakout in palm oil plantations by

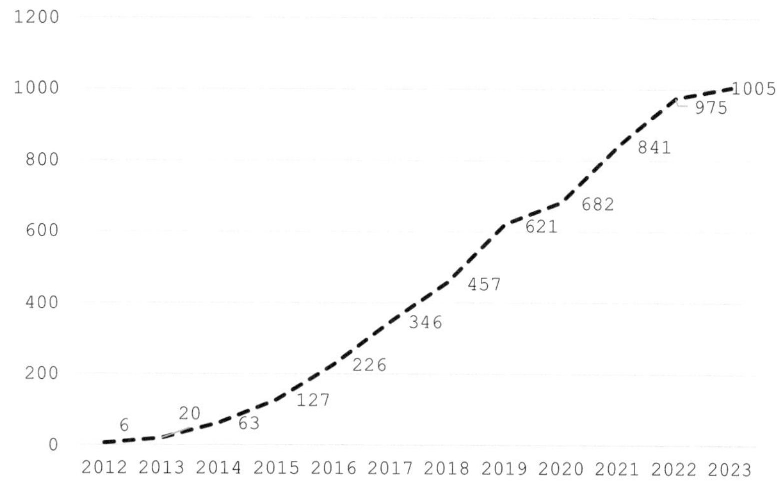

Fig. 5.25 Number of ISPO certifications, Indonesia, 2012–2023. *Source*: GAPKI

calling for forest and land fire alerts; preparation of guidelines for the prevention and control of forest and land fires; and participates in the Weather Modification Technology (WMT) programme.

Conclusions

This chapter examined the development of the agriculture, forestry and fisheries sector of Indonesia, with a focus on its contribution to the national economy, policies to promote it, its performance, the distribution of land under agriculture and efforts to strengthen digitalization and climate resilience. In doing so, the relative emphasis of each of the presidents since Sukarno until Jokowi was detailed.

The country encountered crises over the period 1945–2024, first with the fall of Sukarno, then the global oil crises of 1973–1975 and 1979–1980 followed by the Asian financial crisis in 1997–1998, which resulted in Indonesia defaulting on its debt and balance of payments deficits, and subsequently the global financial crisis of 2008–2009 and the COVID-19 pandemic in 2020–2022. Throughout independent Indonesia, food cultivation in general and rice farming in particular were promoted through the construction of irrigation canals, mechanization, subsidized fertilizer and price stabilization. Indonesia also enjoyed rice aid from the IMF in 1997–2000.

While critics engrossed on the distortionary effects of tariffs and subsidies have been critical of food farming in Indonesia, World Bank (2023) data shows that rice production expanded steadily over the period 1961–2023. Also, rice self-sufficiency levels in 2018–2023 remained over 95%, which is higher than Malaysia's recorded (Annizah & Rasiah, 2025). Rubber and oil palm have been the two major export crops in Indonesia. While the former, largely dominated by smallholder-ownership, has gradually stagnated, oil palm cultivation has expanded strongly. The outer Islands of Indonesia are prime areas where the two tree crops are cultivated. Tuna fishing has become a major export item in which Indonesia enjoyed production of 18% of world output in 2024 (TEMCO.CO, 2024).

Finally, Indonesia's focus on climate change in agriculture started since the turn of the millennium; it has particularly become important from 2014. Arguably one of the major developments on efforts to strengthen climate reliance is the introduction of the POJK, which came into force in 2019. Environmental, social and governance (ESG) reporting has become mandatory in Indonesia and establishments have been given until 2024 to implement ESG practices. In addition to ensuring climate resilience practices, the ESG is also expected to strengthen the position of smallholders and workers against large businesses as fair trade and fair wages are part of the ESG practices.

References

Agung, P. (2023). *7 Start Up Agri-tech Indonesia Bantu Doring Kesahteraan Pertanian.* Hitekno, March 1, 2023 downloaded on August 15, 2024 from https://www.hitekno.com/internet/2023/03/01/194553/7-startup-agritech-indonesia-bantu-dorong-kesejahteraan-sektor-pertanian

Annizah, N.I & Rasiah, R. (2025). Agriculture, Food security and Sustainability in Malaysia, In Rasiah, V. Anter & N.I. Annizah (Eds), *Sustainable Agriculture and Food Security in Southeast Asia: Towards Climate Resilience and Digitalization*, Singapore: Springer Nature.

Arifin, B. (2023). *Reforma Kebijakan Industri Pangan in Akselerasi Menuju Ekonomi Indonesia yang Hijau.* INDEF.

Badan Pangan Nasional. (2023). *Bahan Paparan Presentasi.*

Birowo, A. T. (1974). Masalah Teknologi dan Kesempatan Kerja dalam Pertanian. *Ekonomi Pertanian. Prisma, 3*(6), 1.

Booth, A. (1998). *The indonesian economy in nineteenth and twentieth century.* Macmillan Press.

CEIC. (2023). https://insights.ceicdata.com/Name-your-insight/myseries

Collier, W. L, Soentoro, G. W., & Makali, W. (1974). Sistem Tebasan, Bibit Unggul, Pembaharuan Desa. *Ekonomi Pertanian. Prisma, 3*(6).

FAO. (2008). *An Introduction to the Basic Concepts of Food Security, Rome: Food and Agricultural Organization of the United Nations.* Accessed on September 23, 2023 from https://www.fao.org/3/al936e/al936e00.pdf

GAPKI. (2023). *Pembangunan Kelapa Sawit Berkelanjutan di Indonesia.*

Gie, K. K. (1999). *Ekonomi Indonesia Dalam Krisis dan Transisi Politik.*

Gie, K. K. (2019). *Aku Bermimpi Menjadi Peng-Peng di Republik Banana.*

Glassbuner, B. (2007). *The economy of Indonesia selected readings.* Equinox Publishing.

Hill, H. (1999). *The Indonesian Economy in Crises: Causes, Consequences, and Lessons.* Institute of South East Asian Studies.

KPPU. (2020). *Kajian Efektivitas Progam Kemitraan di Industri kepala Sawit di Indonesia.*

Krisnamurti, B. (2023). *Usaha Stabilitasi Harga Beras in Akselerasi Menuju Ekonomi Indonesia yang Hijau, Inklusif dan Unggul.* INDEF.

Manggi, H., & Siwage, D. N. (2024). *Indonesia's Perennial Rice Self Sufficiency Problem.* April 9, Fulcrum, downloaded on August 14, 2024 from https://fulcrum.sg/indonesias-perennial-rice-self-sufficiency-problem/

McElroy, J. K. (1989). Indonesia's Tuna fishing: Past, present, and future. *Marine Policy, 13*(4), 285–308.

Oswado, I. G. (2023). *Startup Pertanian RI Dapat Pendanaan Awal.* Totalnya Rp 90M, Detikfinance, July 12, downloaded on August 15, 2024 from https://finance.detik.com/berita-ekonomi-bisnis/d-6819181/startup-pertanian-ri-dapat-pendanaan-awal-totalnya-rp-90-m

Rahmadi, S., & Huang, H. W. (2014). Indonesia fisheries development and future strategy. *Marine Policy, 43*(1), 174–183.

Rasiah, R. (2020). Industrial Policy and Industrialization in Southeast Asia, In Arkebe, O C. Cramer, H.J. Chang & R. Kozul-Wright (Eds), *Oxford Handbook of Industrial Policy*, Oxford: Oxford University Press.

Rasiah, R., Cheong, K.C. & Doner, R. (2014). Southeast Asia and the Asian and Financial Crisis, *Journal of Contemporary Asia, 44*(4), 572–580.

Rasiah, R. (2025). Strengthening Agricultural Sustainability and Food Security in Southeast Asia, In R. Rasiah, V. Anter & N.I Annizah (Eds), *Sustainable Agriculture and Food Security in Southeast Asia: Towards Climate Resilience and Digitalization*, Singapore: Springer Nature.

Rasiah, R., & Azmi, S. (2005). *Development of Palm Oil and Related Products in Malaysia and Indonesia.* Downloaded on August 18, 2024 from https://www.researchgate.net/publication/237474157_Development_of_Palm_Oil_and_Related_Products_in_Malaysia_and_Indonesia

Rockmore, W. B., & Jones, F. F. (1996). Business Investment Strategy and Firm Performance: A Comparative Examination of Accounting and Market-Based Measures. *Managerial Finance, 22*(8), 44–56. https://doi.org/10.1108/eb018576

Statista. (2024). *Volume of Total Palm Oil Exports from Indonesia 2014–2023.* March 6, downloaded on August 18 from https://www.statista.com/statistics/1095541/indonesia-palm-oil-export-volume/

Sulaiman, A. A., Herodian, S., Hendriardi, A., Jamal, E., Prabowo, A., Prabowo, A., Mulyantara, L. T., Budihati, U., Syahyuti, S., & Herodian, H. (2018). *Revolusi Mekanisasi Pertanian.* IAARD Press.

Takeuchi, K. (1983). *Export Prospects for Forest Products in Indonesia, 1983–1990.* Division Working Paper No. 1983–1, World Bank, Washington D.C.

TEMCO.CO. (2024). *Indonesia claims 18% of world tuna production, the ministry claims, TEMCO.CO.* June 24 downloaded on August 17, 2024 from https://en.tempo.co/read/1882171/indonesia-corners-18-of-the-worlds-tuna-production-ministry-claims

The Indonesian Financial Services Authority (OJK) rule POJK 33/2014

The Indonesian Financial Services Authority (OJK) rule POJK 51/2017

Thomas, M., & Orden, D. (2004).*Agricultural policies in Indonesia.* MTID discussion papers 78, International Food Policy Research Institute (IFPRI).

University of Michigan. (1998). *Government Policy*. Downloaded on August 4, 2024 from https://websites.umich.edu/~csfound/545/1998/mmaster/govt.html

Van der Meer, C. L. J. (1982). Pembangunan Pertanian dalam Sistem Ekonomi Pasar. *Pangan Dalam Pemerataan Pembangunan. Prisma, 11*(10), 1.

World Bank (2020). *The World Bank in Indonesia*. Downloaded on August 15 from https://www.worldbank.org/en/country/indonesia

World Bank. (2024). *World Development Indicators*. World Bank Institute.

Sustainable Development Challenges in the Agricultural Sector of the Philippines 6

Rene E. Ofreneo

Abstract

The Philippines, host to the International Rice Research Institute (IRRI), is the birthplace of Asia's "Green Revolution" in rice production. In the second half of the 1970s, the country, a rice importer for a century, attained self-sufficiency in rice production. In the 1980s, it reverted to the status of a major rice importer. The agricultural sector has also declined continuously, with the Philippines becoming more and more dependent on imported agricultural products ranging from rice and corn to vegetables and fruits and milk and meat. The sector's share in the GDP has plunged from 25% in 1980 to less than 10% today. The paper argues that this dismal development in Philippine agriculture was due primarily to policy errors, starting with the adoption wholesale by the government of a disastrous "agricultural deregulation" program imposed by the World Bank. This deregulation program has been compounded by the failure of the government to provide a holistic and pro-farmer development blueprint for the sector, on the one hand, and the persistence by the neo-liberal economic technocrats in pursuing a rudderless liberalization program for the sector, on the other.

Keywords

Agriculture · Structural change · Neo-liberal policies · Philippines

Introduction

A Poster Boy for the Green Revolution Turned into an Ailing Agricultural Producer

In the late 1970s, the Philippines was hailed as the miracle rice producer of Asia, the poster boy of the "Green Revolution" program of the International Rice Research Institute (IRRI). Through a supervised and subsidized credit program for the rice farmers dubbed as "Masagana 99" (M99), the martial-law government of then President Ferdinand Marcos Sr. promoted in 1973–1975 the nationwide propagation of the IRRI's high-yielding variety (HYV). In 1976, the government proudly declared the country's success in attaining rice self-sufficiency as a result of the phenomenal rise in average paddy rice harvests, from two tons per hectare in the pre-M99 period to over four tons per hectare under M99 (Ofreneo, 1980). In 1977–1979, the Philippines was able to export a modest amount of rice surplus.

R. E. Ofreneo (✉)
School of Labor and Industrial Relations, University of Philippines, Diliman, Quezon City, Philippines
e-mail: reneofreneo@gmail.com

R. Rasiah et al. (eds.), *Sustainable Agriculture and Food Security in Southeast Asia*, Sustainable Development Goals Series, https://doi.org/10.1007/978-981-95-0477-0_6

There were also production surges in other areas of Philippine agriculture in the 1970s. The traditional sugar and coconut export crops enjoyed relatively stable prices in the global commodity markets in the mid-1970s. The martial-law government also promoted the production of "non-traditional" export crops such as banana, palm oil and sorghum through a partnership program with agribusiness multinationals, e.g., United Fruits and Sumitomo. Thus, in the late 1970s, the Philippines became the Asian rival of Latin American banana producers in the global market for bananas (Ofreneo, 1980).

And then came the collapse of M99 production in the 1980s and the return of the Philippines to the status of a rice-importing country. In 1994–1995, the Philippines also became a net agriculture-importing country because of its growing reliance on imported grains, coffee, vegetables, fruits, garlic, onion, milk, meat and even items such as salt. The trade deficits have become chronic and wider since.

In 2022–2023, the country experienced double-digit food inflation, igniting fierce national debates in the first one and a half years of the Presidency of Ferdinand "Bong Bong" Marcos Jr. Why has the agricultural sector declined precipitously? Why is the sector unable to produce enough to feed the nation?

This chapter is an inquiry on how the Philippine agricultural sector stagnated in the last four decades. Specifically, the chapter inquires on the policy errors and weaknesses that have led to the poor performance of the sector. The chapter is also an inquiry on the lack of a unified vision for agricultural development given the clashing interests of the different politico-economic actors involved in the governance of the sector.

The flow in the chapter discussion is as follows: agricultural decline as reflected in the statistics, the policy shift toward agricultural deregulation in the 1980s, other contributing factors for the weaknesses of the agricultural sector, the unsettled debates on policy directions and conclusions and recommendations.

Decline of Philippine Agriculture Captured by Official Statistics

Indicators of the collapse of Philippine agriculture are not difficult to find. Official statistics indicate rising agricultural imports vis-à-vis agricultural exports, declining share of agriculture in the gross domestic product (GDP) and grinding rural poverty.

On growing import dependence, Table 6.1 shows that in 1993 the Philippines was still a net agricultural exporting country. However, since 1994, the Philippines has become a net agricultural importing country, with the trade deficits widening each year. Agricultural imports from the United States, the traditional destination of the country's major agricultural exports, rose to US$3 billion a year in 2010s while Philippine exports to the United States remained stagnant at around US$1 billion a year. Thus, in the USAID web site in 2014, the US Department of Agriculture (USDA) proudly proclaimed that the U.S. agricultural exports to the Philippines grew by 123% in 2009–12, making it one of the United States' fastest-growing markets. The Philippines ranked ninth among U.S. Customers in fiscal year 2014, with sales of farm and food products reaching a record $2.8 billion. Top U.S. exports to the Philippines include soybean meal, wheat and dairy products.

In 2022, the USDA website reported that the Philippines moved up to number eight in the top importers of American agricultural products, with total imports for 2022 amounting to US$4.15 billion. And the list of imports now includes processed vegetables, various food preparations and meat—poultry, beef and pork.

Ironically, the government campaigned in 1994 for the Senate ratification of Philippine membership in the then newly established World Trade Organization (WTO) based on the argument that agricultural exports would shoot up and that the country would enjoy huge agricultural trade surplus and half a million new jobs a year. And yet, Table 6.1 shows the reverse.

Table 6.1 Philippine agricultural exports and imports, 1990–2014 (in US $ M)

Year	Exports	Imports	Exports minus imports
1990	1683	1665	+ 18
1991	1794	1422	+ 372
1992	1842	1681	+ 161
1993	1916	1887	+ 29
1994	2039	2332	−293
1995	2457	2979	−522
1996	2292	3502	−1210
1997	2299	3557	−1258
1998	2201	3199	−998
1999	1778	3219	−1441
2000	2026	3104	−1078
2001	1958	3087	−1129
2002	2000	3303	−1303
2003	2375	3158	−783
2004	2553	3531	−978
2005	2729	3860	−1131
2006	2830	4073	−1243
2007	3237	4671	−1434
2008	3961	6948	−2987
2009	3174	5601	−2427
2010	4129	6823	−2694
2011	5417	7026	−1609
2012	5075	7240	−2165
2013	6425	7157	−732
2014	6948	8668	−1720
2015	5132	10,966	−5834
2016	5280	12,518	−7238
2017	6580	11,762	−5182
2018	6118	13,141	−7023
2019	6677	13,532	−6855
2020	6200	12,576	−6376
2021	6820	15,707	−8887
2022	7500	19,303	−11,803

Source PSA (2023) http://psa.gov.ph

Table 6.2 Structural change, Philippine, 1980 and 2019

Value-added share (% of GDP)	1980	2009	2019
Agriculture	25.1	13.1	8.8
Industry	38.8	31.7	30.3
Services	36.1	55.2	60.8
Employment Share(% total employment)			
Agriculture	51.8	35.2	22.8
flndustry	15.4	14.5	19.1
Services	32.8	50.3	58

Source PSA website (psa.gov.ph)

Table 6.3 Agricultural rice production and yield, Philippine, 2013–2022

Year	Area planted to rice (million has.)	Total national Production (million tons)	Yield (metric ton per ha.)
2013	4.75	18.44	3.89
2015	4.66	18.15	3.90
2017	4.81	19.28	4.01
2019	4.65	18.81	4.04
2021	4.81	19.96	4.15
2022	4.8	19.76	4.11

Source PSA (2023) website (psa.gov.ph)

The growing trade deficits means a declining agricultural sector, which is also reflected in the shrinking share of agriculture in the country's gross domestic product (GDP). This decline is paralleled by the similar decline in the industrial sector, a phenomenon which has alarmed the Asian Development Bank (ADB). As documented by ADB's Norio Usui (2012), services have been growing while agriculture and industry have been shrinking (see Table 6.2). Agriculture's contribution to the national GDP was reduced by nearly half its share from 1980 to 2009 or a span of three decades. It shrank further to less than 10% of the GDP in 2019.

The sector's contribution to overall employment has also gone down. Of course, the decline in agricultural employment is normal when countries succeed in its industrialization drive and in modernizing its agriculture, after which services expand. But in the case of the Philippines, it has become a services-led economy without going through an industrial revolution and full-scale agricultural modernization. Industrialization and modernization have enabled the developed countries to grow a consumer society and maintain high levels of agricultural productivity despite a shrinking agricultural work force.

In the case of the Philippines, services are sustained largely by the remittances of over ten (10) million overseas Filipino workers (OFWs), composed of short-term migrant contract workers and permanent immigrants. The flow of remittances, over US$30 billion a year (Opiniano &

Ang, 2024), has enabled the country to have a consumption-led growth pattern as reflected in the endless expansion of malls, retailing shops, educational institutions, etc. The problem is that consumption is largely done by the Philippine elite and the OFW families able to receive a steady flow of remittances overseas. Most of the families in the rural areas do not receive any similar remittances because the overseas labor market is highly selective in terms of education and formal urban work experience. Most of the rural poor have not even finished secondary education and have been shut out of the formal labor market.

To meet the food requirements of a growing population, nations with land and other natural resources try to intensify agricultural production through varied productivity-enhancement programs such as the development of high-yielding seeds, better water or irrigation systems, agricultural mechanization and empowerment of the farming population through agrarian reform and skills upgrading. This is not happening in the Philippines, as reflected in the official data on weak growth in agricultural productivity in the last three decades.

This low growth in agricultural productivity is illustrated by rice. The surge in rice production in the 1970s has never been replicated by the succeeding post-Marcos administrations. The Masagana 99 program, an integrated program to promote the Green Revolution via a loan package for the new high-yielding seeds and agricultural inputs, succeeded in transforming the palay deficit crisis in the early 1970s into a situation of self-sufficiency and surplus in the second half of the decade. However, it is all downhill in the succeeding decades.

Note that the rice production surge in the mid-1970s was associated with the Green Revolution or the propagation of the IRRI HYV seeds, which enabled rice farmers to achieve a "Masagana 99" cavans yield or over 4 tons per hectare. As discussed in a subsequent section, the M99 program collapsed in the 1980s and 1990s. It is only in recent years, beginning in 2017, that the average per hectare yield in the Philippines reached 4 tons or more (see Table 6.3). And yet, the average paddy yield in China, Indonesia, South Korea and Vietnam are already over 5 tons per hectare since 2010 (Pabuayon et al, 2013).

Given the foregoing developments, one can only conclude that the Philippines has been under-investing in the sector, meaning no or limited investments on better seeds, agricultural machinery, enabling infrastructures and support services for the farming population, training among others (Mangabat, 2013). Comprehensive data on these investments are difficult to secure. However, an indicator of the country's low investment in agriculture can be seen in the utilization of machinery in various phases of agricultural production. Table 6.4 shows the level of mechanization in select crops in the Philippines. For various activities in the farm, the most common description used by agricultural engineers is "Low", meaning non-usage of agricultural machinery for work is done manually or through the combined efforts of man and animal. From Table 6.4, one can easily conclude that the most advanced use of machinery is in the sugar and pineapple plantations, which adopted mechanization decades ago.

Low means non-use of machines and reliance on man or combined man-animal efforts; high means use of mechanical power operated by man; intermediate is in between the two.

The foregoing observation on low level of mechanization is supported by the low number of tractors and threshers used in the Philippine farms compared to those of its neighbors in Southeast Asia. The Philippines has the lowest rate of agricultural mechanization compared to China and the major agriculture-producing countries in the region. Based on studies commissioned by the United Nations Asian and Pacific Center for Agricultural Engineering and Machinery (UNAPCAEM), the mechanization level in the Philippines in the crops sector is still in the developing stage. It is relatively low compared to other Asian countries like Japan, Korea and China (Peeyush & Ou, 2014). According to the Philippine Center for Postharvest Development and Mechanization (PhilMech), the country's level of mechanization is at 1.23 horsepower per hectare (hp/ha) compared to Japan's 18.87 hp/ha, Korea's 9.38

Table 6.4 State of mechanization, selected crops, Philippine, 2012

Operation	Rice/Corn	Vegetables/legumes	Coconut/Fruits	Sugar/Pineapple
Land preparation	Intermediate to high	Low		Intermediate to high
Planting	Low	Low	Low	Low to intermediate
Crop care	Low	Low	Low	Low to high
Harvesting	Low	Low	Low	Low
Threshing/shelling	Intermediate to high	Low (legumes)		
Cleaning		Low		
Drying	Low	Low (legumes)	Low	
Milling/processing	High	Low	Low	

Source Suministrado (2013)

hp/ha, Thailand's 4.20 hp/ha and India's 2.22 hp/ha (Ranada, 2014).

The Spectacular Collapse of M99 and the World Bank's "Deregulation" Program

Now on deeper reasons for the decline of agriculture through the decades.

First, why the collapse of the M99 or the HYV promotion program in the 1980s? There are two major explanations.

First, government spending on the sector, especially investments on supporting infrastructures and services, ground to a halt. A monstrous debt crisis pushed the Marcos government to the edge of bankruptcy, forcing it to stop all spending on the sector. The national debt crisis coincided with the debt crisis of M99 borrowers who could not pay the loans they incurred in the 1970s, aggravated by the government's withdrawal of debt and farm input subsidies as well as phaseout of price support for harvested rice paddies. In turn, the farmers' debt crisis meant farmers' underinvestments on needed fertilizer and pesticides that the HYV seeds required. (Ofreneo, 1987)

The second explanation was closely related to the first. Non-government spending on support for the agricultural sector and the struggling rice farmers was reinforced by the government's acceptance of the World Bank's policy recommendation for the sector—a shift to "agricultural deregulation". The Department of Agriculture (DA) was resistant to the proposed policy shift given that the government's success in the implementation of M99 in the 1970s was partly due to a web of subsidy programs involving lower interest rates on farmers' loans, cheaper fertilizer and pesticides and support prices for paddy and rice. And yet the DA and the Marcos government had lost the power to resist the World Bank's policy dictation on "deregulation" given the national debt crisis, which triggered an economic recession in 1981–1982 and a full-blown economic depression in 1983–1985 (Ofreneo, 1987).

And yet, the World Bank took a hardline posture on its deregulation program. This was based on the "new" neo-liberal economic thinking that swept the Bank and other aid givers in the 1980s. Accordingly, developing countries would grow faster under "free market" conditions, as farmers, guided by signals in the market, are bound to optimize production by investing on the most profitable uses of the land such as shifting farming in favor of planting high-value crops. Based on this deregulation thinking, the World Bank pressed the crisis-ridden Marcos regime to adopt the following measures: scrapping of the fertilizer and irrigation subsidies for farmers, removal of credit subsidy, "deregulation" of interest rates on agricultural loans, "decontrol" in the prices of agricultural commodities, permission for the private sector to import feed grains and other agricultural products, and dismantling of government support to sugar and coconut industries (Ofreneo, 1987).

In 1986, the Marcos regime was toppled by a "People's Power Revolt". However, the succeeding government of President Corazon C.

Aquino simply continued the agricultural deregulation program. The DA Secretary Ramon Mitra crafted "A Short-Term Recovery Plan for the Rural Sector", which stressed the government's policy of creating "an economic environment for the farmer free from unnecessary and costly government institutional and policy interventions". Per the recovery plan, the support given to the farmers would be in the form of "a market information service" and access to the "new technology". (Ofreneo, 1987)

To reiterate, the rationale given by the World Bank for the deregulation program was the need to boost agricultural efficiency and productivity through a freer market sans government intervention. And yet, the opposite happened—the collapse of M99 and the decline of the agricultural sector as a whole. The indebted farmers had a hard time adjusting to a deregulated environment and the withdrawal of government assistance. In fact, the sector became rudderless while many rural infrastructures, particularly irrigation systems, were neglected. The 1980s became a lost decade for the agricultural sector.

Agricultural Deregulation in the 1990s and National Debates on WTO Membership

Despite the poor performance of agriculture in the 1980s under the SAP's deregulation program, the government pursued in the 1990s a bundle of additional agricultural liberalization programs, all of which reinforced the agricultural deregulation policy in place. These programs included the multilateral trade agreements under the World Trade Organization (WTO) and the regional ASEAN Free Trade Agreement (AFTA) of the Association of Southeast Asian Nations (ASEAN). The thrust of these trade agreements is uniform—open up the economy in general and the agricultural sector in particular.

The most significant among these agreements is the Philippine membership in the WTO. Under the WTO, membership means adherence to the Agreement on Agriculture (AoA), a free trade agreement requiring WTO member states to remove quantitative restrictions on agricultural imports in favor of "tarrification" and to reduce tariffs once they are in place. In short, membership in the WTO means making all agricultural products tradeable (either importables or exportables) under a regime of trade liberalization.

When the government announced to the public its intention to apply for membership in the WTO, various farmer organizations expressed opposition. They registered their position to the Senate, which was given the task of either ratifying or rejecting Philippine membership in the WTO. In the intense debates in the Senate in the last quarter of 1994, government trade negotiators, backed up by neo-liberal economists, argued that membership would be beneficial to the country. They also came out with projections on the huge positive gains accruing to the Philippines (Table 6.5)—half a million new jobs a year created in agriculture, P60 billion GVA to be posted by the sector each year and an agricultural trade surplus amounting to P3.2 billion a year.

After several months of debates, the liberalizers won. The liberalizers argued that Philippine underdevelopment was due to economic inefficiency and low productivity, which the liberalizers blamed on the lack of market competition. They reasoned out that the main benefit of trade liberalization is to promote more efficiency in the economy by encouraging resources to go toward the more efficient sectors, usually the export sector. In the case of agriculture, heavy protection through trade restrictions, high tariffs and subsidies allegedly resulted in inefficiency and wastage of resources. Protection supposedly prevented the inflow of investments, especially in high-value crops. SAP's deregulation program, therefore, allows the economic managers to overcome the boom-and-bust pattern of economic growth (Orbeta, 1996).

As a condition for membership in the WTO, the Philippines committed to tarrify the sector, subscribe to a tariff-reduction schedule, eliminate non-tariff barriers and enforcing WTO trade disciplines (Habito, 1999). The WTO membership became an excuse for the further consolidation of the deregulation program.

Table 6.5 Projected annual gains under the WTO, Philippine, for 1995 onward

Sector	Job creation	Export earnings	GVA	Other gains
Agriculture	500,000 new jobs/ year	P3.2 billion net export surplus/ year	P60 billion/ year	
Industry	500,000 new jobs/ year, rising to 700,000–800/year			More competitive "modernized" industries; "mature" industries adopting "new trading behavior" to avoid collapse"

Source Handouts distributed by department of agriculture and department of trade and industry in the Senate (1994) debates on the ratification of Philippine membership in the WTO, cited in the Fair Trade Alliance (2006)

Government Measures for the Vulnerable: Weak Implementation

As pointed out, farmers' organizations, with the support of a number of civil society organizations, opposed WTO membership of the Philippines. They raised issues such as the lack of readiness of the industrial and farm sectors to global competition, the absence of safeguards and safety nets for home-based producers and their workers, the danger of jobs being drowned out by foreign goods (especially those coming from China), and the high cost of doing business and farming and so on (FairTrade, 2006).

To allay the fears of those opposed to WTO, the government dangled a number of safety nets and competitiveness-enhancing measures to insure the survival of Filipino workers, farmers and industrialists in a globalized environment. The government of President Fidel Ramos set up a Master Plan, also known as a safety net competitiveness enhancement plan to minimize such adverse effects (Philippine Senate, 1994). Subsequently, the Senate (Congress of the Philippines, 1997) came up with the Agricultural Competitiveness Enhancement Program (ACEF) and the Agricultural and Fisheries Modernization Act (AFMA).

And yet, three decades after, Philippine agriculture, with the exception of a few areas, is still hobbled in backwardness and underdevelopment. So why have ACEF and AFMA failed to deliver competitiveness and modernization?

In the case of ACEF, the intent appears good on paper but the fund implementation leaves much to be desired. From 1999 to 2003, the total actual collection amounted to a paltry P5.4 billion, with half or P2.6 billion unallocated as of August 2006 (Dy et al., 2008). ACEF was supposed to help small farmers modernize through special lending program. And yet, the bulk of the borrowers turned out to be the richer sugar producers.

However, the bigger program is AFMA, which was enacted right after the Senate ratification of Philippine membership in the WTO. The AFMA program was designed to increase the productivity of agriculture and fishery sectors and prepare them for globalization. To achieve this, the law mandates a yearly budget of P 17 billion pesos for AFMA. However, the expected additional P20 billion in the first year (1999) and P15 billion a year in 2000–2005 did not materialize. The annual budgetary allocations for agriculture and AFMA even declined in the said period.

Data on the achievements of both ACEF and AFMA were hardly discussed in public because the DA had been niggardly in reporting about the impact of ACEF and AFMA. There are no published official reports on the state of agricultural modernization. One explanation for this failure of DA to report on the accomplishments under ACEF and AFMA is due to the questionable uses of the ACEF and AFMA funds under the various administrations. The problem is further complicated by the revelations in the Congressional hearings that some modernization-related projects of DA, DAR and DENR went to spurious or bogus NGOs and local government units (LGUs), which got the money but did not use the funds for agricultural modernization.

One major study on AFMA was undertaken by a group of researchers from the Center for

Table 6.6 AFMA objectives vs. outcomes

AFMA objectives	CRC findings
1. Poverty alleviation & social equity	Rural poverty remains a challenge; lack of resources/non-compliance with AFMA program
2. Food security	Some improvement
3. Rational use of resources	R&D and marketing not given priority compared to production support
4. Global competitiveness	No significant breakthroughs; same export products
5. Sustainable development	Farm incomes did not improve for coconut farmers, Job creation wanting

Source Dy et al. The AFMA Experience (2008)

Research and Communication (CRC) for the AFMA Congressional Oversight Committee. The executive findings of the group are summarized in Table 6.6.

The foregoing clearly shows a hocus-pocus program of modernization under globalization and liberalization. Aimless and mindless. The biggest problems: no clear vision of development and no one to lead agricultural modernization. There is no Commander-in-Chief to guide the agricultural sector toward the high road to competitiveness and modernization.

Other Missing Safety Nets

An Anti-Dumping Law (RA 7843) was enacted by Congress in 1995, right after the accession of the Philippines to the WTO. However, due to the supposed inconsistencies of its provisions to the WTO's Agreement on Anti-Dumping, it was later amended as RA 8752 and passed in August 1999. Another trade remedy measure against unfair trade practices is the Safeguard Measures Act. This measure was enacted five years after Philippine membership in the WTO, in 2000. This was enacted to impose additional tariff, as a temporary safeguard measure, to give breathing space to a domestic industry that is suffering from serious economic injury or threat. There are two kinds of safeguard measures: (1) General Safeguard Measure, a remedy available to industrial producers, and (2) Special Safeguard Measure (SSM), which is available exclusively to agricultural producers (Fair Trade Alliance, 2006).

Have the government, CSOs and CSOs used the above safeguards in the numerous cases of dumping and smuggling of garlic, onion, sugar and other products? The answer is: hardly. Also, no one in the DA or in the government seems to be on top to insure that local producers are not harmed and that safeguards and fair rules are put in place and implemented decisively.

The damage brought about by the aimless and mindless liberalization in agriculture cannot be more telling than the emasculation of the National Food Authority (NFA), which is originally tasked to intervene in rice trade to provide price support to farmers and ensure available and cheap rice for the population. Since the SAP years, NFA has drastically reduced its palay procurement operations, from as high as 10% of total palay production in the 1980s to as lows as 1–2% in the 1990s and 0.5% in 2005 (Cororaton & Yu, 2019), significantly weakening its price support and stabilization function over the years. Worse, NFA has increasingly relied on rice importation to meet its food security stock and in stabilizing rice price and supply. With minimal budget allocated for its trading operations, NFA resorted to borrowing short-term commercial loans sapping the agency's meager resources.

In 2014, NFA's budget for "food security and stabilization" dramatically rose to P8.9 billion, from its usual P4 billion allocation. But even such amount can only account for a miniscule 1% of the 20 million metric ton palay output during that year. Ironically, these subsidies have gone mostly to the private sector rather than the poor farmers.

The government had been targeting rice self-sufficiency and yet, it had failed to provide NFA sufficient funds that could have allowed it to provide sufficient price support for rice farmers and subsidies for poor consumers. NFA's palay procurement rate has failed to keep pace with the increasing volume of palay production.

Continued Decline of Agriculture at the Turn of the Millennium

With the sector literally left to its own devices and with poor government support and guidance, agriculture continued to decline at the turn of the millennium onward. Some developments from 2000 to 2010s are worth citing here. These include the following:

Drying up of farm credit Since the collapse of M99 in the 1980s, there was hardly any credit assistance flowing to the farms. Most commercial banks have ignored an existing "agri-agra" law requiring banks to set aside at least 25% of their loanable funds for agriculture (Miraflor, 2020). The hesitation of the banks to provide credit to small farmers and small fisherfolk is well known—the former are considered "risky" clients. On the other hand, the voluminous documents needed by the banks dissuade many farmers, especially those that have only gained informal education or little formal education, in applying for loans.

Non-expansion and poor maintenance of the irrigation system Agriculture is dependent on water management, and yet, the government has allowed the irrigation system to deteriorate since the 1980s. Many of the irrigation systems in the Philippines are constructed by individual farmers, especially the construction of shallow tube wells and low-lift pumps. As it is, the irrigated area of the country pales in comparison with the irrigated areas of other Asian countries, particularly with the most competitive agriculture-producing countries, namely Thailand, Vietnam and China.

Slow-moving implementation of agrarian reform In the 1970s, President Marcos Sr. decreed (Presidential Decree No. 27) land reform in the rice and corn areas. He decreed the transfer of lands above the five hectares owned by rich landowning families to landless tenant farmers. He also decreed the transformation of tenancy relations (share-cropping) into more capitalistic lease holding arrangements.

And then President Corazon Aquino signed in 1988 a Comprehensive Agrarian Reform Law (CARL), which extended the land reform program to cover all crops. The CARP coverage included pineapple, banana, sugar, coconut, rubber and other farms.

Today, three decades and a half after, CARP implementation remains unfinished. Many agrarian reform beneficiaries (ARBs) have remained heavily indebted, unable to transform their lands into productive ones. Crisis in farming has induced a growing number of ARBs to mortgage their farms to creditors, lease their farms to rich farm developers and corporations (common in banana plantations), and even sell the lands to new land owners. The program is also hobbled by numerous land cases, meaning unresolved conflicts on land transfer, land titling, compensation to land owners and so on. This is the reason why the Department of Agrarian Reform (DAR) has remained a major line agency of the government.

On the whole, the slow-moving and conflict-ridden implementation of land reform has adversely affected agricultural productivity in large areas of the country. CARP has failed to deliver the promises written in the law: agricultural modernization and transformation of sleepy villages into vibrant communities of producers.

A 2018 study of CARL's implementation by the Asian NGO Coalition for Agrarian Reform and Rural Development (Quizon et al., 2018) documented the following troubling concerns in CARP implementation:

- DAR reported success in covering 90% of CARP-able lands. However, the remaining 10% (over 560,000 hectares) are the most difficult because landlord resistance in these areas are most intense. Hence, the study estimates CARP completion to take another 19 years.
- There are 200,000 hectares of private lands that are deemed "CARP-able" which have escaped DAR's notices of coverage (NOCs).
- Around 1.4 million ARBs in large plantations and who were given collective "Certificates of Land Ownership" were still waiting for the parcellization of land as of 2017.
- A large number of ARBs who have entered into "grower's contract" or leasehold arrangements with big agribusiness corporations have

lost control of their lands and have been piling up debts.

- Many ARBs are facing exhausting and expensive legal cases such as non-installation on their awarded lands, overlapping land rights, absence of legal contracts and so on.
- Many ARBs have also pawned or sold their lands.

Declining farm hectarage The crisis in farming arising from the aimless deregulation program and weaknesses in government support programs and agrarian reform implementation has given rise to a dangerous but largely un-monitored trend in Rural Philippines—the declining farm hectarage (Project Development Institute, 1999). This, of course, is also partly due to pressures of urbanization. But the selling of farms and their conversions into subdivisions and other commercial purposes are undoubtedly rooted in the crisis of farming in general. Farms are disappearing in Luzon, Visayas and Mindanao. Declining investments in farming means declining productivity, declining farm income and declining interest in farming. This is fueled to a great extent by the neglect of the government of agriculture under SAP/deregulation and the entry of cheaper agricultural products from other countries, compounded by rampant smuggling. Hence, many farmers are giving up farming, selling or renting out lands or converting their farms into non-farm uses.

In turn, all of this is contributing to another serious and un-monitored problem—the lack of interest among the youth in Rural Philippines to continue the culture of farming because they do not see any bright future in farming. This is why agricultural schools in the country are having less and less students enrolled in agronomy or related agricultural disciplines. What is happening instead is that these agricultural schools have been reengineering themselves by offering non-agriculture courses such as mass communications and HRM because majority of the enrollees do not want to go into farming.

Tariffication of Rice: Two and Half Decades of Debates

The agricultural deregulation program, which the World Bank imposed on the debt-saddled Philippines in the 1980s, was hailed by neo-liberal economists and technocrats as a needed pill to spur growth in the countryside. However, some of these economists and technocrats found the early package of policies under deregulation as still insufficient or lacking.

One major policy debate revolves around the tariffication of rice and the role of the government's NFA in the importation of rice. Arsenio Balisacan, the present Director General of the National Economic and Development Authority (NEDA), wrote in 1998 that the "deregulation of agriculture has been substantially incomplete" because it failed to include the abolition of "NFA monopoly of international trade and domestic market operations in rice and corn" (Balisacan, 1998). He also lamented the government's enactment in 1992 of the "Magna Carta of Small Farmers" (RA 7607), which prohibits importation of agricultural products "that are produced locally in sufficient quantity". Balisacan also criticized the CARP, which he wrote "diminished the collateral value of agricultural lands".

It is clear from the foregoing that Balisacan's position was that of a free trade advocate. On the other hand, farmer leaders had remained steadfast in their opposition to the deregulation program and the diminution of the role of the government in the rice market. Responding to the collective pressures of the organized farmers, the government applied to the WTO for the exemption of rice, a political and economic staple, from the coverage of agricultural tariffication.

The rice tariffication exemption was granted and renewed after 10 years. The exemption lasted until 2019 or two and a half decades. This is a reflection of the intensity of the debates between the pro- and anti-liberalizers, especially with regard to the rice commodity.

Nevertheless, there were continuing debates between the pros and cons of rice tariffication and agricultural deregulation. The Rice Watch and Action Network organized a formal debate in 2006 on the positions held by both sides.

Arze Glipo of the Integrated Rural Development Foundation gave the following reasons advanced by the pro-exemption camp: (a) markets are not free and are dominated by monopolies; (b) free trade punishes the weak; (c) comparative advantage is an illusion; (d) elimination of government intervention does not increase efficiency; and (e) the so-called playing field is good only for big capital. Ramon Clarete, a liberal economist, and Jessica Cantos of Action for Economic Reforms gave the following arguments in favor of tariffication: (a) consumers will pay less and farmers will receive more; (b) quantitative restrictions are uncertain, non-transparent and not subject to political lobbying; (c) tariff revenues can be used as subsidy to farmers; (d) tariffs instill discipline among the players in the market; and (e) the NFA and the traders create a supply driven traders' market to the detriment of the farmers and consumers (Clarete, 1992; Rice Watch and Action Network, 2006).

The foregoing outline of the debates went on until 2019 when the government enacted the Rice Tariffication Law (RTL) despite the fierce lobbying against the proposed RTL by the farmer organizations and their supporters in the civil society movement. The immediate reason for the passage of the law was the rice crisis of 2018. The NFA was unable to import enough on time, a situation that resulted in skyrocketing prices of rice. On the other hand, the NFA blamed the crisis to the failure of another government agency advising the NFA on when to import or not, the NFA Advisory Council headed by NEDA, a proponent of deregulation and diminution of the role of NFA in the import and domestic market.

After the RTL passage, the domestic market was flooded with imported rice brought into the market by the private importers-distributors. The rice shortage in the market disappeared. And the NFA's role was effectively whittled down, with the agency limited to the job of maintaining "buffer stock" good for 15 days during harvest period and 30 days during lean months, with the proviso that NFA rice sourcing should be based on the procurement of rice from local producers.

The problem is that the RTL enactment, contrary to the projections made by its proponents, resulted in a double whammy for the nation: continuous rise in the price of rice and decline of the farm-gate price for local rice harvest. According to some RTL supporters, this scissor phenomenon was temporary. But the scissor phenomenon continued from February 2019 (passage of the law) throughout 2019, 2020, 2021, 2022 and 2023.

Thus, not surprisingly, the farmer organizations and CSOs opposed to the RTL have maintained their unity in continuing their criticisms of the RTL and in opposing a new "enemy"—the Regional Economic Partnership Agreement (RCEP). Their united opposition led to the postponement by the Senate of its ratification of Philippine membership in the RCEP by more than a year. RCEP was ratified in February 2023 after the Senate came up with a list of support measures for the strengthening of local agriculture. But like the 1994 Senate and DA promises in relation to WTO membership, the new Senate package of commitments to the Philippine farming sector under another layer of trade liberalization is seen by many organizations as illusory.

The Climate Emergency Deepens Agricultural Crisis

Another major stumbling block to Philippine agricultural development is climate change (CC). The risks CC creates for the country are well chronicled by global and local media. CC devastations come in the form of typhoons or cyclones (around 20 a year), floods, drought, farm disruptions, soil erosion, damages in infrastructures, community displacements, sea rise, river and lake siltation and so on (Ofreneo, 2012).

Thus as pursued by almost all countries in Southeast Asia and in the world (see Rasiah, 2025), the Philippines has launched master plans to address global warming and the challenge of instituting climate resiliency programs, including the promotion of "climate-smart agriculture" or CSA.

The climate projections undertaken by the Philippine Atmospheric, Geophysical and Astronomical Services Administration (PAGASA) for 2020 and 2050 indicate the whole of the Philippines will get warmer each year. The observed mean temperature rose by 0.01 °C per year from 1951 to 2010.

In addition, there is evidence to show that the cyclone's passage over the Islands of Luzon, Visayas and Mindanao over a 30-year mean shows that there has been a slight increase in the number of cyclones in the Visayas during the 1971–2000 period when compared to the period 1951–1980 and 1960–1990 (PAGASA, 2011). The climate projections also show that, generally, there shall be a fall in rainfall in most parts of the country during the summer (MAM) season, though there is a likely increase in rainfall during the southwest monsoon season in June, July and August (JJA) until the transition months of September, October and November (SON) in most areas of Luzon and Visayas. Increase in rainfall is also likely during the northeast monsoon months of December, January and February (DJF), particularly in provinces/areas characterized as Type I climate. There is, however, a generally decreasing trend in rainfall in Mindanao, especially by 2050 (PAGASA, 2011). These changes are expected to dampen agricultural productivity in the Philippines with negative ramifications for the country's food security. To make matters worse, the Philippines has yet to submit its carbon net-zero targets to the United Nations Convention for Climate Change (Climate Change Tracker, 2023; Rasiah & Gopi Krishnan, 2025).

Government Climate Response:

Digitalization and Climate-Smart Agriculture

In response to the existential climate threat, the Philippines launched the National Climate Change Action Plan (NCCAP) 2011–2028 under the Administration of President Benigno Aquino II. The NCCAP prioritizes food security, water sufficiency, ecological and environmental stability, human security, climate-smart industries and services, sustainable energy, and knowledge and capacity development as the strategic direction for 2011 to 2028 (Philippines, 2011).

The Philippines Office of the Executive Director for Research and Development of the Department of Science and Technology emphasized in 2023 the government's digitalization as an instrument to drive innovation and efficiency to transform agriculture (Open Gov, 2023). Among the prioritized strategies earmarked include precision agriculture that benefits from digitization for farmers to raise yields, and optimize agricultural planting, irrigation, and pest management by deploying global positing system (GPS) and Geographic Information System (GIS) technologies, drones, robots and satellite imagery. The range of precision and smart agriculture practices to be promoted includes drip irrigation, controlled-release of fertilizers to reduce waste, water conservation, instant access to market prices on inputs and output, and farming methods that do not negatively impact on the environment. These developments have been planned with support from the Philippine Rice Research Institute (PhilRice) and the International Rice Research Institute (IRRI) (Open Gov, 2023).

Meanwhile, the Balik Scientist Programme focuses on bringing back Filipino scientists to contribute to the nation's growth by targeting the harnessing of global science for sustainable development and innovation, including digitalization, artificial intelligence and robotics (Open Gov, 2023).

However, the Climate Change Tracker (2023) considers the Philippine initiatives on climate change and digitalization as still falling short in achieving their desired consequences (Open Gov, 2023). Besides, the Philippines, while experiencing a rising share of non-renewable energy use in its total energy consumption, remains greatly dependent on coal and gas to fuel its electricity requirements. In fact, the Climate Change Tracker (2023) questions the Philippines' ability to achieve its National Renewable Energy Program (NREP) 2020–2040 target of

35% renewable energy share in power generation by 2030, and 50% by 2040.

In general, the Philippines has a good body of environmental laws and an outline of needed policy measures and programs to build up resiliency in agriculture and other economic sectors. The problem, however, is in the weak implementation of the laws and programs as reflected in the failure of the country to fulfill its commitments under the Paris Agreement (Ofreneo, 2012).

Conclusion: Agricultural Collapse, Yet Renewed Call for Agricultural Recovery

In summary, it is abundantly clear that the SAP-deregulation decades—1980s, 1990s, 2000s and the present—are lost decades for Philippine agriculture. Neo-liberal economists and government technocrats have been purveying the idea that agricultural deregulation is the key to unlocking the wealth in the countryside. This has not happened based on the actual progress or lack of it in the Philippine farming sector.

But why has the theory of deregulation failed the Philippines. First, the agricultural deregulation model assumes the existence of an agricultural market where the producers are fully informed and capacitated with land rights and access to credit, technology and market. As it turned out, most of the farmers (and even officials of the DA) were not even fully informed of the agricultural deregulation program or what it meant, for there were no consultations with the stakeholders and no information drives on SAP and the agricultural deregulation program.

In rice farming, agricultural deregulation had little chance of success because the small farmers were unable to influence the market or shift to the so-called high-value crops due to the following reasons: farmers' immediate need for cash after harvest, prevalence of tied sales, farmer dependence on the middlemen, dominance by the big traders, inability of heavily indebted farmers to make sound investments, inability of small farmers to switch to new crops because the infrastructures in the community were designed for rice farming, and the unresolved land tenure issues (Ofreneo, 1987).

As things turned out, the biggest winners in the RTL and deregulation programs are the big private traders-distributors who have managed to control the buying of local products and the importation of the same, allowing them to dictate the prices at the consumers' level and the prices at the farm-gate level. In fact, this is what the numerous Congressional investigations on the calamitous swings in prices of rice, onion, salt and so on are showing.

As to the initiatives to strengthen climate resilience and smartification of agriculture to raise yields and food security, the impact has yet to show tangible improvements. Indeed, critics remain unconvinced that there is policy traction between smartification efforts and improvements in agricultural productivity and food security.

And yet, policy makers have remained hopeful. In early 2024, President Ferdinand Marcos, in response to the twin agricultural disasters in 2023 (high food inflation and decline of rice farm output), launched a sweeping funding program for agricultural revival. The government re-strengthened the capacity of the NFA to buy palay and import rice, allocated additional funds for irrigation and farm mechanization and tried to revive key features of the Masagana 99 program of the 1970s such as increased assistance to farmers in terms of farm credit and procurement of needed inputs.

In addition, he appointed as Francisco Tiu Laurel, a successful fishery businessman, as Agriculture Secretary. Immediately, Tiu Laurel launched a series of dialogues with farmer and CSO leaders who have been criticizing the government's neo-liberal program in agriculture, including the rice tariffication law and the country's participation in the RCEP. In a dramatic manifestation of the government's sincerity in the dialogue process, Tiu Laurel recruited and hired some of these farmer leaders to take care of the rice sector and agricultural planning departments of the Department. The Philippine Council for Agricultural and Fishery is also co-chaired by a CSO leader.

So will the Marcos administration be able to stop the decline of the Philippine agricultural sector? Will it be able to overhaul the SAP-deregulation economic model that has been in place, with horrifying impact on the Filipino rural masses as well as the consuming public?

References

Balisacan, A. (1998). Policy Reforms and Agricultural Development in the Philippines. *ASEAN Economic Bulletin, 15*(1).

Clarete, R. (1992). *Structural Adjustment and Agriculture: Developing a Research Analytic Framework*. PIDS.

Climate Action Tracker. (2023). *Philippines, Country Summary*. June 5, 2023 downloaded on July 3, 2024 from https://climateactiontracker.org/countries/philippines/

Congress of the Philippines. (1997). *Agriculture and Fisheries Modernization Act of 1997*. Senate Library.

Cororaton, C., & Yu, K. (2019). Assessing the Poverty and Distributional Impact of Alternative Rice Policies in the Philippines. *DLSU Business & Economics Review, 28*(2).

Department of Agriculture. (1994). *Action Plan for the GATT-UR Adjustment Measures for the Agricultural Sector*. DA.

Dy, R. et al. (2008). *Modernizing Philippine Agriculture and Fisheries: The AFMA Implementation Experience*. University of Asia and the Pacific.

Fair Trade Alliance. (2006). *Nationalist Development Agenda: A Roadmap for Economic Revival, Growth and Sustainability*. FairTrade Alliance.

Open Gov. (2023). *Digitalization Boosts Philippine Agriculture*. September30, 2023. downloaded on July 3, 2024 from https://opengovasia.com/2023/09/30/digitalisation-boots-philippine-agriculture/

Habito, C., et al. (1999). *Farms, Food and Foreign Trade: The World Trade Organization and Philippine Agriculture*. AGILE.

Mangabat, M., Sanguyo, E., & Sombilla, M. (2013). *Productivity in the Rice Sector in Philippine Agriculture*. SEARCA and DA Philrice.

Miraflor, M. (2020, August 4). A Decade of Non-Compliance of Agri-Agra Act Hoped to Be Finally Addressed. *Manila Bulletin*.

National Economic Development Authority. (2004). *Medium Term Philippine Development Plan 2004–2010*. NEDA

Ofreneo, R. (1980). *Capitalism in Philippine Agriculture*. Foundation for Nationalist Studies.

Ofreneo, R. (1987). *Deregulation and the Agrarian Crisis*. Institute of Industrial Relations, UP.

Ofreneo, R. (October 2005). "Promises and Realities", Newsbreak (Vol. 5 No. 20). Newsbreak Publishing, Inc.

Ofreneo, R. (2012). *Green Jobs and Green Skills in a Brown Philippine Economy*. University of the Philippines Press.

Opiniano, J. & Ang, A. (2024). *The Philippines' Landmark Labor Export and Development Policy Enters the Next Generation*. Downloaded on July 3, 2024 from https://www.migration.policy.org

Orbeta, A. (1996). *Structural Adjustment and Poverty Alleviation in the Philippines*. Philippine Institute of Development Studies.

Pabuayon, I., Catelo, S., Rola, A., & Paris, T. (2013). *Agricultural Policy*. University of the Philippine Press.

PAGASA. (2011). *Climate Change in the Philippines*. February, downloaded on July 4, 2024 from https://pubfiles.pagasa.dost.gov.ph/iaas/ClimateChange_in_the_Philippines_MDGF_Report_2011.pdf

Peeyush, S., & Ou, Y. (2014). Philippines. In *Agricultural Mechanization at a Glance: Selected Country Studies in Asia on Agricultural Machinery Development* (p. 142). United Nations Asian and Pacific Center for Agricultural Engineering and Machinery (UNAPCAEM).

Philippine Tariff Commission. (2006). *Annual Report 2006*. Tariff Commission.

Philippines. (2011). *National Climate Change Action Plan 2011–2028*. downloaded on July 3, 204 from https://climate.emb.gov.ph/wp-content/uploads/2016/06/NCCAP-1.pdf

Project Development Institute. (1999). *The Disappearing Rice Industry of Central Luzon*. Project Development Institute.

Quizon, A., Marzan, A., De Vera, D., & Rodriguez, M. (2018). *State of Land and Resource Tenure Reform in the Philippines 2018*. ANGOC.

Ranada, P. (2014). *Machines on PH farms: Catching up with ASEAN integratio*n. Rappler, online news magazine downloadable at www.rappler.com

Rasiah, R., & Gopi Krishnan, V. (2025). Greening in Southeast Asia. In J. Mathews & O. Arkebe (Eds.), *The Oxford Handbook of Greening Economic Development*. Oxford: Oxford University Press.

Rasiah, R. (2025). Building Agricultural Sustainability and Food Security in Southeast Asia. In R. Rasiah, A. Venus & N. I. Annizah (Eds.), *Sustainable Agriculture and Food Security in Southeast Asia: Towards Climate Resilience and Digitalization, Singapore: Springer Nature.*

Rice Watch and Action Network. (2006). *Development Prospects for the Rice Industry: Proceedings for a Series of Round Table Discussions on Rice*. Rice Watch and Action Network.

Philippine Senate. (December 1994). *Committee Report No. 72. Report of the Committee of the Whole on the General Agreement on Tariffs and Trade*. Senate.

Philippine Statistical Authority (PSA). (2023). Various years for select agricultural data.

Suministrado, D. (2013). "Status of Agricultural Mechanization in the Philippines", presented in Regional Forum on Sustainable Agricultural Mechanization in Asia and the Pacific, organized by UN CSAM in Quingdao, China, October 25–27. (unpublished)

USDA website. (2014). *2014 Annual Exporters Guide Philippines.* https://www.fas.usda.gov/regions/philippines

USDA website. (2023). *2022 Annual Exporters Guide Philippines.* https://www.fas.usda.gov/regions/philippines

Usui, N. (2012). *Taking the Right Road to Inclusive Growth: Industrial Upgrading and Diversification in the Philippines.* Asian Development Bank.

Rene E. Ofreneo is a Professor Emeritus at the University of the Philippines and former Dean of the University of Philippines School of Labor and Industrial Relations. He is one of the leading scholars on labor relations and labor market studies in the Philippines and in the Asia–Pacific. He writes a column, "Laborem Exercens", for Philippine Business Mirror. He is President of three civil society organizations: Freedom from Debt Coalition, Integrated Rural Development Foundation and Philippine Movement for Climate Justice.

Agriculture, Food Security, and Economic Development in Thailand

7

Yot Amornkitvikai and Teerawat Charoenrat

Abstract

This chapter begins by discussing the growing importance of agriculture to the Thailand economy, including the policies the government launched to promote the sector. While Thailand enjoys food security if the measure is confined to trade measurements, it does raise a number of security issues facing especially small rural farmers. Thailand's agriculture sector has often sheltered the national economy whenever external shocks destabilize economic development, but is now facing serious challenges. While historically a major contributor to exports and rural livelihoods, the sector has been grappling with price volatility, resource depletion, and farmer indebtedness. Government policies, though evolving towards sustainability and commercialization, have often provided inadequate support, leading to declining economic contribution and persistent poverty among especially the small farmers. Recent initiatives, such as digital agriculture and climate resilience, hold promise but have yet to fully address the sector's deep-rooted issues. Overcoming challenges like small farm size, landlessness, ageing farmers, and limited market access requires a comprehensive approach focused on sustainable practices, value chain development, and farmer empowerment.

Keyword

Agricultural policies · Food security · Smart farming · Poverty · Thailand

Introduction: State of Agriculture

Despite industrialization, agriculture remains essential to Thailand's economic and social development, particularly in rural regions. Given Thailand's tropical climate and abundance of biodiversity, agriculture has historically been the backbone of the Thai economy and a significant source of income for Thai people in rural areas. Thailand is recognized as a global leader in various agricultural exports, such as rice, cassava products, and sugar. Nationally, agriculture contributes to almost one-third of total employment and remains a critical source of livelihood for Thailand's rural population; more than two-thirds of the Thai rural

Y. Amornkitvikai
College of Population Studies, Chulalongkorn University, Bangkok, Thailand
e-mail: yot.a@chula.ac.th

T. Charoenrat (✉)
Faculty of Economics, Khon Kaen University, Khon Kaen, Thailand
e-mail: teerch@kku.ac.th

R. Rasiah et al. (eds.), *Sustainable Agriculture and Food Security in Southeast Asia*, Sustainable Development Goals Series, https://doi.org/10.1007/978-981-95-0477-0_7

population are engaged in agricultural activities, covering around 6 million households.

The agriculture sector in Thailand has remained a crucial contributor to economic development, serving as the primary source for both internal consumption and exports to other nations worldwide. Simultaneously, it has also functioned as a significant sector in employing excess workers throughout various economic crises (Office of Agricultural Economics, 2017). Thailand has greatly benefited in terms of its food security and export profits as a result of this. However, the Thai agriculture industry continues to encounter several issues. Throughout decades, the Thai agricultural sector has continued to face numerous challenges, including volatile product pricing and inefficient utilization of production resources, the degradation of natural resources, and the persistent issue of excessive debt levels among agricultural households. While the government has consistently implemented rules and procedures to offer aid, the majority of these initiatives are temporary and need to address the present crisis with long-term remedies.

This chapter examines the development of Thai agriculture, focusing on inter- and intra-structural changes, policies for poverty alleviation and agricultural growth, as well as the integration of Industry 4.0 technologies and climate resilience strategies.

Structural Change

Thailand's economic structure has undergone significant transformations, both within and across sectors. Inter-sectoral structural change involves shifts in resources, like labour and capital, between different economic sectors, such as from agriculture to manufacturing and services, which have been influenced by factors like technological advancements, globalization, and government policies. Meanwhile, intra-agricultural structural change has occurred within the agricultural sector, reflecting alterations in production methods, for instance, a transition from subsistence to commercial farming and from the use of traditional to modern technologies.

Thailand's economic restructuring has been marked by rapid industrialization and urbanization at the expense of agriculture. This inter-sectoral shift, initiated by the 1961 National Economic and Social Development Plan, has redirected labour from rural to industrial activity (Temsumrit & Sriket, 2023). However, while this transformation has spurred economic growth, it has also exacerbated rural poverty and income inequality due to stagnant labour productivity.

While government policies can assist agriculture, Thailand's experience highlights a challenge. From the 1950s, the Thai government relied on export taxes and quotas on agricultural commodities as a primary revenue source, which had a limited impact on production due to inelastic supply of agricultural commodities. Although agricultural distortion policies ended in 1985, subsequent economic strategies had unintended negative consequences for the sector. The 1980s focus on industrialization through protectionism and import substitution inadvertently increased input costs for farmers, such as fertilizer prices thereby reducing profitability and labour productivity (Temsumrit & Sriket, 2023). Also, the agricultural sector, which employs a large portion of the workforce, has suffered from decades of government underinvestment.

Hence, the inter-sectoral and intra-agricultural structural change in Thailand is a complex and multifaceted process. It is crucial to consider the various factors driving this change and its potential impact on the economy and society. The following sub-section discusses inter-sectoral and intra-agricultural structural change in Thailand, with a focus on changes in value-added, employment, productivity, exports, imports, and trade balance.

Changes in GDP Structure

Figure 7.1 shows that agricultural GDP sector has increased marginally compared to non-agricultural GDP in 1993–2022. More specifically, the agricultural GDP increased from THB 426.985 billion in 1997, which was at the end of the Seventh National Development Plan, to

THB 1,531.121 billion in 2022, which was at the end of the Twelfth Plan. However, agricultural GDP declined between the eleventh and twelfth plans owing to climate change causing natural disasters in the form of continuous droughts and dwindling rainfall from 2013 to 2015. Early Mortality Syndrome (EMS) disease caused significant harm to various vital crops, as well as a substantial fall in farmed marine shrimp output (Office of Agricultural Economics, 2017). Furthermore, Thailand's exports of agricultural goods and other commodities were negatively affected by falling price due to economic depression in Thailand's key trade partners, as well as severe competition in the global market. Nevertheless, the percentage of agricultural GDP to total GDP increased from 9.0% during the seventh plan to 10.0% during the eleventh plan (Table 7.1).

Between 2021 and 2022, agricultural GDP increased from 1.0% to 2.4%, following a fall by 0.2% in 2019 to 3.6% in 2020. Specifically, the GDP growth of the agricultural sector in Thailand has been driven by capital inputs (Table 7.2). When analysing the source of GDP growth, the capital element plays the most significant role in generating GDP growth in the agriculture sector, with capital generating an average positive growth of 0.9% between 2018 and 2022. Furthermore, between 2018 and 2022, labour and land contributed only 0.04% and 0.01% of the country's GDP growth, respectively. However, total factor productivity (TFP) was negative, causing a 0.05% loss in GDP growth between 2018 and 2022.

Besides its contribution to the country's GDP growth rate, the agricultural sector continues to play an essential role in overall development. This is because agriculture and commodity exports continue to provide sizeable foreign currency for the Thai government. Indeed, Thailand's agriculture sector has laid a significant foundation for Thai and global food security. It is also a major source of raw materials for the industrial sector, which includes the rubber, food processing, and animal feed industries. Furthermore, agriculture is still the primary source of income for the vast majority of Thai people.

Despite contributing only 8.8% of Thailand's GDP in 2021 (Table 7.1), agriculture remains a vital sector, employing 31.6% of the workforce (Fig. 7.2) and demonstrating steady growth with an average GDP expansion of 1.0% between 2018 and 2022 (Table 7.2).

Agricultural Prices

The average price of agrarian items increased in 2013–2015 and 2019–2022 while the income index had a negative growth in the former period before rising from 1.5% in 2016 to 12.9% in 2022 (Fig. 7.3).

The rise in agricultural prices could be attributed to higher prices for farm products, such as oil crops, fruit trees, vegetables, livestock, and fisheries with the exception of cut flowers and perennial crops, whose price index has remained low compared to other crops (Fig. 7.4). The price index of oil crops, fruit trees, vegetables, and livestock increased substantially from 2013 to 2023 compared to other agricultural products.

Trade Performance

Agricultural exports grew from 1,406,428 million baht in 2018 to 1,816,433 million baht in 2022 (Table 7.3). Furthermore, the value of agricultural product imports climbed from 514,720 million baht in 2018 to 719,403 million baht in 2022. The agriculture sector's trade balance was in surplus, increasing from 892,437 million baht in 2018 to 1,097,594 million baht in 2022. The top agricultural exports in 2022 were natural rubber (2.8%), fruits and products (2.7%), rice and products (1.7%), cassava (1.5%), check meat and products (1.4%) sugar and products (1.2%), and fish and products (1.2%), shrimps and products (0.5%), vegetables and products (0.3%), and residues and waste from the food industries and prepared animal fodder (0.3%) (Table 7.4).

Thailand exported agricultural products mainly to China, accounting for 27.9% of total agricultural export in 2022, followed by Japan (9.9%), USA (9.6%), Malaysia, (4.6%), Indonesia (3.5%),

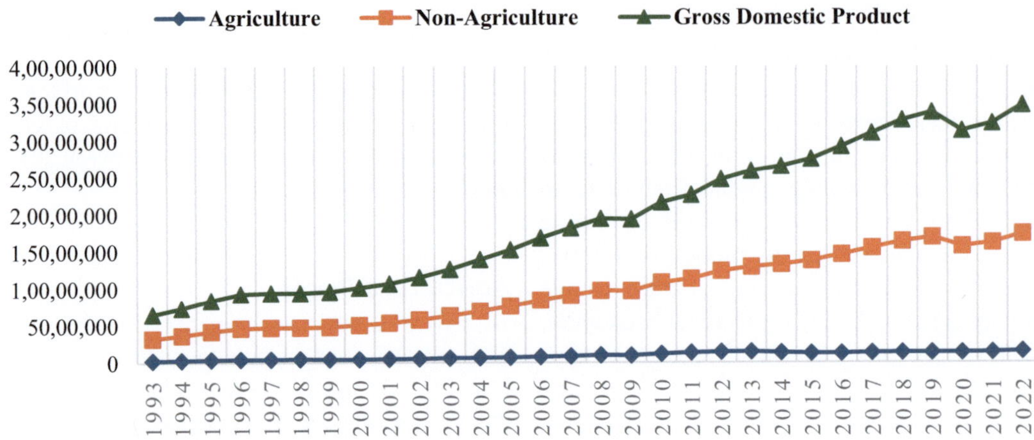

Fig. 7.1 GDP structure, Thailand, 1993–2022 (million baht). *Source* NESDC (2023c)

India (3.2%), South Korea (3.0%), Cambodia (2.8%), Vietnam (2.6%), and Myanmar (2.4%) (Table 7.4).

The top import of agricultural products in 2022 were fish and products, accounting for 1.1% of total agricultural import value, followed by residues and waste from industries and prepared animal fodder (0.8%), food crops and products (0.7%), oil plants (0.7%), fruits and products (0.5%), other food products (0.4%), milk and products (0.3%), vegetables and products (0.3%), cotton (0.2%), and beverages (0.2%) (FOA, 2022). Imports came mainly from Brazil, accounting for 16.3% of total agricultural import values in 2022, followed by China (13.0%), USA (7.9%), Australia (5.9%), India (4.8%), Vietnam (4.8%), Myanmar (4.1%), Indonesia (3.9%), New Zealand (3.5%), and Malaysia (3.0%) (Table 7.5).

Figure 7.5 presents the percentage distribution of Thailand's agricultural exports by destination country between 2018 and 2022. China consistently ranked as the top export market, rising notably from 20.37% in 2018 to a peak of 29.56% in 2021, before a slight decline in 2022. Japan and the United States remained stable but showed modest decreases in recent years. ASEAN neighbors such as Malaysia, Vietnam, and Indonesia held moderate shares with minimal variation. The declining share of the "Others" category from 35.51% to 30.45%, suggests increasing export concentration, underscoring Thailand's growing dependence on a few major markets, particularly China. This figure highlights Thailand's increasing reliance on China as a primary export destination for agricultural goods, which raises important implications regarding export diversification and market vulnerability in the long run.

Agricultural Land Use and Ownership

Thailand's agriculture land amounted to 149.7 million Rai in 2021 which accounted for 46.7% of total land in Thailand in that year (Fig. 7.6). The main plantation crops grown are rubber, oil palm, sugarcane, cassava, shallots, potatoes, garlic, and onions. Thailand was ranked among the top ten rice-producing countries in the world over the last few decades and is also one of ASEAN 's most geographically fertile countries with forests accounting for one-third of the country's total land area.

The breakdown by principal crops shows that 65.0 million Rai were in the northeastern region, 31.9 million Rai were in the northern region, 29.0 million Rai were in the central region, and another 23.9 million Rai were in the southern region (Fig. 7.7).

Paddy cultivation comprises the largest proportion of agricultural land in Thailand (43.7%) in

Table 7.1 GDP structure, Thailand, 1993–2023 (million baht)

Year	National development plans	Agriculture	%	Non-agriculture	%	Gross domestic product
1993	7	261,917	8.03	3,001,509	91.97	3,263,426
1994	7	318,278	8.63	3,370,812	91.37	3,689,090
1995	7	383,011	9.08	3,834,603	90.92	4,217,614
1996	7	420,416	9.06	4,218,188	90.94	4,638,604
1997	7	426,985	9.06	4,283,325	90.94	4,710,310
1998	8	482,343	10.26	4,219,210	89.74	4,701,553
1999	8	425,959	8.89	4,363,867	91.11	4,789,826
2000	8	430,927	8.50	4,638,893	91.50	5,069,820
2001	8	458,563	8.58	4,886,438	91.42	5,345,001
2002	8	501,502	8.69	5,268,076	91.31	5,769,578
2003	9	596,439	9.44	5,720,864	90.56	6,317,303
2004	9	646,130	9.29	6,308,152	90.71	6,954,282
2005	9	700,096	9.19	6,914,315	90.81	7,614,411
2006	9	789,828	9.40	7,610,816	90.60	8,400,644
2007	9	848,424	9.35	8,227,877	90.65	9,076,301
2008	10	977,724	10.07	8,729,205	89.93	9,706,929
2009	10	945,297	9.79	8,713,370	90.21	9,658,667
2010	10	1,137,252	10.52	9,670,893	89.48	10,808,145
2011	10	1,310,673	11.59	9,996,233	88.41	11,306,906
2012	10	1,421,601	11.50	10,935,741	88.50	12,357,342
2013	11	1,462,282	11.32	11,452,876	88.68	12,915,158
2014	11	1,334,795	10.09	11,895,509	89.91	13,230,304
2015	11	1,219,421	8.87	12,524,059	91.13	13,743,480
2016	11	1,236,980	8.48	13,353,357	91.52	14,590,337
2017	11	1,302,041	8.41	14,186,623	91.59	15,488,664
2018	12	1,342,912	8.20	15,030,431	91.80	16,373,343
2019	12	1,372,848	8.13	15,516,326	91.87	16,889,174
2020	12	1,362,953	8.70	14,298,193	91.30	15,661,146
2021	12	1,407,848	8.71	14,758,749	91.29	16,166,597
2022	12	1,531,121	8.81	15,839,115	91.19	17,370,236
2023	13					
Q1		419,399	9.28	4,099,405	90.72	4,518,804
Q2		383,626	8.86	3,946,691	91.14	4,330,317

Source NESDC (2023c)

2021, followed by fruit trees and perennial trees (26.3%), upland field cereals (20.6%), and vegetables, cut flowers, and ornamental plants (0.7%) (Table 7.6). Paddy cultivation is most concentrated in the northeastern region (60.9%) of agricultural land use (Table 7.7). Only 48.3% of the entire agricultural land was effectively utilized by the landowners (Table 7.8). Mortgage, redemption, and free usage accounted for 32.3% of overall agricultural land utilization, while leased land accounted for 19.4% of overall agricultural land utilization. Most Thai farmers are smallholders

Table 7.2 Sources of GDP growth, Thailand, 2018–2022(%)

	2018	2019	2020	2021	2022	(2018–2022)
GDP Growth, All Sectors	4.1	2.35	−6.1	1.53	2.62	0.9
Labour	0.37	−0.22	0.06	0.07	0.58	0.172
Land	0	0	0	0	0	0
Capital	2.01	0.92	−0.63	0.41	0.38	0.618
TFP	1.76	1.66	−5.53	1.05	1.65	0.118
GDP Growth, Agriculture	5.1	−0.16	−3.57	0.98	2.4	0.95
Labour	0.35	−0.44	−0.01	0.29	0.03	0.044
Land	0	0	−0.01	0	0.04	0.006
Capital	2.48	1.04	−1.1	1.16	1.14	0.944
TFP	2.25	−0.76	−2.45	−0.47	1.19	−0.048
GDP Growth, Industrial	2.7	−0.03	−5.94	3.4	−0.5	−0.074
Labour	0.33	−0.63	−0.72	−0.45	0.54	−0.186
Land	0					0
Capital	2.53	0.13	−0.18	1.14	0.04	0.732
TFP	−0.17	0.47	−5.03	2.72	−1.08	−0.618
GDP Growth, Services, and Others	4.8	3.99	−6.52	0.67	4.3	1.448
Labour	−0.18	0.42	0.43	−0.12	1.03	0.316
Land	0.00	0.00	0.00	0.00	0.00	0
Capital	1.72	1.2	−0.75	0.06	0.42	0.53
TFP	3.26	2.36	−6.2	0.73	2.85	0.6

Source NESDC (2018–2022)

with four out of ten classified as poor, living below the poverty line in 2017. Indeed, around 50% of agricultural households held 1.6 hectares or less land ([Craig Keating, 2022]).

Agricultural Policies

Since the 1960s, the Thai government has attempted to strengthen the agricultural sector and socio-economic infrastructure through national economic and social development plans (Mahathanaseth & Pensupar, 2014). Although the performance was good during the first plan (1961–1965), several governmental projects experienced delays due to technical issues and challenges in drafting applications for foreign loans. The construction of the Bhumibol Dam was finished in 1964. The implementation of distribution ditches and dikes within the Greater Chao Phraya commenced in 1963 with the purpose of supplying water to a land area spanning 700,000 Rai. The second plan (1967–1971) focused on extending irrigation, livestock and forest management, and land development programmes, and fisheries projects, which helped boost productivity and diversification to achieve balanced agricultural production. In addition, the agrarian lending programme through the new Bank for Agriculture and Cooperatives and farmer associations helped promote companies' groupings.

The third plan (1972–1976) enhanced Thailand's agricultural products for export while the fourth plan (1977–1981) emphasized commercial agriculture but at the same time emphasized forest and waterway conservation through crop diversification (i.e. dry season paddy, maize, beans, sorghum, sugarcane, kenaf, tobacco, cotton, rubber, coconuts, sericulture, oil palm, livestock, and coastal fishing). The fifth plan (1982–1986) boosted agricultural production through

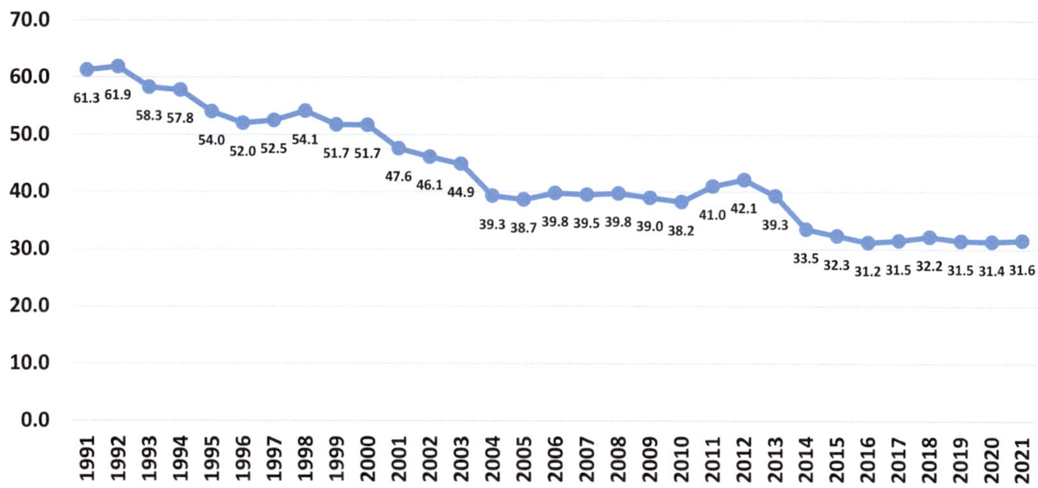

Fig. 7.2 Agricultural employment in total employment, Thailand, 1991–2021(%) *Note* ILO estimates *Source* World Bank (2021)

raising yield rather than an extension of cultivated land targeted at maintaining forest and water resources through reforming the production process to enhance land, water, forestry resources, and fishing efficiency, and agricultural marketing and pricing system. Also, this plan supported restructuring to stabilize prices, quickening the distribution of land ownership and the provision of land titles, and provided monetary incentives to encourage government's financial institutions and commercial banks to expand credit to farmers and agricultural organizations. The sixth plan (1987–1991) emphasized the further diversification of agriculture with middle-income farmers made the primary target group. Additionally, agro-industries were established to stimulate economic growth and create jobs in both rural and urban areas.

The seventh to thirteen plans (1992–2027) emphasized sustainable agriculture to protect the environment with a focus on minimizing the use of nonrenewable resources (see Table 7.9). The ninth (2002–2006), tenth (2007–2012), and eleventh (2012–2016) plans promoted sustainable agriculture, including organic farming, integrated farming, and new theory agriculture to strengthen household food security and energy security.

Table 7.9 National economic and development plans, Thailand, 1961–2027

Plans	Period
The first plan	(1961–1966)
The second plan	(1967–1971)
The third plan	(1972–1976)
The fourth plan	(1977–1981)
The fifth plan	(1982–1986)
The sixth plan	(1987–1991)
The seventh plan	(1992–1996)
The eight Plan	(1997–2001)
The ninth plan	(2002–2006)
The tenth plan	(2007–2011)
The eleventh plan	(2012–2016)
The twelfth plan	(2017–2022)
The thirteenth plan	(2023–2027)

Source Compiled by authors from NESDC (2023b)

The twelfth plan (2017–2022) initiated the first five years of the implementation of the 20-year National Strategy (2017–2036), which aimed to strengthen competitiveness in the production and service sectors to establish and strengthen value chain connectivity between the agricultural, industrial, service, trade, and investment sectors. Specifically, this plan aimed to (i) strengthen

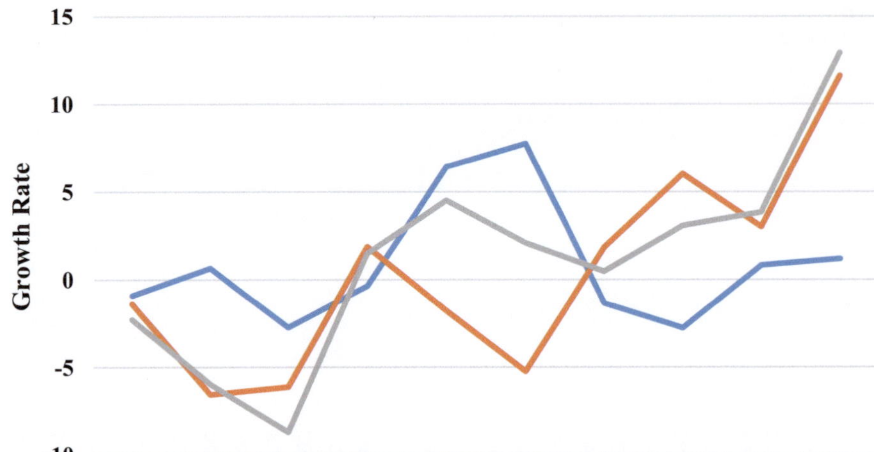

	2013	2014	2015	2016	2017	2018	2019	2020	2021	2022
Production Index	-0.92	0.64	-2.74	-0.37	6.41	7.74	-1.36	-2.77	0.82	1.18
Price Index	-1.38	-6.56	-6.12	1.86	-1.79	-5.25	1.84	6.01	3.00	11.60
Income Index	-2.28	-5.95	-8.69	1.48	4.51	2.08	0.46	3.07	3.84	12.92

Fig. 7.3 Annual growth of agricultural economic index, Thailand, 2013–2022. *Source* Office of Agriculture Economics (2022)

sustainable agricultural production through developing and maintaining water storage systems for agriculture to match available water resources, (ii) build and transfer technical knowledge, science, technology, innovation, and co-creation to support the use of bioresources, (iii) upgrade agricultural and food production to meet international health standards, respond to market demands in food production, (iv) strengthen the production capacity of the agro-industry value chain, and (v) develop enabling factors in agriculture to support new generation farmers (NESDB, 2017).

The thirteenth plan (2023–2027) attempts to elevate the agriculture sector towards the production of high-value agricultural products through the use of technology to improve productivity, reduce reliance on natural resources, and add value to agricultural produce with a strong focus on digitalization and climate resilience (NESDC, 2023b).[1]

Despite Thailand's successful promotion of agriculture, including food for export, several problems remain. Government expenditures have remained controversial owing to the rice-pledging scheme, which started in 2011 that expanded rapidly through direct government purchase of un-milled rice at prices 50% above the market price. The amount of un-milled rice stockpiled in warehouses at the end of the third quarter of 2013 was double the annual volume of global commercial rice (18 million tons). The government has been unable to auction the paddy stock as rice prices were 12 to 15% lower at the end of 2013 than they were at the beginning of 2011. In its first year of operation, the subsidy cost the government \$12.7 billion, accounting for 3.5% of GDP, which increased the fiscal deficit to 4.4% of GDP in 2012 from 1.7% in 2011. In 2013, the scheme cost the government an additional \$13.9 billion, accounting for 3.6% of the GDP (World Bank, 2014).

Also, despite the Thai government promoting aggressively agriculture through the supply chain, there are still limitations, such as 83% of agricultural lands are situated outside irrigated zones; 30% of farmland is unsuitable for farming and has unsuitable soil quality; land holdings of farmers have shrunk so that they only owned 48% of agricultural land in 2022; open burning

[1] See Rasiah et al. (2023) on the development implications of digitalization.

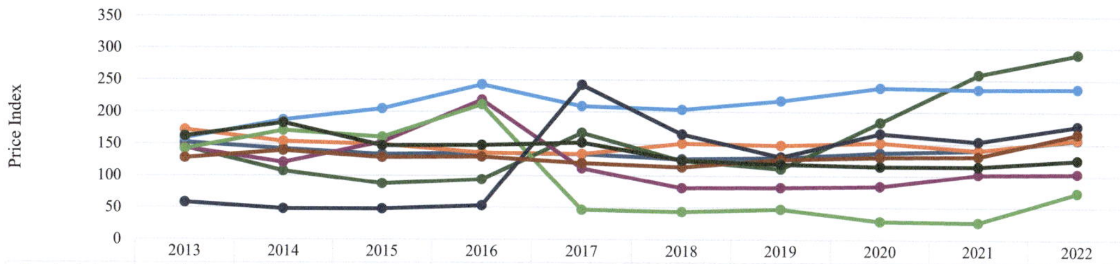

	2013	2014	2015	2016	2017	2018	2019	2020	2021	2022
Agricultures	151.84	141.88	133.2	135.68	133.39	126.17	128.49	136.22	140.3	156.58
Food crops	171.56	153.45	148.89	135.79	133.9	150.39	147.7	152.0	140.01	157.24
Oil crops	144.31	107.27	87.22	93.83	167.3	123.49	110.89	183.86	258.62	289.99
Fruit trees	158.98	186.78	204.53	242.47	208.85	203.55	217.46	237.69	235.49	236.31
Perennial	143.27	120.41	152.56	218.25	111.52	80.48	81.42	83.22	101.98	103.51
Cut flowers	141.83	170.73	160.14	211.15	46.21	43.28	47.42	28.45	26.83	73.42
Vegetables	57.72	47.77	47.53	52.96	241.81	164.99	129.61	166.44	154.09	178.35
Livestock	127.59	139.03	128.43	129.55	119.79	113.32	125.54	129.35	130.41	165.87
Fishery	161.63	182.69	147.2	147.66	151.96	124.16	118.34	114.98	114.97	124.99

Fig. 7.4 Agricultural price index, Thailand, 2013–2022. *Source* Office of Agriculture Economics (2022)

Table 7.3 Value of exports, imports, and balance of trade, Thailand, 2018–2022

Unit: Million Baht	2018	2019	2020	2021	2022
Total Exports	8,085,563	7,618,889	7,173,166	8,536,631	9,933,553
Agricultural products	1,406,428	1,317,937	1,290,048	1,503,024	1,816,433
–Food and agricultural products	*1,156,469*	*1,103,210*	*1,084,262*	*1,228,652*	*1,501,749*
–Agricultural products for agro-industry	*249,959*	*214,727*	*205,786*	*274,372*	*314,684*
Non-agricultural products	6,679,135	6,300,952	5,883,118	7,033,607	8,117,120
Total Re-Exports	7,878	9,512	5,328	5,471	10,764
Agricultural products	729	680	432	596	564
–Food and agricultural products	*728*	*667*	*427*	*593*	*564*
–Agricultural products for agro-industry	*1*	*13*	*5*	*3*	
Non-agricultural products	7,149	8,832	4,896	4,875	10,200
Total Imports	8,098,098	7,425,649	6,502,407	8,549,082	10,646,953
Agricultural products	514,720	510,925	525,184	604,194	719,403
–Food and agricultural products	*492,230*	*493,177*	*513,631*	*587,586*	*694,286*
–Agricultural products for agro-industry	*22,490*	*17,748*	*11,553*	*16,608*	*25,117*
Non-agricultural products	7,583,378	6,914,724	5,977,223	7,944,888	9,927,550
Balance of Trade	−4,657	202,752	676,087	−6,980	−702,636
Agricultural products	892,437	807,692	765,296	899,426	1,097,594
–Food and agricultural products	*664,967*	*610,700*	*571,058*	*641,659*	*808,027*
–Agricultural products for agro-industry	*227,470*	*196,992*	*194,238*	*257,767*	*289,567*
Non-agricultural products	−897,094	−604,940	−89,209	−906,406	−1,800,230

Source Office of Agricultural Economics (2022)

Table 7.4 Agricultural exports by country, Thailand, 2018–2022 (million baht)

	2018	%	2019	%	2020	%	2021	%	2022	%
Total	1,406,428		1,317,937		1,289,656		1,503,024		1,816,433	
China	286,496	20.37%	281,577	21.36%	315,324	24.45%	444,365	29.56%	507,085	27.92%
Japan	162,824	11.58%	155,585	11.81%	145,324	11.27%	154,028	10.25%	179,013	9.86%
USA	128,573	9.14%	128,478	9.75%	142,390	11.04%	150,393	10.01%	173,853	9.57%
Malaysia	61,462	4.37%	56,204	4.26%	52,266	4.05%	72,135	4.80%	84,207	4.64%
Indonesia	63,740	4.53%	51,232	3.89%	36,498	2.83%	34,371	2.29%	63,082	3.47%
India	17,796	1.27%	16,359	1.24%	13,908	1.08%	36,364	2.42%	57,521	3.17%
South Korea	33,532	2.38%	34,189	2.59%	27,499	2.13%	34,436	2.29%	54,669	3.01%
Cambodia	39,230	2.79%	44,105	3.35%	48,329	3.75%	45,273	3.01%	51,577	2.84%
Vietnam	71,305	5.07%	62,604	4.75%	60,850	4.72%	48,589	3.23%	47,970	2.64%
Myanmar	42,103	2.99%	34,310	2.60%	33,705	2.61%	41,098	2.73%	44,329	2.44%
Others	499,367	35.51%	453,294	34.39%	413,563	32.07%	441,972	29.41%	553,127	30.45%

Source Office of Agriculture Economics (2022)

Table 7.5 Agricultural imports by country, Thailand, 2018–2022 (million baht)

	2018	%	2019	%	2020	%	2021	%	2022	%
	514,720		510,925		525,184		604,194		719,403	
Brazil	49,378	9.59	48,195	9.43	62,614	11.92	96,174	15.92	117,270	16.30
China	65,463	12.72	68,613	13.43%	69,423	13.22	73,339	12.14	93,687	13.02
USA	77,679	15.09	67,931	13.30	58,137	11.07	58,280	9.65	56,629	7.87
Australia	18,605	3.61	18,924	3.70	21,893	4.17	36,497	6.04	42,722	5.94
India	22,807	4.43	19,063	3.73	15,873	3.02	24,029	3.98	34,595	4.81
Vietnam	23,545	4.57	22,269	4.36	24,459	4.66	25,469	4.22	34,573	4.81
Myanmar	10,273	2.00	14,872	2.91	19,079	3.63	24,431	4.04	29,407	4.09
Indonesia	20,705	4.02	18,665	3.65	24,944	4.75	26,779	4.43	28,005	3.89
New Zealand	14,960	2.91	16,030	3.14	16,254	3.09	18,390	3.04	24,924	3.46
Malaysia	15,910	3.09	15,868	3.11	17,416	3.32	18,314	3.03	21,737	3.02
Others	195,395	37.96	200,495	39.24	195,092	37.15	202,492	33.51	235,854	32.78%

Source Office of Agriculture Economics (2022)

of crop residues remains a problem; cluster link between agricultural organizations and farms is weak; out-migration from farms has raised the average age of the agricultural workforce (NESDC, 2023a).

To meet its ambitious objective of achieving income parity between farmers and non-farm workers by 2037, the thirteenth plan (2022–2026) aims to increase the competitiveness of the Thai agricultural sector by raising the value-added, developing infrastructure and management systems for food quality and security and agricultural sector sustainability, and enhancing the potential role of farm entrepreneurs as economic participants in the supply chain to receive a proportionate share of the benefits, which is consistent with the country's 20-year master plan, which seeks to utilize the uniqueness and distinct agricultural products and the biodiversity of each region to increase value-added of farm products by using new technologies and innovation to raise agricultural productivity.

Food Safety and Security

Since 2008, Thailand (and ASEAN) has been facing increasing challenges on food security, caused by the escalating global and national food prices. The ASEAN Ministers of Agriculture and Forestry introduced the ASEAN Integrated Food Security Framework and Strategic Plan of Action in Food Security (2013–2020) to address this problem in the area (Apichart Pongsrihadulchai, 2019). Thailand launched the Food Security Strategic Plan (2017–2021) on 3 May 2017 to support food security initiatives with a focus on providing the Thai population with adequate access to nutritious food of high quality, even during emergencies, while preserving natural resources for agricultural production and promoting the well-being of farmers. The food and beverage industry is also the country's third largest industry, contributing 21% of Gross Domestic Product (GDP) in 2021 (NESDC, 2023c).

Thailand's food production is controlled for processing, storage, and distribution to consumers, such as Food Safety, Good Agricultural Practice (GAP), Good Manufacturing Practice (GMP), Hazard Analysis, Critical Control Point System (HACCP), and Organic Thailand (see Fig. 7.8). Traditional farming methods have also been increasingly promoted. Government agencies supervise food farming and processing, such as agencies from the Ministry of Industry, Ministry of Agriculture and Agricultural Cooperatives, and Ministry of Public Health.

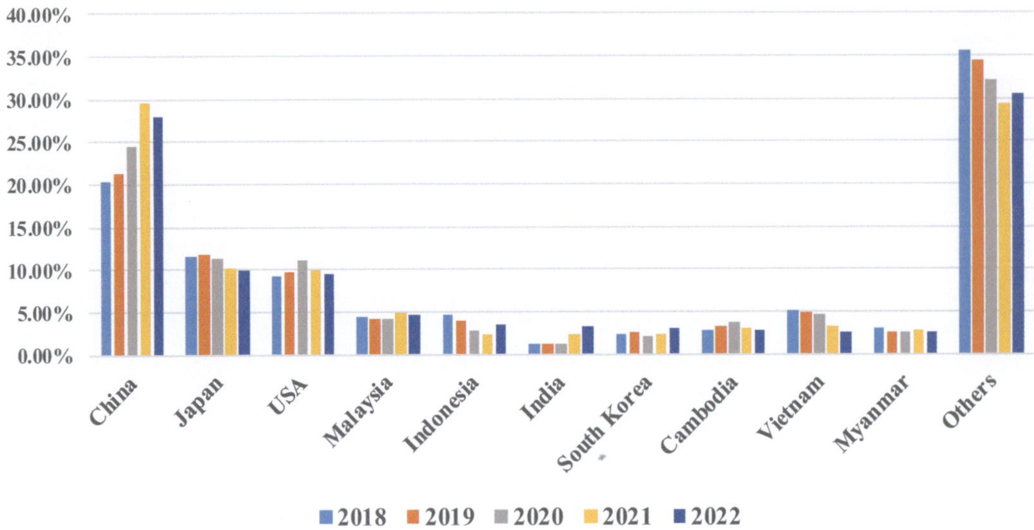

Fig. 7.5 Thailand's agricultural exports by destination country, 2018–2022 (%). *Source* office of agriculture economics (2022)

Entrepreneurs must request permission from both the Department of Industrial Works and the Food and Drug Administration (FDA) set-up factories to operate the business. In addition, food processing operators must register with the FDA and display food labels that meet standardized laws, which is a measure to protect consumers (Bureau Veritas Thailand, 2023; Thai Food and Drug Administration (FDA), 2023):

According to the UN (2023), food security and nutrition and sustainable agriculture have been designated as the second goal of Sustainable Development Goals (SDGs), aimed at eradicating hunger, achieving food security, improving nutrition, and promoting sustainable agriculture. The UN has reaffirmed everyone's right to have access to safe and nutritious food. Food security consists of four main factors, i.e. food availability, access to food, food utilization, and food stability (FAO, 2008).

However adequate domestic consumption in Thailand only comes to 75% (NESDC, 2023b). Hence, although Thailand is often the world's largest rice exporter, 260,000 Thai people in 2023 were undernourished (NESDC, 2023c). The GFSI (2022) ranked Thailand 51 out of 113 countries 9th among Asia-Pacific countries, and 15th among the upper middle-income countries in 2022 on food security based on: affordability, availability, quality and safety, and natural resources. In addition, NESDC (2023c) expressed that food insecurity mainly occurs in rural areas where drought and transportation are serious problems. Thai farmers' ability to subsist on the food is declining leaving many to require government measures to make food more affordable.

Unlike Malaysia (Annizah & Rasiah, 2025), Thailand has remained one of the largest net food exporting countries in the world (Office of the Board of Investment [BOI], 2021). Yet, although the Thai food industry contributed around 23% of the country's GDP in 2021 (NESDC, 2023c), land ownership in the country is highly concentrated. The data from 399 land offices in Thailand show that around 21 million people Thais, and approximately 1 million legal entities own small land of no more than four rai per person with a title deed. At the same time, large landholders numbered 4613 persons and 2205 juristic persons in 2018. Therefore, strict enforcement of land and building taxes can be a mechanism that helps support the distribution of land ownership, which is an essential resource for creating opportunities for poor people in Thailand (The Secretariat of the Senate, 2014).

Moreover, the production of food and energy crops has been affected by the energy crisis and

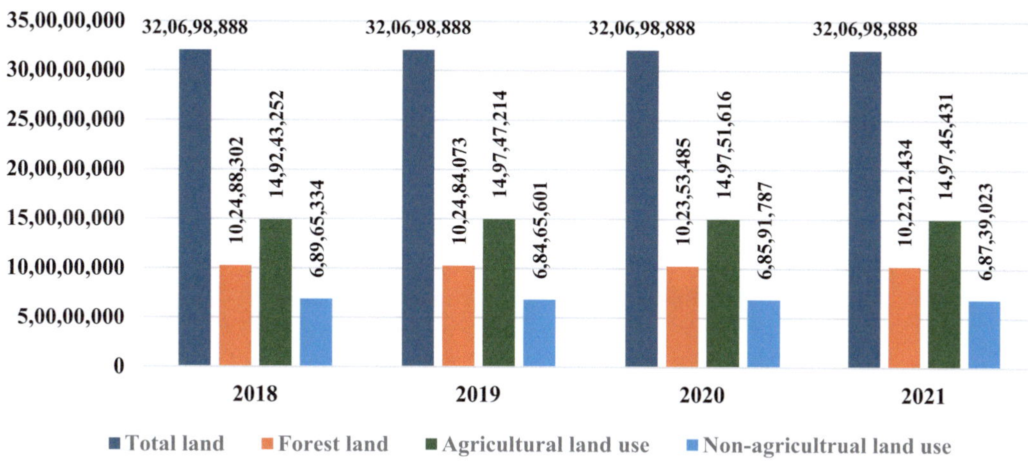

Fig. 7.6 Land utilization, Thailand, 2018–2021 (Rai) Note: 1rai = 0.395 Aker. *Source* Office of Agriculture Economics (2022)

rising oil prices since 2022, which drove the Thai government to develop biofuels and biomass, such as ethanol and biodiesel, which is a product of plants, cassava, sugarcane, and oil palm (NESDC, 2023c). As with other countries, the deterioration of natural and soil resources that are being improperly exploited has caused problems of saline soil, acidic soil, reduction of forest reserves, and erosion of fertile soil from forest destruction (NESDC, 2023c).

NESDC (2023c) reported that the majority of people with low incomes are in the agricultural sector, up to around 2.8 million people who accounted for 68.5% of total working poor people (4.1 million people). Approximately 660,000 of poor people in 2022 did not have any arable land, and hence have to rent land for farming. In addition, over half of Thai farmers are debt-ridden, accounting for 59.9% of them. Besides, even though the national workforce has increased, workers in the agriculture sector have declined, and the average age of farmers has continued to rise (NESDC, 2023c).

Agricultural products in Thailand face highly volatile fluctuation in prices due to a lack of production planning, as well as the dominance of small-scale agriculture who tend to be vulnerable to the dictates of buyers. The logistics system lacks systematic management, which affects the quality of products and operating costs. There is also an encroachment of imported goods from the opening of free trade zones, especially the ASEAN Free Trade Areas (AFTA) (NESDC, 2023c; The Secretariat of the Senate, 2014).

The Global Food Security Index ranked Thailand 64th out of 113 countries in 2022 (Corteva, 2022). Thailand's score was 60.1% out of 100% in 2022, representing an increase of 4% in food security compared to 2012. Although progress has been made in some areas in 2022 compared to 2012, such as affordability scores (+7.2%), availability (+4.4%), and sustainability and adaptation scores (+172%), there remains scope to improve the quality and safety of food (−10.8%), particularly in nutritional standards (−48.3%), food safety (−4.1%), dietary diversity (−0.6%), and micronutrient availability (−0.7%) (Table 7.10).

Despite its glorious achievements, Thai agriculture sector is one of the most labour-intensive among food exporting countries, which calls for greater efficiency and productivity (Food and Agriculture Organization of the United Nations (FAO), 2023; UN, 2023). Also, rural farmers still live in poverty owing to low value added per worker, while the majority of Thai farmers are smallholders with around 40% of them landless (FAO, 2023; GSMA, 2022). Approximately ten million farmers obtained emergency relief

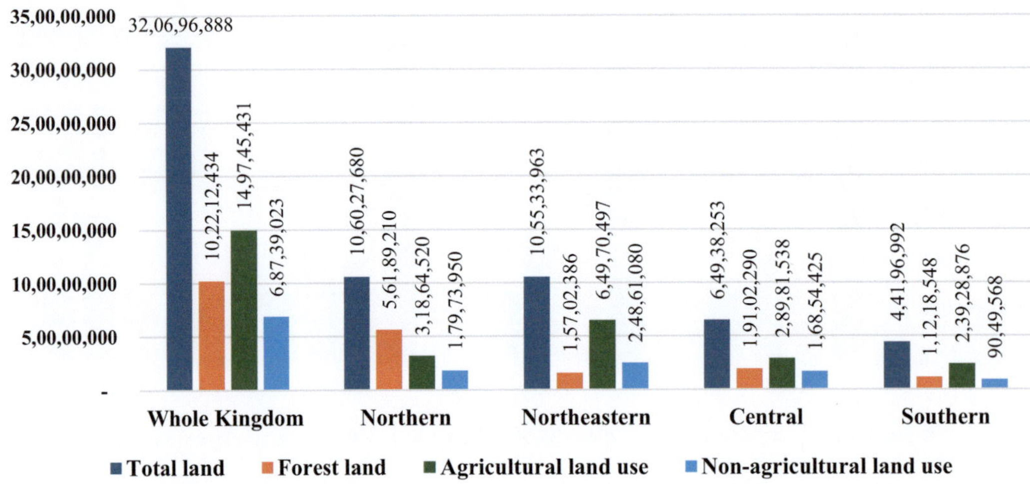

Fig. 7.7 Land utilization by region, Thailand, 2021 (Rai) *Note* 1rai = 0.395 Aker. *Source* Office of Agriculture Economics (2022)

Table 7.6 Agricultural land, Thailand, 2018–2021 (Rai)

	Total land	Paddy land	%	Upland field crop	%	Fruit trees and perennial trees	%	Vegetables, cut flowers, and ornamental plants	%	Others	%
2018	149,243,252	69,066,573	46.28	31,279,222	20.96	38,521,616	25.81	1,571,479	1.05	8,804,361	5.90
2019	149,747,215	65,408,870	43.68	30,889,123	20.63	39,377,672	26.30	1,113,158	0.74	12,958,391	8.65
2020	149,751,617	65,410,147	43.68	30,889,608	20.63	39,376,962	26.29	1,114,548	0.74	12,960,351	8.65
2021	149,745,432	65,406,318	43.68	30,885,302	20.63	39,377,681	26.30	1,114,787	0.74	12,961,343	8.66

Note 1rai = 0.395 Aker
Source Office of Agriculture Economics (2022)

packages to deal with low-income and financial problems with the debt profiles in the agricultural sector rising by more than 25% during the COVID-19 pandemic in Thailand (NESDC, 2023c).

Thai farmers operating in the agricultural sector face problems ranging from small size, lack of land ownership, limited networking, and access to information to obtain management operations, an ageing workforce, rising production costs, lack of access to water, increased frequency of extreme weather events, and rising debt (Food and Agriculture Organization of the United Nations [FAO], 2023; GSMA, 2022; NESDC, 2023c; Table 7.11).

Digitalisation and Smartification of Agriculture

The MOAC's Digital Agriculture Strategy (2017–2021) launched by the Ministry of Agriculture and Cooperatives (MOAC) encompasses three primary objectives, i.e. developing an agricultural information system for proactive management, using suitable digital technologies within the agricultural sector, and providing assistance for sustainable agriculture. However, most smallholder farmers are still at a basic level of digitization and have yet to apply digital agricultural solutions (GSMA, 2022). According to a

Table 7.7 Agricultural land by region, Thailand, 2021 (Rai)

	Total land	Paddy land	%	Upland field crop	%	Fruit trees and perennial trees	%	Vegetables, cut flowers, and ornamental plants	%	Others	%
Whole Kingdom	149,745,431	65,406,318	43.68	30,885,302	20.63	39,377,681	26.30	1,114,787	0.74	12,961,343	8.66
Northern	31,864,520	15,099,244	47.39	10,373,639	32.56	3,921,287	12.31	419,599	1.32	2,050,751	6.44
Northeastern	64,970,497	39,595,675	60.94	12,710,674	19.56	7,333,302	11.29	154,587	0.24	5,176,259	7.97
Central	28,981,538	9,740,276	33.61	7,785,019	26.86	6,666,793	23.00	490,844	1.69	4,298,606	14.83
Southern	23,928,876	971,123	4.06	15,970	0.07	21,456,299	89.67	49,757	0.21	1,435,727	6.00%

Note 1rai = 0.395 Aker
Source Office of Agriculture Economics (2022)

Table 7.8 Ownership mode, agricultural regions, Thailand, 2021 (Rai)

	Total land	Own	%	Rent	%	Mortgage/ redemption/free use	%
Whole Kingdom	149,745,431	72,354,962	48.32	29,009,336	19.37	48,381,133	32.31
Northern	31,864,520	12,025,349	37.74	9,731,772	30.54	10,107,399	31.72
Northeastern	64,970,497	37,598,766	57.87	7,734,746	11.91	19,636,985	30.22
Central	28,981,538	10,068,770	34.74	10,976,083	37.87	7,936,685	27.39
Southern	23,928,876	12,662,077	52.92	566,735	2.37	10,700,064	44.72

Note 1rai = 0.395 Aker
Source Office of Agriculture Economics (2022)

Fig. 7.8 Thai food safety standards. *Source* Bureau Veritas Thailand (2023)

survey conducted by Thailand's National Statistical Office, only 23% of farmers use ICT-related technology in their operations. Firm-level adoption hurdles, agritech provider barriers, and ecosystem difficulties can all be used to explain weak digital adoption.

Digital solutions are crucial to the development of the agricultural sector to provide opportunities for Thai small-scale farmers, including better access to new and profitable domestic and international markets (see Table 7.12). Adapting farm operations through the utilization of digital technologies can have a positive impact on smallholder farmer earnings. Some of these agriculture-related digital solutions for smallholder farmers were presented by GSMA (GSMA, 2022): mobile applications that can expand the reach of extension services to farmers in rural areas; e-commerce platforms that allow smallholder farmers to sell and buy their products directly; platforms that connect smallholder farmers with affordable agricultural and farm equipment rentals; smallholder data portrait that is collected through Internet of Things (IoT) to support government advisory services and to understand the needs of farmers; multi-functional stakeholder information system and database platforms that can facilitate knowledge sharing and exchange, integration, and access to innovation and technology; utilization of artificial intelligence (AI), such as GPS, sensor technology, information and communications technology (ICT), drones, and robots for precision and smart farming; and data set from remote sensors on satellites, combined with big data and digital footprints that can support financial service providers to better assess risk profiles of smallholder farmers to improve their ability to access appropriate and affordable financing.

The enabling infrastructure for digital uptake, such as smartphone ownership and mobile broadband network coverage, has been steadily strengthening in Thailand, and innovative agritech solutions for more intelligent agriculture are emerging and happening in practice, supported by significant Thai government efforts (GSMA, 2022). According to the Digital Economy Promotion Agency (DEPA), there were more than 100 digital agriculture solutions that were available through smartphone applications in 2020. However, most local agritech start-ups need help

Table 7.10 Food security environment, Thailand

Affordability			Availability			Quality and Safety			Sustainability and Adaptation		
	Score	2012–2020 change		Score	2012–2020 change		Score	2012–2020 change		Score	2012–2020 change
Overall	83.7	+ 7.2	Overall	52.9	+ 4.4	Overall	45.3	−10.8	Overall	51.6	+ 17.2
1.1 Change in average food costs	100.0	+ 29.5	2.1 Access to agricultural inputs	56.2	+ 6.2	3.1 Dietary diversity	36.0	−0.6	4.1 Exposure	62.0	< − >
1.2% under global poverty line	99.7	+ 2.1	2.2 Agricultural research & development	33.0	+ 4.7	3.2 Nutri standards	20.2	−48.3	4.2 Water	25.0	< − >
1.3 Inequality-adjusted income index	59.7	+ 10.5	2.3 Farm Infrastructure	84.6	+ 42.8	3.3 Micronutrient availability	39.3	−0.7	4.3 Land	62.6	−3.0
1.4 Agricultural trade	51.1	−10.6	2.4 Volatility of agricultural production	44.2	−27.2	3.4 Protein quality	59.1	+ 0.3	4.4 Oceans, rivers and lakes	29.1	−1.5
1.5 Food safety net programmes	100.0	< − >	2.5 Food loss	87.6	+ 4.4	3.5 Food safety	71.4	−4.1	4.5 Political commitment to adaptation	52.9	+ 30.8
			2.6 Supply chain infrastructure	55.9	+ 14.3				4.6 Disaster risk management	77.4	+ 77.4
			2.7 Sufficiency of supply	71.2	+ 50.6						
			2.8 Political and social barriers to access	56.1	+ 9.7						
			2.9 Food security and access policy commitments	0.0	−52.5						

Source Corteva (2022)

Table 7.11 Challenges facing agriculture in the supply chain, Thailand

Inputs→	Production→	Harvest→	Processing→	Market
• Higher labour costs due to labour shortages • Higher wage rates • Ageing farmers • Lack of land ownership • High input costs • Absence of good management • Lack of access to water and irrigation	• Small-scale production leading to diseconomies of scale • Poor harvest and high production costs • Inefficient production • Damages plants, crops, insects, and pests • Lack of information and knowledge • Inadequate Agricultural practices	• Inadequate storage facilities • Quality degradation during the drying stage • Degradation of soil and water quality • High transportation costs • Delays in transportation	• Poor quality of raw material supply • Obsolete processing and ancillary equipment • Poor quality of processed products • Poorly trained personnel • Inappropriate packaging materials	• Poor marketing skills • Weak market development • Low market bargaining power • Insufficient market channels • Inefficient trade platforms • Inadequate expertise to create branding strategies to attract new markets

Source GSMA (2022) andUN (2023)

Table 7.12 Digital solutions for smallholder farmers

Access to services			Access to markets	Access to assets	
Farm management and advisory	Finance		Supply chain management	Market networks	Precision and smart farming
Precision agricultural advisory	Payments		Procurement	Mechanization access services	Smart crop management
Farmer information services	Insurance		Traceability and certification	Value chain integration	Smart livestock management
Participatory advisory	Credit and loans	Savings	Quality assurance and counterfeiting	E-commerce services	
Record keeping	Credit scoring	Crowdfunding	Logistics and Production network		

Source GSMA (2022)

to reach their intended users and smallholder farmers, which indicate that there are systemic issues that are stalling wider adoption and deployment of these solutions (DEPA, 2020; GSMA, 2022).

While digital technologies are proliferating in Thailand, a review of the digital platforms identified by DEPA implies that there are still supply-side gaps (DEPA, 2020). Are few agritech solutions exist to help support agricultural processing activities, distribution services, and support functions, such as farm management and advisory services and digital payments (GSMA, 2022), NSO (2023). Most Thai smallholders remained at a basic level of digitalization and technology in 2022. Utilization of ICT-related tools by smallholder farmers revealed that only 23% of Thai farmers use some form of ICT-related devices in their agricultural activities.

The primary device of business in Thailand is the smartphone, which is used to share and exchange information throughout social media channels (Line app and Facebook) with smallholder farmer groups (NSO, 2023). Thai smallholder farmers who have adopted digital agriculture solutions are likely to be those who engage in formal supply chains and grow high-value crops (HVC), such as fruits, flowers, spices, vegetables, aromatic plants, and organic food (GSMA, 2022). Aspects of traditional farming methods have also be included in such agricultural schemes. These Thai farmers have recognized the benefits of ICT across supply chains of agricultural products, from production to distribution of fruits and vegetables to the markets. However, most Thai smallholder farmers typically cultivate staple crops (e.g. potatoes, cassava, yams, and taro), which are not HVCs, and therefore, are unlikely to realize the full benefits of digital transformation in agricultural operations (GSMA, 2022; NSO, 2023b).

Even though there is a widening digital gap among Thai smallholder farmers, mobile phones, smartphones, and ICT provide an opportunity to deliver essential farm support services in agricultural activities to support participation in HVCs. However, it is important to emphasize what digital adoption challenges among are for Thai smallholder farmers need to be addressed. The GSMA (2022) interviews of smallholder farmers, ICT suppliers, and government officials in Thailand show that the following three critical barriers are being address by the government:

Small Farmer Problems

Trust and Behavioural

Smallholder farmers are typically conservative and averse to risks and uncertainties, which can hinder them from trying innovations and technologies that cause them to stick with traditional farming solutions. Hence, Thai farmers require tangible evidence of successful agricultural cases before they opt to adopt specific innovations and technologies in their operations. Many smallholder farmers are part-time farmers who have other sources of income, while some of them only grow agricultural products to obtain government incentives. Hence, they need motivation and creativity to be innovative and competitive (GSMA, 2022).

Digital Competence

The digital literacy of Thai farmers is still too low. Also, ICT adoption tends to decline significantly as the age of most Thai farmers increases. Most Thai farmers have a strong preference for highly simplified digital tools (NSO, 2023).

Finance

Thai farmers need more resources to invest in innovations, technologies, and processes to appropriate uncertain economic benefits. Thai farmers consider investment in innovation and technology solutions not worthwhile due to high costs and small farm size. Many Thai farmers seek financial loans and many of them are in debt (GSMA, 2022).

Nformation Gap

Smallholder farmers lack a clear understanding of the value of digital agriculture solutions, which explains why ICT adoption in their operations is low (NSO, 2023). Such information gaps provide limited scope for them to adopt ICT tools (NSO, 2023).

Aggrotech Barriers

Visibility

Online and digital marketing has not been effective in Thai agricultural businesses, and not all online marketing efforts have been successful. Thai suppliers often do in-person marketing, showcases, and events to meet with smallholder farmers, which is expensive and requires skilled workers for one-on-one conversations, it is time consuming, and their solutions are often characterized by ambiguity (GSMA, 2022; National Innovation Agency, 2021).

Research and Development (R&D)

When Thai agrotech start-ups conduct pilot tests on new products, they are only able to obtain data and feedback from a minimal number of users, which raises concerns concern about appropriateness, especially when agrotech start-ups need to fully consider the diversity of the needs of Thai small farmers (GSMA, 2022; National Innovation Agency, 2021).

Finances

The promotion of agrotech also requires access to finance (GSMA, 2022). Thai small farmers who can access government funds face delayed disbursement of funds that create cash flow issues for service providers in Thailand (National Innovation Agency, 2021). Also, Thai smallholder farmers always complain about the high cost of digital agriculture tools, which only provide basic features and instructions. While adapting ICT solutions can help Thai farmers efficiently, it is only sometimes cost-effective for suppliers (GSMA, 2022; National Innovation Agency, 2021).

Ecosystem Barriers

Public Extension Services

The Department of Agricultural Extension (DAE), Ministry of Agriculture and Cooperatives (MOAC), is the central government organization responsible for supporting and promoting community businesses (The Department of Agricultural Extension (DAE), 2023). DAE has regional offices in all 76 provinces in Thailand, which help assist in registering and providing technical advisory services, training programmes, financial assistance, and other services to Thai agriculture. However, the quality of DAE services is faced with low coverage of villages and rural communities. There is also a great deal of emphasis on "ticking the box," for instance, registering a high number of attendees in agricultural training programmes rather than focusing on the impact these programmes have on farmers' operations and activities (GSMA, 2022).

Infrastructure

The requisite infrastructure to appropriate digital synergies needs to be expanded, such as big data platforms (cloud-based services) in the supply chain (GSMA, 2022). Also, delays in issuing guidelines and regulations for innovations and technologies discourage, which also necessitates public investments in R&D to widen the spread of innovative digital solutions (GSMA, 2022).

Digitalization Cases

To extend the geographical reach of Thai agricultural extension service through digital solutions deployed via public-private partnerships (PPPs), the private sector has been engaged to develop and implement a mobile app-enabled remote participatory advisory service featuring feedback loops between content providers and end users to increase the number of Thai smallholder farmers receiving advice and service, as well as offer greater interactivity with the service providers and possibly a role for farmers to create advisory content (GSMA, 2022). The Kaset Go app was developed by *Total Access Communication Public Company Limited (DTAC),* a leading mobile providers in Thailand, in partnership with Yara Thailand. The Kaset Go app is an online community platform for Thai farmers to exchange information on a range of topics. It consists of a number of question-and-answer communities where peer farmers and experts from across the country provide support by responding to and answering queries from Thai farmers. Once the experts verify a Q&A thread to be accurate and valuable, the Kaset Go app will sort all the responses in an extensive database and file them in a knowledge directory that Thai farmers can access and filter responses by topic (GSMA, 2022).

In addition, The Kaset Go app has a section that utilizes infographics, videos, and other formats to represent trends and insights on Thai agricultural data in an easy-to-understand way. The Kaset Go app became available for smartphone download in August 2020 and has about 200,000 users. The developers are currently working on additional features and instructions, such as information on different crop types, weather warning data, pests, insects, and disease alerts, and market price information. A longer-term plan is to include ancillary services (electric power), including farm equipment rentals, agri-insurance, and digital financial services (GSMA, 2022).

Thai Innovative Farming Technology

To increase awareness about the business value of emerging low-cost agrotech digital solutions in Thailand, a collaborative private-sector digital agriculture solution for farmers was developed throughout a broad network of community-based digital road shows and pilot farms in Thailand. This Thai innovative farming technology exhibition project MOAC was established in partnership with DEPA and NIA to develop an intelligent farming project involving 150 maize farmers in Nakhon Sawan Province and Sukhothai Province with 65 farmers in 2019. The project aimed to exhibit a number of advanced digital technologies with practical demonstrations that include utilizing drones to spray crop pests and insects or through uncrewed aerial vehicles (UAV) to spray liquid supplements on crops. While it took only three minutes for the drone to spray pilot farms, it would have taken farm labourers an hour to pour over one Rai. Although the success of these demonstrations is unknown in Thailand, it offers scope for scaling them nationally so that Thai farmers who need to be made aware of these digital agriculture solutions can realize the benefits first-hand. Notably, implementing these innovative farming technologies would require more than conducting demonstrations or pilots. In order to deploy the solutions by scale and sustainably, Thai farmers need to be shown how to use and operate them efficiently (GSMA, 2022).

Furthermore, multistakeholder collaboration is required to drive the digital transformation of Thai smallholder farm operations. Table 7.13 shows a roadmap to improve Thai smallholder farmers' utilization of agrotech solutions to strengthen sustainable farming.

Conclusions and Policy Implementations

Thailand's agriculture sector remains a vital component of its economy, particularly supporting the livelihood of the rural people. Benefiting from a tropical climate and rich biodiversity, agriculture has been a historical mainstay, providing livelihoods for millions and driving exports of rice, cassava, and sugar. The sector employs nearly a third of the workforce and supports over two-thirds of the rural population. Thailand boasts abundant agricultural land and resources, ranking among the world's top rice producers. However, the agricultural sector's contribution to GDP has declined due to inefficiencies and low productivity. Thailand also lags behind ASEAN peers in rice yield, with most farmers operating small, often unprofitable holdings. In addition, Thailand's agricultural sector faces significant challenges hindering its growth and farmer prosperity. Small farms, limited resources, ageing farmers, and climate change impacts have become critical obstacles.

To address these problems, the government pursued various agricultural development plans since the 1960s. The early strategies focused on infrastructure and cultivation of non-commercial crops, shifting later towards commercialization and sustainability. However, Thailand's controversial rice-pledging scheme that was initiated in 2011 resulted in skyrocketing government expenditures due to direct rice purchases at prices 50% above market rates. Nevertheless, the 13th development plan emphasizes water management, quality standards, farmer support, sustainability through the deployment of climate resilience, and digitalization strategies to modernize the sector.

As with countries in general, the agricultural sector's contribution to GDP has fallen in trend terms, but even so the livelihood of the small rural farmers have not improved tangibly as owing to small farm size and landlessness. While the digitalization and sustainability initiatives hold promise, rural households have remained disadvantaged. Importantly, the agricultural sector's vulnerabilities have been further exacerbated by inadequate research and development, limited marketing capabilities, and insufficient government support. In order to confront these obstacles, a comprehensive strategy is required to improve the quality of life of especially small rural farmers by raising farm productivity, and strengthening digitalization and climate resilience. The

Table 7.13 Stakeholder mapping for fast-tracking digital adoption by Thai farmers

Action	Lead partner	Supporting stakeholders	Expected outcome
Digital training to help farmers become more open and willing to try new digital solutions	MOAC	–PPPs under the agricultural extension scheme	Improved digital competence to utilize technology tools to transform farming operations
Restructuring the agricultural extension system to incorporate PPPs for greater reach and impact		–Department of agricultural extension (DAE) –Agritech companies, particularly those with digital advisory expertise	Thai Farmers have adequate support from government organizations
Encouraging digital uptake via community networks that can share costs and make it more affordable for farmer members		–Community enterprise promotion division, MOAC –Community organizations development institute –Local agricultural cooperatives and community enterprises –Agritech companies	Digital agriculture solutions become more affordable and accessible when costs are shared
Extending support to providers of digital solutions, to showcase demonstrations and exhibitions of existing technologies, to show benefits, and how to implement solutions		–DEPA –NIA –Bank for agriculture and agricultural cooperatives (BAAC) –Agritech start-ups –Start-up Thailand	Improved attitudes of Thai farmers towards adopting digital agriculture solutions
Infrastructure investments in a National agricultural data aggregation platform that farmers involved in agriculture can access, share, and use real-time agricultural data		–Office of agricultural economics (OAE) –National agricultural big data center (NABC) –Digital government development agency (DGA) –Data protection commission –Open government data of Thailand –BAAC –Large private-sector agriculture aggregators –Cooperatives	Available information for planning, decision-making, networking, and marketing

Source GSMA (2022)

following should be the primary focus of Thailand's digital agriculture strategy (GSMA, 2022): fostering precision farming; investing in research and development to develop innovative agricultural systems that optimize resource use, production costs, and market outcomes through digital technologies; increasing farmer capabilities to provide farmers with the knowledge and skills necessary to adopt and benefit from digital tools, sustainable practices, and data-driven decision-making; increasing technology access to provide farmers with affordable digital agriculture technologies and subsidies to assist them in transitioning to demand-driven production models; and providing assistance to farmer organizations by introducing e-commerce guidance to enhance market access and profitability.

Thailand launched its Food Security Strategic Plan in 2017, aiming to ensure food access, quality, and safety for its population while safeguarding agricultural resources while prioritizing digital agriculture to enhance smallholder farmers' livelihoods. The government is investing in research, technology development, and farmer training to promote precision agriculture and data-driven farming practices to develop innovative farming systems targeted at improving farmers' access to technology and information, providing affordable digital tools, and support farmer organizations to leverage e-commerce.

Finally, it is important to recognize that behind the glorious agricultural export story, Thai farmers confront a multitude of challenges affecting their livelihoods, which include small-scale farming, land scarcity, ageing farmers, rising costs, water shortages, extreme weather conditions, and debt burdens. Low productivity, limited research, and inadequate technology have raised questions over its sustainability over the long run. A holistic approach to the agricultural sector is essential to revitalize the agricultural sector, improve farmer livelihoods, and foster sustainability, which must address the interconnectedness of economic, social, and environmental factors. Consequently, government policy must consider ecological stewardship by prioritizing soil health, biodiversity, and water management through practices such as agro-ecology; farmer empowerment by providing farmers with resources, technology, training, and fair market access; strengthening value chains by fostering collaboration among agricultural stakeholders to increase efficiency and farmers' incomes; and strengthening climate resilience by developing strategies to mitigate climate change and global warming.

References

Annizah, N. I., & Rasiah, R. (2025). Agricultural Sustainability and Food Security in Malaysia, Rasiah. R., Anter, V. & Annizah, N.I. (eds), *Sustainable Agriculture and Food Security in Southeast Asia: Towards Climate Resilience and Digitalization*, Singapore: Springer Nature.

Apichart Pongsrihadulchai. (2019). Thailand agricultural policies and development strategies. *FFTC Agricultural Policy Platform (FFTC-AP).* https://ap.fftc.org.tw/system/files/field/file/article/980_1.pdf

Bureau Veritas Thailand. (2023). https://www.bureauveritas.co.th/en

Corteva. (2022). *Global Food Security Index 2022: Thailand country report.* https://impact.economist.com/sustainability/project/food-security-index/reports/Economist_Impact_GFSI_2022_Thailand_country_report_Sep_2022.pdf

Craig Keating. (2022). *Thailand's Agricultural Policies: A Hand Out, Not a Hand Up.* The University of Melbourne. https://asialink.unimelb.edu.au/insights/thailands-agricultural-policies-a-hand-out,-not-a-hand-up

DEPA. (2020). *Agriculture Landscape in Thailand.* https://www.depa.or.th/storage/app/media/file/investment-bulletin.pdf.

Food and Agriculture Organization of the United Nations (FAO). (2008). An introduction to the basic concepts of food security. https://www.fao.org/3/al936e/al936e00.pdf

Food and Agriculture Organization of the United Nations (FAO). (2023). Country Case Studies: Thailand. https://www.fao.org/3/ac158e/ac158e08.htm

GSMA. (2022). *Digitalizing Rural MSMEs: Thailand's Agriculture and Tourism Sectors.* https://www.gsma.com/mobilefordevelopment/wp-content/uploads/2022/06/GSMA_CIU_Thailand-MSME-Digitalisation_Jun2022.pdf.

Mahathanaseth, I., & Pensupar, K. (2014). Thai Agricultural Policies: The Rice Pledging Scheme. *Taipei (TW): FFTC Agricultural Policy Platform.*

National Innovation Agency. (2021). *Thailand Agtech Startup Ecosystem Development White Paper.* https://www.nia.or.th/2022/Thailand%20AgTech%20Startup%20Ecosystem%20Development%20White%20Paper

NESDB. (2017). *The Twelfth National Economic and Social Development Plan (2017–2021).* Office of the National Economic and Social Development Board. https://www.nesdc.go.th/ewt_dl_link.php?nid=9640

NESDC. (2023a). *The 13th National Economic and Social Development Plan (2023–2027)* Office of the National Economic and Social Development Council, Office of the Prime Minister. https://www.nesdc.go.th/article_attach/article_file_20230615134223.pdf

NESDC. (2023b). *Office of the National Economic and Social Development Council (NESDC).* Gross Domestic Product (GDP). https://www.nesdc.go.th/nesdb_en/article_attach/article_file_20230821091724.pdf

NESDC. (2023c). *Office of the National Economic and Social Development Council (NESDC).* NESDC Economic Report. https://www.nesdc.go.th/nesdb_en/article_attach/article_file_20221206142518.pdf

NSO. (2023). *National Statistical Office of Thailand, The Information Communication and Technologies Using of Population and Household in Thailand*. https://ittdashboard.nso.go.th/preview2en.php?id_project=138

Office of Agricultural Economics. (2017). *The Twenty-Year Agricultural and Cooperative Strategy (2017–2036) and the Five-Year Agricultural Development Plan under the Twelfth National Economic and Social Development Plan (2017–2021)*

Office of Agricultural Economics. (2022). *Agricultural Statistics of Thailand 2022*. Office of Agricultural Economics. https://www.oae.go.th/assets/portals/1/files/jounal/2566/yearbook2565.pdf

Office of the Board of Investment (BOI). (2021). *Thailand Food Industry*. https://www.boi.go.th/upload/content/Food%20industry_5aa7b40bd758b.pdf

Rasiah, R., Low, W. Y., & Kamarudin, N. (2023). *Digitalization and Development: Ecosystem for Stimulating Industry 4.0 Technologies*. Routledge.

Temsumrit, N., & Sriket, H. (2023). The Structural Transformation of Thailand: The Role of Policy Distortion. *Asian Development Review, 40*(01), 203–245.

Thai Food and Drug Administration (FDA). (2023). https://en.fda.moph.go.th/cat2-health-products

The Department of Agricultural Extension (DAE). (2023). The Ministry of Agriculture and Cooperatives (MOAC). https://www.doae.go.th/en/home1-english/

The Secretariat of the Senate. (2014). *Thailand Food Security*. https://library.senate.go.th/document/Ext7091/7091777_0002.PDF.

UN. (2023). *Food Security and Nutrition and Sustainable Agriculture*. https://sdgs.un.org/topics/food-security-and-nutrition-and-sustainable-agriculture

World Bank. (2014). *Preserving Stability and Promoting Growth*. World Bank. https://documents1.worldbank.org/curated/en/474401468027633486/pdf/Preserving-stability-and-promoting-growth.pdf

World Bank. (2021). *Employment in Agriculture (% of total employment) (modeled ILO estimate)—Thailand*. World Bank. https://data.worldbank.org/indicator/SL.AGR.EMPL.ZS?locations=TH&most_recent_year_desc=true&view=map

Yot Amornkitvikai is Associate Professor of Economics at the College of Population Studies, Chulalongkorn University. He has published extensively in journals, including Gender, Technology and Development, Journal of Asian Business and Economic Studies, Journal of Economic Studies, and Gender in Management, as well as book chapters by the Institute of Southeast Asian Studies and the Asian Development Bank Institute.

Teerawat Charoenrat is Associate Professor of Economics at Khon Kaen University, Thailand. He obtained the Royal Thai Government Scholarship Fund for pursuing a Ph.D. degree. His research interests include enterprise performance, firm productivity and technical efficiency in micro, small and medium enterprises in Thailand. He has published in refereed journals, such as Economic Modelling, Journal of Asian Economics, Asian Economic Papers, Journal of Southeast Asian Economies, Journal of Entrepreneurship in Emerging Economies. He has undertaken research projects for the World Bank, UNCTAD, ADB, ADBI, ITD and ERIA.

Agriculture Development and Food Security in Vietnam

8

Bach Tan Sinh, Dang Kim Khôi, Nguyen Thanh Bình, Nguyen Minh Tu, Le Van Thuy Tien, and Nguyen Hong Quan

Abstract

Vietnam's agriculture has made significant progress since the 1990s, which has helped strengthen food security in the country, as well as enable strong exports. However, it is facing many challenges in a changing ecological and social context. Drawing on documents and secondary data over the period 1975-2022, this chapter analyzes agricultural development and food security in Vietnam. Firstly, the chapter examines agricultural development and food security using indicators, such as changes in share of agriculture in GDP and employment in overall employment, farm acreage, production and yields of key crops and livestock; food farming; food security and consumption; and rural livelihood. Secondly, it discusses government policies to promote sustainable agriculture. Thirdly, the chapter assesses climate-resilient smart agricultural practices, such as organic farming, circular agriculture, digital transformation, and the focus on science, technology, and innovation in agriculture. The chapter finishes with an analytical proposal toward accelerating green transition in agriculture.

Keywords

Agriculture · Food security · Sustainable Development · Vietnam

B. T. Sinh (✉) · N. M. Tu · N. H. Quan
Vietnam National University, Ho Chi Mnh, Vietnam
e-mail: sinhbt@gmail.com

N. M. Tu
e-mail: nm.tu@iced.org.vn

N. H. Quan
e-mail: nh.quan@iced.org.vn

D. K. Khôi
vietnam national university of agriculture, Hanoi, Vietnam
e-mail: khoidk@gmail.com

N. T. Bình · L. V. T. Tien
Can Tho University, Can Tho, Vietnam
e-mail: ntbinh02@ctu.edu.vn

L. V. T. Tien
e-mail: lvttien@ctu.edu.vn

Introduction

Since the 1990s, agricultural development has changed the socioeconomic status of Vietnam toward improving food security, boosting agricultural exports, and strengthening the livelihood of people. The agriculture sector plays a crucial role in Viet Nam's economy and society. Nevertheless, typical of rapidly industrializing and urbaning countries, the contribution of agriculture to Vietnam's GDP has fallen from 35 percent in

R. Rasiah et al. (eds.), *Sustainable Agriculture and Food Security in Southeast Asia*, Sustainable Development Goals Series, https://doi.org/10.1007/978-981-95-0477-0_8

1986 to 11 percent in 2022, while its share in overall employment shrunk from 84 percent in 1986 to 28 percent in 2022. Despite a contracting share in both GDP and employment, agriculture remains a cornerstone in sustaining the stability of Vietnam's economy and political infrastructure. The robust agricultural base ensures the country's continued access to low-cost agricultural commodities. This resilience becomes particularly evident during external shocks, when the crop cultivation and livestock farming subsectors emerge as stabilizing forces, anchoring the growth of the broader agriculture, forestry, and fisheries sectors, and by extension, underpinning the populace's livelihoods.

The chapter aims to provide Vietnam experience in agricultural development and food security. A qualitative interpretative method is used to systematically review the relevant literature and desk studies pertaining sustainable agriculture development, food security, and enabling agricultural policies in Vietnam. In addition, the chapter uses for analysis data from the Statistical Yearbooks, Rural, Agricultural and Fishery Census, and Household Living Standard Surveys. The chapter is organized as the followings. Firstly, the chapter examines agricultural development and food security using such indicators as changes in share of agriculture in GDP and employment, agricultural trade. Secondly, it presents government enabling policies to promote sustainable agriculture and food security. Thirdly, the chapter assesses climate-resilient smart agricultural practices, such as organic farming, circular agriculture, integrated ecological farming, digital transformation, and the role of science, technology, and innovation in agriculture. The chapter proposes the proposal toward accelerating green transition in agriculture.

Agricultural Development and Food Security

Between 1976 and 1986, agriculture served as the backbone for the expansion of heavy industry in the centrally planned economy. This era saw the organization of agricultural production through cooperatives and state farms, while state-owned enterprises monopolized the provision of inputs and the regulation of output markets. Despite incremental reforms aimed at boosting production, agricultural outputs consistently fell short of state goals, precipitating food shortages and catalyzing a significant economic downturn (OECD 2015).

In response, the government initiated a series of sweeping reforms in the mid-1980s, collectively termed *Doi Moi* (Renovation). These reforms pivoted the Vietnamese economy from a centrally planned system to a more market-oriented paradigm. Over the subsequent 35 years, this strategic shift has redefined Vietnam's economic narrative, propelling it from one of the world's poorest nations, trailing behind countries like Laos and Kenya in 1986, to a lower-middle-income country by 2022. This remarkable economic upturn has been paralleled by a profound realignment in both GDP composition and employment patterns, with a pronounced migration of economic activities from agriculture to services and manufacturing. Notably, the non-agricultural economy has surged at a significantly higher rate during this period, relegating agriculture's contribution to both GDP and overall employment. While agriculture's share in GDP and employment in overall employment has shown a trend fall between 1986 and 2022 (Fig. 8.1), as the chapter will show it continues to play a major role in the development of Vietnam.

The shift in Vietnam's workforce from agriculture to the secondary and tertiary sectors of manufacturing and services aligns with established theory in development economics, which posits that for impoverished nations to ascend the economic ladder, they must undergo structural change from primary to the secondary and tertiary sectors. This evolution involves the redistribution of labor from low-yield sectors to high-yield sectors (Laborde et al., 2019).

Despite a contracting share in GDP and overall employment, agriculture remains a cornerstone in sustaining stability in Vietnam's economy and political infrastructure. The sector persisted as the primary source of income for 28 percent of the

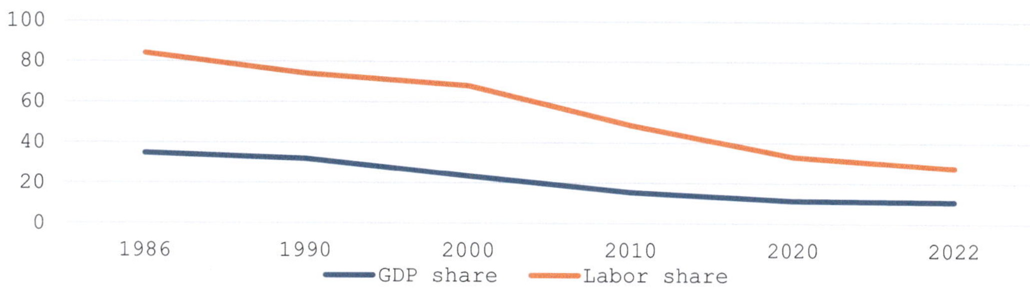

Fig. 8.1 GDP composition and employment share, agriculture, Vietnam, 1986–2022 (%). *Source* Compiled from Laborde et al. (2019) and GSO (2023)

workforce. Furthermore, a robust agricultural base ensures the country's continued access to low-cost agricultural commodities, which becomes particularly evident during external shocks, when the crop cultivation and livestock farming subsectors emerge as stabilizers, anchoring the growth of the broader agriculture, forestry, and fisheries sectors. Figure 8.2 illustrates this resilience, highlighting the sector's performance during the 2008 Global economic crisis and the 2020–2021 COVID-19 pandemic. While these crises inflicted contractions in the growth rates of industrial and service sectors, agriculture has demonstrated remarkable stability. This steadiness not only buffered the sector but also played a pivotal role in anchoring Vietnam's economy amid these tumultuous times.

The evolution of Vietnam's agriculture from 1986 to 2019 can be segmented into four distinct phases: reform, expansion, consolidation, and reorientation (OECD, 2015). The initial phase was marked by transformative reform policies, including Resolution 10, which spearheaded decollectivization, relaxation of government-mandated price controls, and the pioneering land law. These reforms invigorated the agricultural sector, reversing a -1.1 percent downturn in 1986–1987 to an upswing of 6.9 percent in 1991–1992, culminating in an average growth rate of 2.9 percent over the 1986–1992 bracket.

Having stabilized national food supplies, Vietnam embarked on legislative initiatives such as the new cooperative law and liberalization of rice exports, coupled with a substantial escalation in agricultural budgetary allocations. These strategic moves, aimed at bolstering food production for export during 1994–2002, propelled the average growth rate to 3.8 percent.

The dawn of the new millennium saw the promulgation of Decree 5 in 2001, geared toward catalyzing agricultural modernization, rural industrialization, and a ten-year development blueprint spanning 2001–2010. Concurrently, Vietnam's global economic integration deepened, underscored by the successful negotiation of the US-Vietnam bilateral trade agreement and ascension talks with the WTO. These concerted policy efforts sustained an average GDP growth rate of approximately 3.1 percent during this period.

Despite noteworthy achievements, there was growing apprehension within the government regarding the unsustainable trajectory of agricultural development. Addressing these concerns, the Central Party Committee promulgated Resolution 26 in 2008 to focus on farmers, and rural areas (collectively referred to as "Tam Nong"), which was solidified in 2013 with the approval of the National Agricultural Restructuring Programme, an initiative aimed at augmenting value added and promoting sustainable development.

However, these strategic efforts did not arrest the declining trend of agriculture in 2010–2019, which was exacerbated when it grew by 2.9 percent during the COVID-19 pandemic (Fig. 8.3). Nevertheless, growth rebounded significantly to 4.4 percent in 2022.

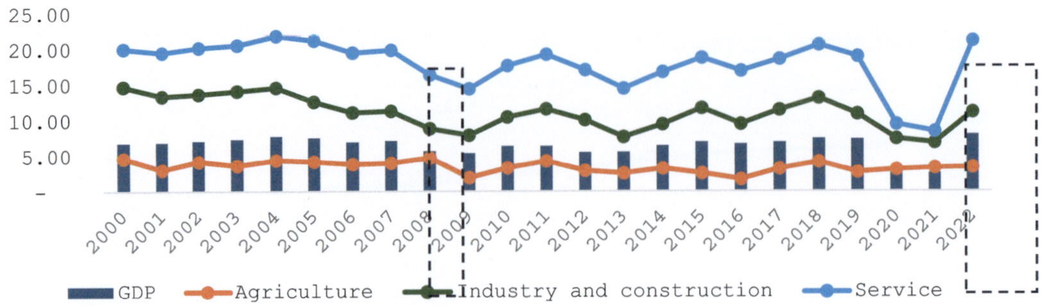

Fig. 8.2 Sectoral GDP growth, Vietnam, 2000–2022 (%). *Source* Author's calculation based on GSO data (2023)

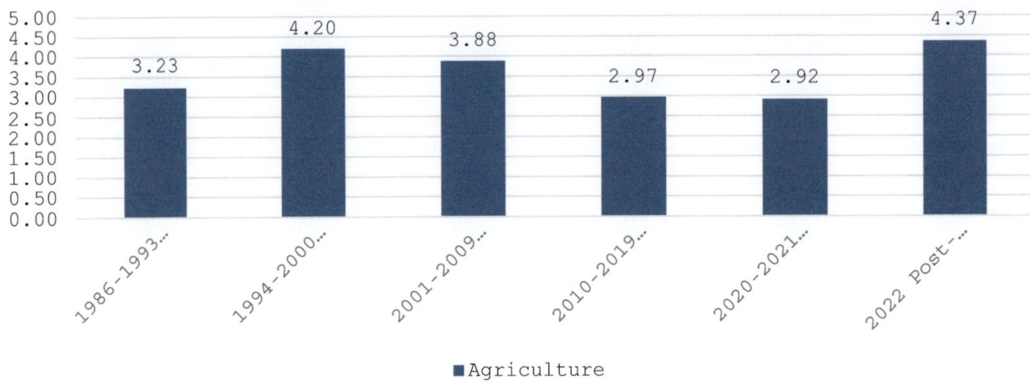

Fig. 8.3 Value added growth, agriculture, Vietnam, 1986–2022 (%). *Source* Calculated by author based on GSO data (2023)

Livestock and plantation subsectors have steadfastly held the lion's share of agricultural contribution to GDP since the mid-1980s, peaking at 73 percent in 2022, which, however, has waned since evidenced by a 10 percent decline in their GDP share from 1986 to 2022. Conversely, the fisheries' subsector has expanded its economic footprint, rising from a modest 7 percent share in the initial phase to a substantial 21 percent in 2022, which saw the subsector grow by approximately 9 percent in its early stages, though it decelerated markedly in 2010–2019 (Fig. 8.4).

The forestry subsector has enjoyed a resurgence in growth, accelerating from a paltry 1 percent at the onset of the reforms to almost 8 percent in 2022. Amid these shifting dynamics, the growth rates for livestock and plantations have remained relatively resilient, hovering around 3–4 percent.

International Trade

Economic integration has played a pivotal role in expanding and diversifying Vietnam's agricultural, forestry, and fishery import-export markets, facilitating access to a multitude of high-value domains. The export turnover of agricultural commodities has witnessed sustained growth, particularly since Vietnam's accession to the WTO, a milestone that precipitated the establishment of agricultural trade relations with an expansive network of approximately 220 countries and territories worldwide. By 2022, Vietnam's agricultural exports had made inroads into about 180 countries and territories, with no fewer than 24 markets registering transaction values exceeding $100 million each. The landscape of Vietnam's agricultural exports has been dominated by five key market groups: China, Europe, the United

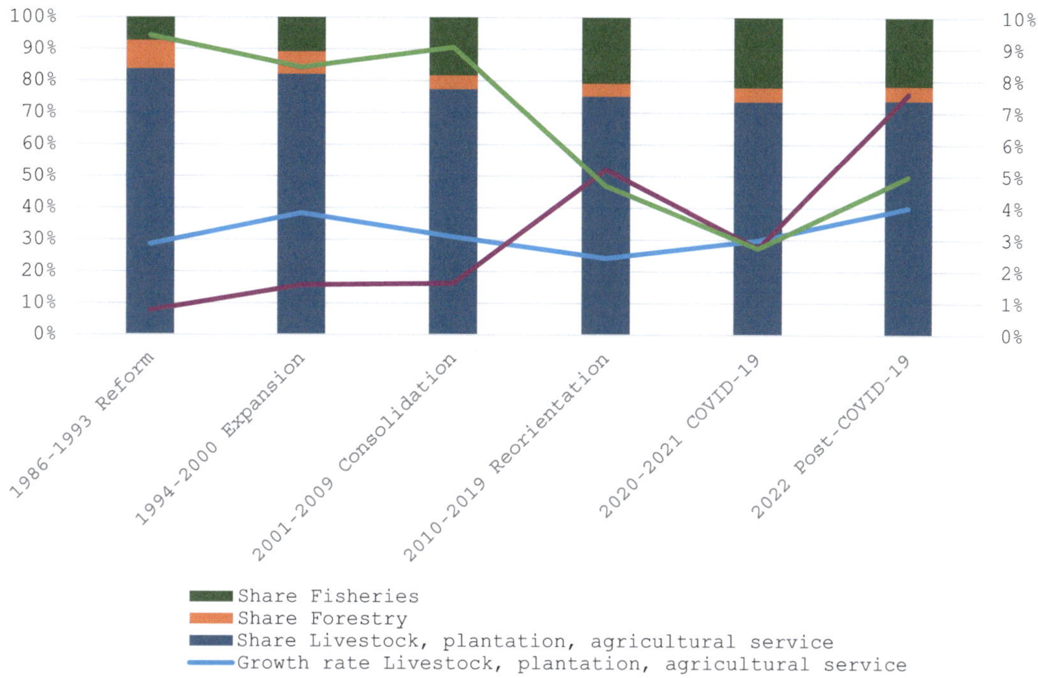

Fig. 8.4 GDP share and growth, agricultural subsectors, Vietnam, 1986–2022. *Source* Calculated by author based on GSO data (2023)

States, Japan-Korea, and ASEAN. Together, these markets accounted for 84.8% of the agricultural export turnover in 2022, which rose from $4.2 billion in 2000 to $53.2 billion in 2022, while the trade surplus in the commensurate period surged from $2.9 billion to $8.5 billion (Fig. 8.5).

By 2023, Vietnam had become the largest exporter of cashews and black pepper, the second-largest exporter of rubber, and the third-largest exporter of rice and fishery products. Indeed, Vietnam enjoyed exports of ten commodities worth over USD 1 billion in 2023. Figure 8.6 shows a dramatic surge in Vietnam's exports since 1995.

Government Policies to Promote Sustainable Agriculture.

The most notable agricultural policy to show traction with economic development in Vietnam between 2010 and 2018 is the *Restructuring the Agricultural Sector Towards Increasing Value and Sustainable Development*, which laid out clear directives for overhauling the agricultural sector. It addressed restructuring by (i) forming an integral part of the broader economic realignment geared toward economic, social, and environmental sustainability; (ii) realignment with markets, pivoting from extensive expansion to a concerted emphasis on quality enhancement; (iii) accentuating the state's role in providing support, fostering an enabling environment, and amplifying the engagement of diverse economic and social sectors (including farmers and enterprises) through Public-Private Partnerships (PPP) and community collaboration; and (iv) constituting a long-haul endeavor necessitating cohesive strategies from pertinent ministries, alongside robust monitoring and evaluation systems for learning and refinement.

In particular, the key goals of the plan include: i) maintaining growth, improving efficiency and competitiveness through raising productivity, quality and added value; better meet the needs and tastes of domestic consumers and to promote

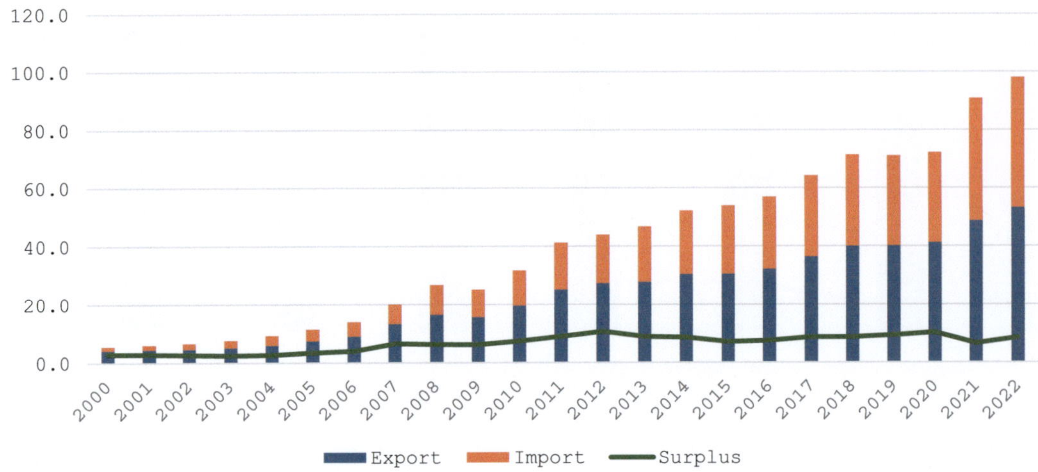

Fig. 8.5 Agricultural trade, Vietnam, 2000–2022 (USD billion). *Source* Author's calculation based on MARD data (2022)

exports. Strive to achieve an average industry-wide GDP growth rate of 2.6–3.0 percent in the 2011–2015, and 3.5–4.0 percent/year in the period 2016 2020; ii) raise income and improve living standards of rural residents, ensuring food security (including nutritional security) in both the immediate and long term, contributing to reducing poverty rates. Rural household income was targeted to increase by 2.5 times from 2008 to 2020. The number of communes meeting this new rural criteria constituted 20 percent by 2015 and 50 percent by 2020. It also seeks to strengthen natural resource management, reduce greenhouse gas emissions and other negative impact on environment to exploit environmental benefits, improve risk management capacity, prevent natural disasters, and increase national forest coverage rate to 42 percent by 2020 to contribute to strengthen climate resilience.

Land Reform

In accordance with the provisions of the Constitution, land in Vietnam is regarded as the collective property of *the entire people*, and its allocation or lease is conducted by the state, extending to various entities, such as organizations, households, and individuals. The state issues land use right certificates, which confer the right to engage in activities such as selling, leasing, exchanging, mortgaging, and inheriting land. The government initiated the allocation of land use rights to farmers in the 1980s, marking a pivotal development in the country's land tenure system. This undertaking represents one of the most significant efforts in the developing world, both in terms of scale—nearly 11 million registrations were issued expeditiously to rural households by year 2000 (OECD 2015). Approximately 75 percent of the total land area in Vietnam has been meticulously mapped by 2012 with land use right certificates issued to approximately 85 percent of agricultural land, which has significantly contributed to agricultural and rural development in Vietnam during the period 1990 to 2000 (Kompas et al., 2012).

The Land Law in Vietnam underwent several revisions since 1986 (including 1988, 1993, 1998, 2003, and 2013), with the overarching objective of facilitating the development of the land market. In a concerted effort to promote the land market while simultaneously upholding principles of land equity, significant amendment was introduced in 2003 entailing an amendment to establish a structured land price framework, the regulation of land use, and the introduction of land area quotas.

In the most recent revision of the Land Law in 2013, further adjustments were made to increase

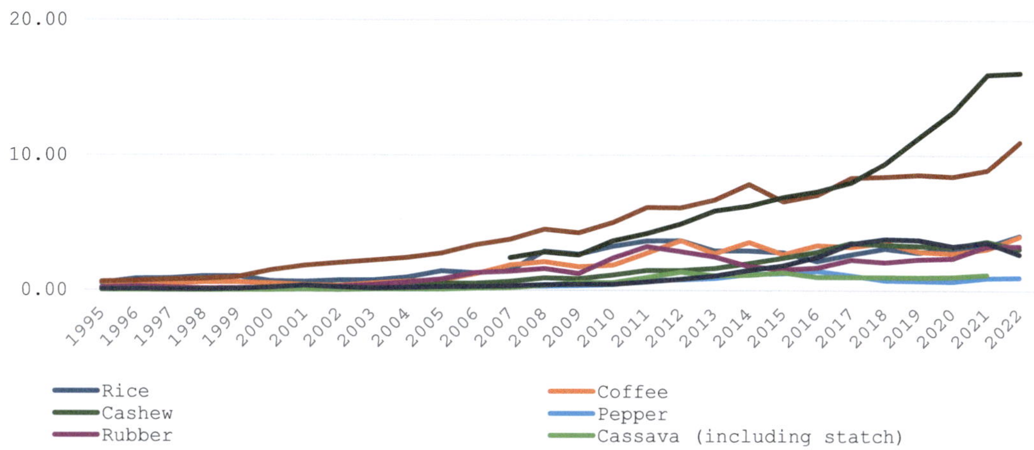

Fig. 8.6 Top agricultural exports, Vietnam, 1995–2022 (USD billion). *Source* Author's calculation based on MARD data (2022)

the limits on land use and area, in addition to instituting a negotiation mechanism for land acquisition to incentivize the establishment of large-scale farms or material zones. Under the current regulations, the annual crop land area allocated to each household or individual directly engaged in agricultural production must not exceed 3 hectares in centrally managed provinces and cities within the Southeast and the Mekong Delta regions. The limit is set at 2 hectares in other centrally managed provinces and cities. For perennial crops, the area must not exceed 10 hectares in the plains and 30 hectares in the midland and mountainous regions. Furthermore, protection[1] or production forests are limited to a maximum of 30 hectares per household. However, these limits are subject to ad hoc adjustments, as the law affords some flexibility. For instance, the law permits a ten-fold increase in allocation limits under unspecified conditions and stipulates that limits can be altered "in accordance with the specific conditions of each locality and period."

[1] Protection forests are primarily used for water source protection, soil conservation, erosion control, landslide prevention, flash flood prevention, combating desertification, mitigating natural disasters, regulating climate, contributing to environmental protection, national defence, and security, while production forests are primarily used to supply forest products, for the combined production and business activities of forestry, agriculture, and aquaculture (Article 5, Forestry Law 2017).

In addition to the Land Law, the Central Government and National Assembly of Vietnam actively implement directives and resolutions issued by the National Central Party and the Party's Politburo, which serve as strategic frameworks for government initiatives aimed at enhancing the capacity of rural communities in response to the increasingly intricate socioeconomic and environmental challenges facing Vietnam. For instance, in 2008, the Party issued Resolution 26-NQ/TW on Agriculture, Farmer, and Rural Areas, which advocates inclusive and sustainable agricultural and rural development. The National Assembly approved the National Target Program on New Rural Development in 2013, which places a strong emphasis on improving the livelihoods of farmers and the living conditions of rural households. It is executed through a collaborative effort between the government and local communities with a significant focus on enhancing rural infrastructure, including roads, markets, schools, and irrigation systems. Furthermore, the Party has also provided policy guidelines for economic development in the Vietnam Mekong Delta (VMD), such as Resolution 21 (2001–2010) and Conclusion 28 (2011–2020 period). The government has subsequently developed policies related to land use, including initiatives to reshape agricultural production structures, preserve rice-growing land to ensure food security, and establish large agricultural material zones to promote collaboration, production, and consumption within the agricultural sector (Nguyen et al. 2020).

Rice First Policy

The Rice First Policy has been maintained by the government since the reform in 1986.

Rice Land Safeguard

Recognizing rice's critical role in national food security, the government enacted policies to safeguard rice land. Following significant global rice price fluctuations in 2008, the Politburo's Conclusion No. 53-KL/TW on August 5, 2009, and Resolution 63/2009/NQ-CP on December 23, 2009, mandated the preservation of 3.8 million hectares for rice cultivation, including a minimum of 3.2 million hectares explicitly dedicated to rice farming, which guarantees an annual rice output of 32–41 million tons, sufficient to satisfy domestic needs and enable the export of approximately four million tons per year.

Under Resolution 63, rice land areas are to be stringently protected, with specific allocations for each locality and detailed land use status maps down to the household level. This land conservation directive further emphasized the National Assembly's Resolution No. 13/2011/QH13 concerning land use planning through 2020. The resolution explicitly stipulates that the rice land area would encompass 3.95 million hectares in 2011–2015, which was projected 3.8 million hectares in 2020.

The government rolled out initiatives aimed at ensuring profitability for producers, such as Decree 42/2012/ND-CP and Circular 205/2012/TT-BTC, which provide direct support for rice producers. Further, the Official Dispatch 430/TTg-KTN (2010) was instituted to guarantee a 30 percent profit margin for producers, while Circular 171/2010/TTLT outlines the methodology for determining production costs.

Irrigation

Since the early 1980s, the government is developing and maintaining irrigation systems to mainly raise rice production. Vietnam's irrigation infrastructure is categorized based on scale, encompassing 1.6 million hectares serviced by small-scale systems (less than 5000 hectares each), 1.2 million hectares by medium-sized systems (ranging from 5000 to 50,000 hectares), and 1.7 million hectares by large-scale systems (exceeding 50,000 hectares each). Approximately half of the total irrigated land is situated in the Mekong Delta, while the Red River Delta (RRD) accounts for over 16 percent. Supporting this extensive irrigation network are roughly 5600 reservoirs, strategically positioned to supply water as required that complement irrigation water sourced directly from rivers. In regions where field-level water ingress is impeded due to low water levels, an infrastructure comprising more than 11,500 water pumps is deployed to facilitate irrigation across about 2.1 million hectares of farmland (OECD 2015).

Rice storage, Reserve, Price Stabilization, Licensing, Quotas and Export taxes

To ensure a stable supply chain, the state has encouraged the expansion of rice storage capacity of enterprises following Decision 3242/QD-BNN-CB (2010), which endorsed plans for a storage network capable of handling four million tons of rice from the Mekong Delta. Concurrently, measures were instituted to curb post-harvest losses, as outlined in Decision 68/2013/QD-TTg (superseding Decisions 63/2010-TTg and 65/2011/QD-TTg) and Circular 13/2014/TT-NHNN, which provides guidelines for loans aimed at mitigating post-harvest wastage. Significantly, Decision 68 has streamlined businesses' access to credit facilities intended to diminish post-harvest losses by eliminating previously enforced regional restrictions. This strategic shift fosters a conducive environment that incentivizes businesses to collaborate closely with farmers over product distribution, thereby reinforcing the entire agricultural supply chain.

Strategy for State Reserve out to 2020 has set the target of 500 thousand tonnes of paddy and rice and 10 thousand tonnes of rice seed. At the moment, the annual quantity of paddy/rice to be brought into the reserve is set by Prime Ministerial decision. This can be brought from farmers or traders, with a maximum price set by Ministry of Finance.

The PSF was established in 1993 with the objective of regulating and stabilizing domestic prices. An abnormal fluctuation for rice was considered to be rise in the retail price of 25% or more over 30 days.

Export quotas on rice exports were introduced in 1992 to ensure food security and price stability. Total quantity of rice exports as well as the allocation of export quotas was determined by the government in two phases on the basis of rice production and consumption estimates. The rice export quota system was formally abandoned in 2021. Enterprises were from then on permitted to export rice provided they had obtained general business licenses to trade in rice. A flexible control system has developed in its place and formalized in Decree 109/2010/ND-CP. Accordingly, MARD is responsible for forecasting the quantity of rice available for commercial export, while MOIT for seeking markets and negotiating government-to-government (G2G) food exporting agreements, and Vietnam Food Association is in charge of operating contracting registration of rice exporters and allocating G2G contracts among them.

During the period of sharply rising international prices for rice in 2008, the government imposed export taxes on rice for a six-month period as part of bread range of policy measures to limit price increase on the domestic market. The taxes ranged from USD30 to USD60 per tonne of rice. When domestic price returned to pre-crisis levels, this export tax was removed in November 2009.

Expansion in Agricultural Cultivation.

Vietnam has witnessed a remarkable expansion in agricultural land area since the early 1990s. Around two-thirds of the expansion occurred during the 1990s, followed by a modest increase after 2000 (Fig. 8.7). By the year 2012, agricultural land spanned approximately 10.8 million hectares, which constituted 35 percent of the total land area. It encompassed 6.5 million hectares dedicated to arable cultivation (60%), 3.7 million hectares designated for perennial crops (34%), and the remaining 0.6 million hectares (6%) allocated for pasture and livestock farming.

Rice cultivation area (which is concentrated in the Mekong and Red River Deltas) has exhibited stability since 2000, accounting for 55 percent of the total agricultural land in 2021. Corn is Vietnam's second-largest short-term crop by cultivation area, accounting for around 8 percent of agricultural land area, which is grown in less fertile lands, with the bulk of it targeted for animal feed. Vietnam is also the world's largest producer of Robusta coffee, which grew in importance before 2000 with an annual growth rate of 10.7 percent before dropping to 1.6 percent per year after 2000. The majority of Vietnamese coffee production is concentrated in the Central Highlands region.

The 2016 Agricultural Census showed 8.6 million households in Vietnam were still engaged in agriculture, forestry, and fisheries, using about 90 percent of agricultural land. Most agricultural households are small with less than 0.5 hectare of agricultural land. Households with over 2 hectares of land accounted for less than 6 percent of agricultural land (Fig. 8.7). The average cultivated area per agricultural household in Vietnam is only 0.3 hectares, which is about half (0.6–0.8 times) compared to Cambodia, Myanmar, and the Philippines, and much lower than Thailand. In 2013, Thailand had about 25 percent of farming households with less than 1 hectare of agricultural land, 40 percent from 1 to 3 hectares and 35 percent over 3 hectares. About 70 percent owned over

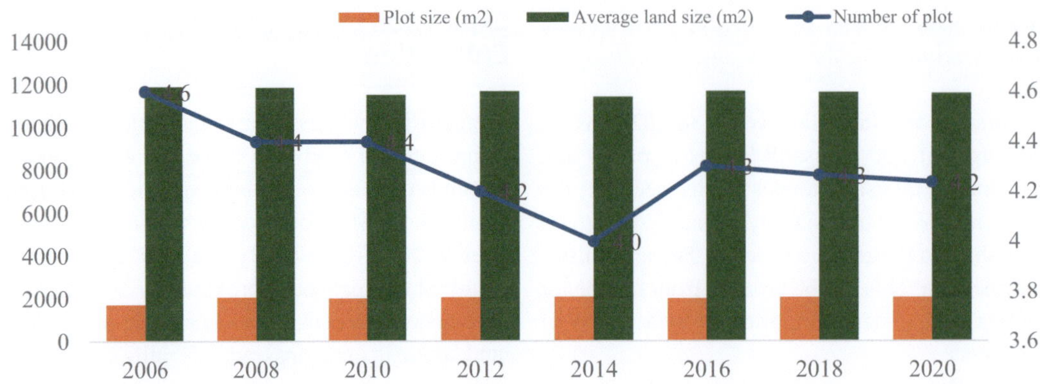

Fig. 8.7 Agricultural land accumulation and consolidation, Vietnam, 2006–2014. *Source* Calculated by author using VHLSS data (2023)

3 hectares arable land in Thailand (World Bank, 2016).

Although Vietnam has had many policies to promote horizontal linkages between farmers, vertical linkages between farmers and businesses, and multi-stakeholder linkages between farmers, businesses, government, and scientists, these linkages have grown slowly. Of the approximately 11 thousand service cooperatives operating in 2016, 84 percent were located in production areas with a low export intensity as they remain dominated by old-style commune-wide cooperatives that operate ineffectively. Less than 1 percent of small-scale coffee growers participate in new-style cooperatives, about 10 percent of farmers are affiliated with businesses in rice production, and 16 percent are engaged in livestock production (MARD (Ministry of Agriculture and Rural Development), 2018).

In the context of commercialization and processing in Vietnam's agricultural sector, private trading companies, including both multinational corporations and small enterprises, have played significant roles, particularly in coffee and animal feed industries. Small private farms and firms have ventured into cashew nuts and seafood processing for export. State-owned enterprises still dominate the import and export of essential commodities, such as rice, rubber, fertilizers, cashew nuts, pepper, and cassava.

Vietnam is currently in the early stages of modernizing its retail industry, with a significant portion of food goods still being primarily consumed through traditional channels, such as small stores and local markets. Nonetheless, the market share of modern retail has risen to 4–20 percent in 2013. Large retail stores are concentrated in major cities like Ho Chi Minh City, Hanoi, Da Nang, Can Tho, and provincial urban centers.

Following accession to the World Trade Organization (WTO) in 2009, the retail industry opened up fully to foreign retailers. In fact, foreign companies now dominate the majority of major supermarket chains in Vietnam, which include their strong supply chain networks, efficient logistics, professional management practices, and high-quality customer service, all of which provide them with a significant advantage over domestic retailers (OECD 2015).

Food Security

Since the 1990s, Vietnam has bolstered national food security with rice output soaring from 19.6 million tons in 1990 to 42.6 million tons in 2022 (Fig. 8.8). Concurrently, the fraction of the population experiencing perennial food insecurity has shrunk to under 10% by 2022. Looking ahead to

2030, even as supply pressures mount, Vietnam can ensure continued food security for the 2020s and 2030s by conserving three million hectares of rice land, though this strategy hinges on the adoption of advanced technology to curtail post-harvest losses, and amplify productivity.

Data derived from the VHLSS underscores a marked enhancement in the protein content of meals consumed by the Vietnamese population. The average monthly per capita consumption of eggs from 2.9 eggs in 2008 to 4.6 eggs in 2022 (Fig. 8.9), and from 1.4 kg of meat to 2.3 kg of meat. This trend aligns with research undertaken by the FAO in 2010, which highlighted a consistent annual growth of 1.8 percent in daily per capita energy intake, which grew from 2311 kcal in the mid-1990s to 2769 kcal in 2005. Notably, this level of energy intake surpasses figures recorded in India (2300 kcal) and several Southeast Asian nations, including Indonesia (2538 kcal), Thailand (2530 kcal), and the Philippines (2518 kcal). However, it lags behind most OECD nations and certain Asian countries, such as China (2974 kcal) and Malaysia (2908 kcal).

Climate- Resilient Smart Agriculture.

Nature-based Solutions (NbS) are defined as actions to address societal challenges through the protection, sustainable management, and restoration of ecosystems, benefiting both biodiversity and human well-being. NbS have significant, but currently underutilized potential to help address global challenges such as climate change, human health, food and water security, natural disasters, and biodiversity loss (IUCN, 2020). Recently, the Vietnamese government pays more attention on NbS. For example, in 2017 the government released the Resolution 120 on "sustainable and climate-resilient development of the Mekong Delta" (The Government, 2017). The viewpoints of the Resolution 120 emphasized that any interventions should "respect natural laws and avoid violent interference with nature; select development models adaptive to natural conditions and friendly to the environment." The Prime Minister issued Decision 150/QD-TTg in 2022 to approve the sustainable agriculture and rural development strategies for the period 2020–2030 with a vision toward 2050. The Decision stated that "agricultural production must be carried out responsibly, modernly, effectively and sustainably; ecological, organic, circular agriculture shall be developed, carbon emissions must be low, friendly with the environment and adaptable to climate change." In practice, there are many farming systems that apply nature-based solutions; for instance, rice-shrimp and mangrove-shrimp in the coastal areas or garden-pond-livestock integrated farming system (well known as VAC in Vietnamese) in the freshwater zone (Binh et al., 2021; Tin et al., 2022; Binh et al., 2023; Binh and Nhan, 2024).

Transformational polices promoting green transition and decarbonization in Vietnam agriculture over the period 2011–2020 is described in Fig. 8.10. These efforts at the national and local levels by the Party and Government reflect the strong political and administrative commitment to pursue the development in agriculture toward greening, low carbon and circular. However, the effective implementation of these policies is considered the key challenge. For example, while the government wants to reduce the level of intensification for rice farming in the Vietnam Mekong Delta (VMD), the intensive-based rice production in this region continues to increase from 23.6 million in 2017 to 23.8 million in 2020 (GSO, 2021). Binh et al. (2021 and 2023) identified five key drivers influencing the transition process in the VMD that includes price, place (location), producer, policy, and project. The last driver "Project" means the actual innovation initiative at the local level to implement the development directions at the national level toward agreeing transition through a particular policy to promote the VMD development with the principle of "living with nature" ("Thuận Thiên") at district level where local authority and community are co-designing and co-implementing the transition from rice-intensive mono-culture farming to integrated rice-shrimp farming system.

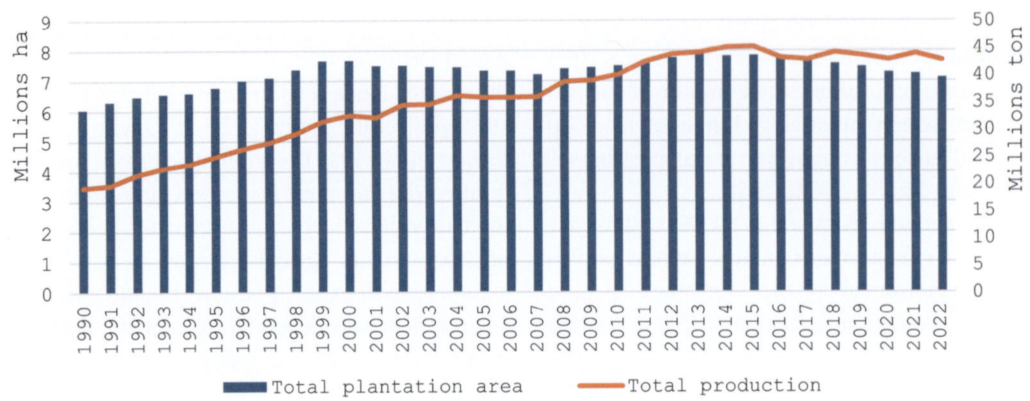

Fig. 8.8 Rice plantation area and production, Vietnam, 1990–2022. *Source* GSO (2022)

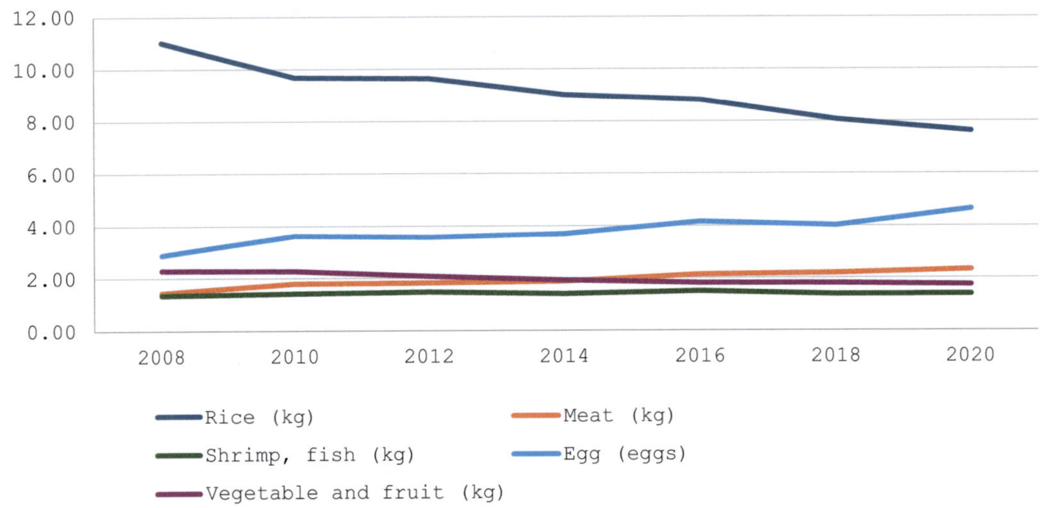

Fig. 8.9 Consumption of key food items monthly per person, Vietnam, 2008–2020. *Source* Author's calculations based on VHLSS

In sync with the goals of the UN's SDGs, this section examines the expansion of climate-resilient smart technologies (including digitalization and digitization) in Vietnam (see also Rasiah, 2025).

Organic Farming

Globally, organic products generated a total value of almost 90 billion US dollars in 2016. As of 2016, nearly 58 million hectares of land was under organic farming systems in 2016, which accounted for 1.2 percent of total farmland in the world (Table 8.1). Countries in Asia share almost 4.9 million hectares, with China and India constituting the largest shares of 2.3 million and 1.5 million ha, respectively (Willer & Lernoud, 2018). Vietnam accounted for 53,348 ha of organic agriculture, which constituted 0.5 percent of the total farmland of the country in 2016. Most organic products in Vietnam are exported (77 million EUR), and only a small portion (18 million EUR) is domestically consumed (Willer & Lernoud, 2018). Nevertheless, concerns for food safety and environmental protection have led to increasing demand for organic products in Vietnam. Using a dichotomous choice

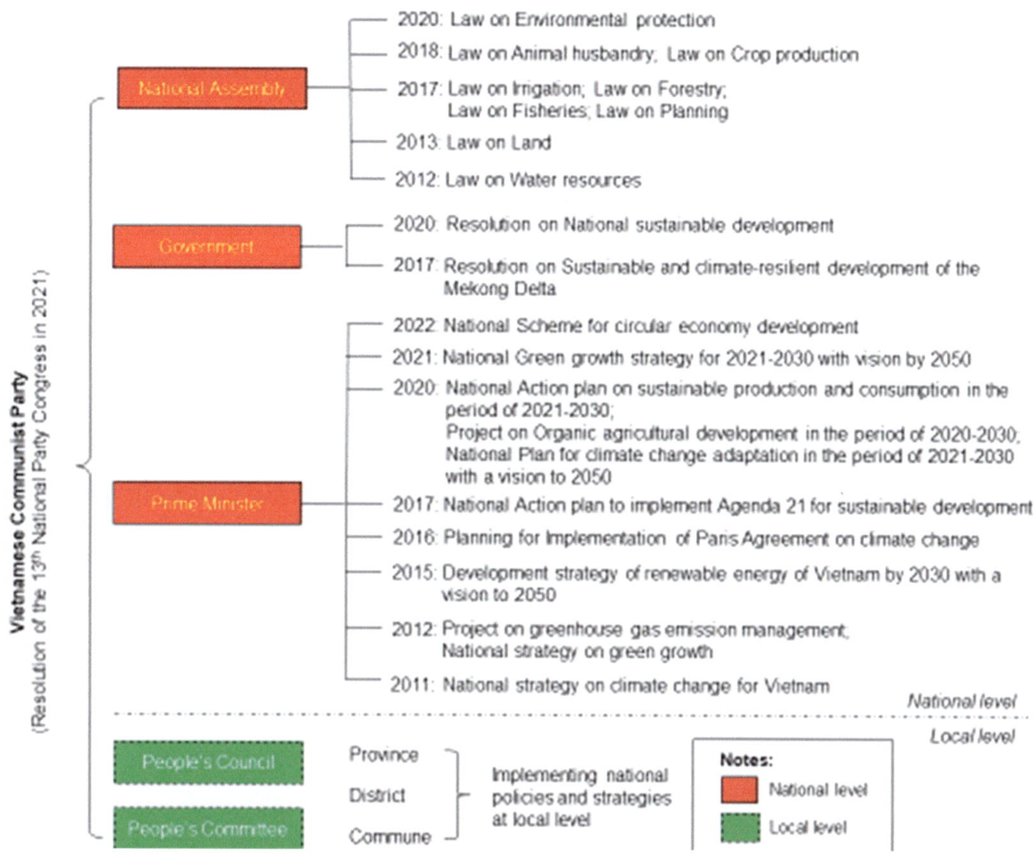

Fig. 8.10 Policy framework to accelerate green transition, Vietnam, 2011–2020. *Source* Binh, NT et al. (2023)

Table 8.1 Statistics of organic agriculture in Vietnam

	Land (ha)	Farmland Share (%)	Retail (million EURO)
World	57,816,759	1.2	84,698
Asia	4,897,837	0.3	7343
Vietnam	53,348	0.5	18

Source Willer & Lernoud (2018)

contingent valuation model, Khai (2015) found that customers in Vietnam are willing to pay 59 percent premium prices for organic vegetables. The government released the national standards for organic agriculture in Vietnam (TCVN 11041–5) in 2018, which are applied on agricultural crops (organic crop production in general, organic rice, organic tea), aquaculture (organic shrimp), and animal husbandry (organic milk).

Several studies have also examined the potential for organic rice farming in Vietnam based on economic analyses. Thanh et al. (2016) found that organic rice-shrimp systems generate higher benefits to farmers compared to traditional rice-shrimp systems. However, Long (2014) demonstrated that the production costs of organic rice in Vietnam are higher than traditional rice with lower yields. The cost-benefit ratio of organic rice, though organic products fetch premium prices. The barriers to adoption of organic rice include increases in pests and insects during crop maturity, lower productivity compared to traditional practices, and lack of awareness and trust (Long, 2014). The establishment of third-party organizations that support the certification, registration, or adaptation of international certification schemes on organic farming in Vietnam (including standards, including standards with lower requirements, such as VietGAP) is evolving in Vietnam.

Circular Agriculture

The transition to circular agriculture is known to climate resilience. Agriculture production in the country currently relies heavily on fertilizers, pesticides, and inputs (Braun et al., 2019), which generates a large amount of residues and waste, which is released into the environment or underused in the production of energy or value-added products (Tu et al., 2022). Pesticide bags and bottles are thrown away, whereas the recycling of these materials is currently limited (Braun et al., 2019). The *circular economy* is gaining traction in the political agenda of Vietnam, and the government has issued the national plan for circular economy development (687/QĐ-TTg) in 2022 (Kiem & Huong, 2023).

Several circular economy frameworks have been developed and applied in Vietnam for a while now, e.g., the Garden-Pond-Animal husbandry (VAC) system, the valorization of waste and by-products to produce bio-fertilizer, biogas, or agroforestry systems (such as integrated garden-wooden tree-animal husbandry, rice-shrimp, and mangrove-shrimp systems). These practices usually operate within the farm scale but links between different systems across farms are limited. Extraction of agricultural by-products or waste (e.g., shrimp by-products or rice residues) for the production of high-value co-products is being undertaken by large commercial companies. Technical solutions to recycle and close the material loops are favored presently in Vietnam, whereas the use of biological resources for optimizing input use and valorization of by-products is less preferred (Haas et al., 2020; Navare et al., 2021).

Integrated Ecological Farming System

Integrated ecological farming systems like rice-shrimp or mangrove-shrimp farming help strengthen climate resilience in agriculture (Binh et al., 2021; Tin et al., 2022; Binh et al., 2023). In fact, intensive rice farming in the Mekong Delta has recently faced many challenges and risks related to saltwater intrusion, climate change, and changes in the flow of the Mekong River. Consequently, many rice-growing areas by the coast have switched to integrated rice-shrimp farming, in which farmers undertake aquaculture in the dry season and cultivate rice in the wet season (Fig. 8.11).

The results of household interviews by Binh et al. (2021) show that 100 percent of farmers agreed to shift from the intensive rice framework to the rice-shrimp framework because of ecological and social efficiency. From an ecological perspective, the rice-shrimp framework generates less environmental pollution, which is suitable for changing environmental conditions, local soil and water. From a social perspective, this model provides higher income, lower investment costs, and easier-to-sell products, creates more job opportunities, reduces conflicts between shrimp farmers and rice growers, and is beneficial for the public.

The shrimp-mangrove farming system is a sustainable livelihood model that adapts to mitigate climate change in the coastal provinces of the Mekong Delta, which accounted for an estimated 200,000 hectares, of which the water surface area covered about 40,000 hectares in 2023. According to Tin et al. (2022), the shrimp-mangrove model has its own strengths, weaknesses, opportunities, and threats (SWOT).

- Strengths: large farming area, many sources of income in the model, low investment costs compared to intensive farming, taking advantage of family labor and products that are ecological, clean, and safe.
- Weaknesses: shrimp productivity is still low, pond design is not appropriate (low water level), rapid sedimentation increases dredging costs, farmers' farming techniques are limited, and poor transportation infrastructure develops.
- Opportunities: diverse natural food sources for aquatic species, application of ecological and

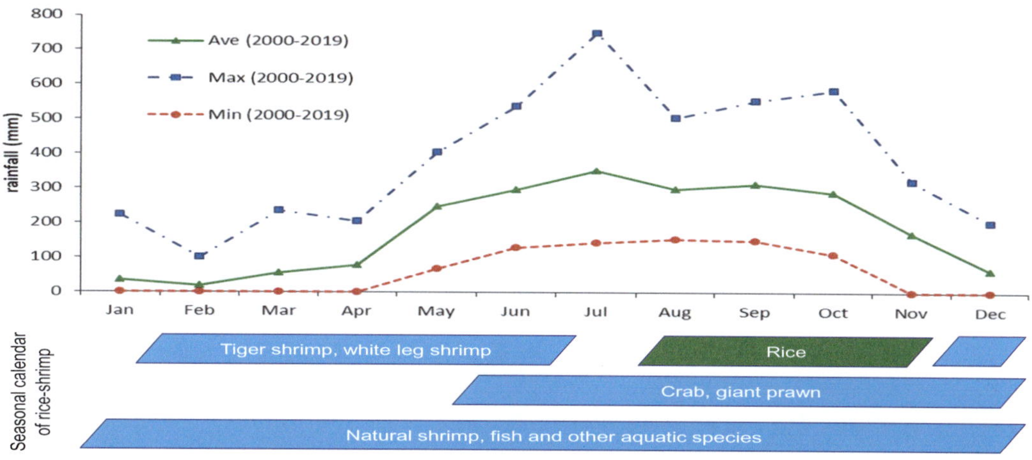

Fig. 8.11 Seasonal changes in rice-shrimp ecological farming system, Vietnam. *Source* Binh, NT et al. (2023)

organic certification standards, stable shrimp and crab prices, consistent with sustainable development trends.

- Threats: effects of climate change, environmental pollution, unstable seed quality, risk of disease, forest land allocation policy changes, and not yet created conditions for exploiting existing resources.

Transformation to Digital Agriculture

Agricultural digital transformation is the process of applying digital technologies from production to processing, distribution, and consumption of agricultural products. The basic difference between digital agriculture and traditional agriculture is the application of digital technologies (big data, cloud computing, and internet of things) activities to smartify agriculture. The benefits of digital transformation in agriculture include: (i) minimizing risks and damages due to climate change; (ii) helping farmers connect directly with consumers; (iii) improving labor productivity; and (iv) enhancing the quality of agricultural products. Digital transformation of the agricultural industry includes basic activities such as: (i) applying modern technology in farming (e.g., IoT, sensors in the field, machine learning and analytics, and drones) and in crop monitoring; (ii) linking components of the agricultural ecosystem in the value chain; and (iii) changing the way businesses operations are managed.

Digital transformation solutions in agriculture have been applied in fields, such as farming, animal husbandry, forestry, and fisheries, which has helped connect nine million farming households with processing and trading enterprises and 100 million consumers across the country, and billions of users around the world. Digitalization is changing Vietnam's agricultural industry from a small and inefficient model to a new model with value chain links. A typical example of successful digital transformation of the agricultural industry is the Vinamilk company, which has applied IoT technology to livestock monitoring to all stages of care that is carefully monitored according to smart agriculture standards to produce up to 23 liters/cow/day of milk. The farm has also been certified as an organic farm according to European standards. Another example is that FPT Corporation has combined with Fujitsu, the Fruit and Vegetable Institute, and Japanese experts to build a new vegetable growing model where Akisai technology is applied to connect and control elements on the farm remotely in which the greenhouse environment is monitored and managed by computers to create the best environment for tomato and lettuce plants to grow.

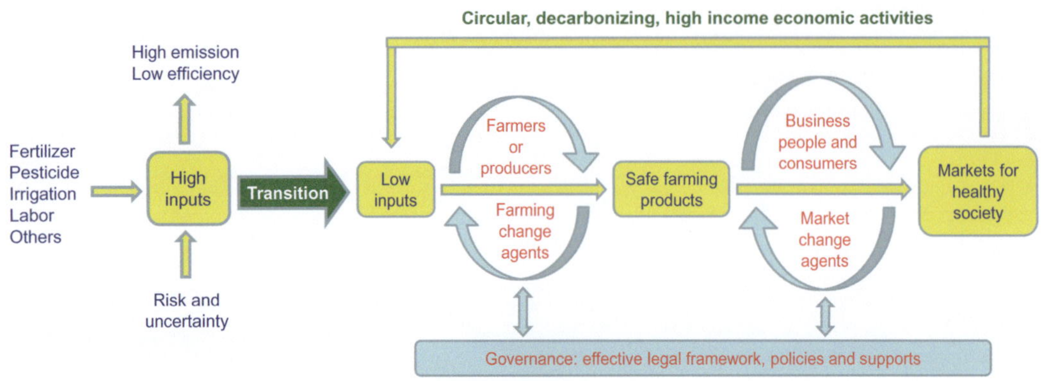

Fig. 8.12 Agro-ecological farming framework to accelerate green transition. *Source* Binh, NT et al. (2023)

Science, Technology, and Innovation

The government recognizes that scientific and technological achievements related to production technology, e-commerce, and digital transformation are key toward transforming agricultural production, distribution, and consumption in the country (TW 2021; Gautam et al., 2022). However, support for science and technology and innovation in Vietnamese agriculture is still low, accounting for only 0.3% of agricultural production value, compared to the mean of 1.1 percent among the OECD (2022) countries. For example, the VMD is still a low-lying region, which requires significance assistance from scientific and technological advancement focused on strengthening its climate resilience (VCCI, 2020, 2021), especially disaster-related problems, such as riverbank erosion, climate change mitigation and adaptation, water scarcity, migration, and ecosystem degradation (Hwangbo & Park, 2021). Thong et al. (2019) focused on indigenous knowledge as a farmer-led innovation process to support flood and salinization mitigation. Hoang et al. (2021) examined lotus farming as a potential flood-based adaptive innovation model to cope with floods and to improve the ecological system by washing away acid sulfate, diversify livelihoods, and contribute to cultural enhancement as lotus is Vietnam's national flower.

In addition, high-tech solutions using artificial intelligence (AI) are increasing to monitor environmental quality, production processes, and traceability, thereby contributing to improving the efficiency of natural resource use and reducing environmental pollution. The government is increasingly focusing on STI to extend the agricultural value chains beyond raw agricultural exports to include processing and extraction of essences (for example, chitin, chitosan, and silicates) for use in production of chemicals, medicine, and pharmaceuticals. Some companies such as Loc Troi (rice), VNfood (shrimp), and Vinh Hoan (catfish industry) have already started implementing green technology and innovation (Tu et al., 2022).

Proposal for Green Technology Transition.

To address the multiple barriers and challenges that have come up with poverty alleviation and promoting climate-resilient smart technologies it is worth considering the agro-ecological farming framework proposed by Binh et al. (2023) to accelerate green transition in agriculture. This is the Green-TIA framework, which emphasizes a shift from high to low-input farming systems (Fig. 8.12). As described, the Green-TIA framework is an attempt to examine the dynamic transformative greening transition from a high-input agricultural development mode associated with increased use of fertilizer and pesticides, immigration labor, high emission, low efficiency, high risk, and uncertainties to a low-input integrated agricultural development model associated with low input, safe farming products, and markets for a healthy society, which targets the future agriculture to focus on circular, de-carbonizing, and high-income economic activities.

Conclusions

Typical of rapidly industrializing and urbanizing nations, the contribution of agriculture to its national economy has gradually fallen. However, unlike countries such a Malaysia which has increasingly faced net food trade since the early 1990s, Vietnam has enjoyed strong exports of both cash crops, such as coffee and cashew nuts but also exports of food crops, such as rice. In recognition of food security, the government has implemented agricultural policies not only to address sufficient food production to feed the domestic population but also to enable exports. Land reforms to ensure adequate distribution of farmland, irrigation, supply of fertilizers, and initiatives to check desalination and floods in both the Mekong River Delta and the Red River Delta and effective distribution have ensured effective allocation of farmland for rural households. The focus on strong productivity and output has also helped support the industrial and service workers with cheap food prices.

Since especially the introduction of the national climate strategy in 2011 and a series of other agricultural policies, which include climate-resilient smart agriculture policies, the government has strengthened farming in Vietnam to be increasingly green. With the support from both traditional sources of knowledge and modern knowhow from R&D activity, the government has raised both farm productivity and ensured increasing climate resilience in the sector. Examples include the expansion of rice-shrimp integrated farming, and increased emphasis on circular, de-carbonizing, and high-income economic activities. Indeed, we are recommending the government to adopt the Binh et al. (2023) Green-TIA framework to accelerate green transition in Vietnam's agricultural sector. The rising focus on smart agriculture should help Vietnam manage agricultural development to meet both poverty alleviation goals, as well as achieve its carbon net-zero goals. Importantly, such an emphasis should allow the country to manage the transition of Vietnam's development to accommodate smoothly the changing needs of its population as living standards rise with rapid economic growth.

References

Binh, N.T., and Nhan, D.K. (2024). Principles of agricultural systems. Can Tho University Publishing House, Can Tho, Vietnam.

Binh, N. T., Sinh, B. T., & Tien, L. V. T. (2023). Transformational policies and strategies framework accelerating green transition—The case of agriculture in Vietnam. *Environmental Progress & Sustainable Energy, 42*(2023), e14108. https://doi.org/10.1002/ep.14108

Binh, N. T., Tien, L. V. T., Thuy, N. T. T., Minh, N. A., & Sinh, B. T. (2021). The rice-shrimp farming system in the Mekong Delta and its policy implications for the socio-ecological transformation of agriculture. In D. T. Truong & P. Degenhardt (Eds.), *Agricultural production in Vietnam from the socio-ecological transformation approach: Opportunities, challenges and policy implications* (pp. 262–277). Labor Publishing House.

Braun, G., Braun, M., Kruse, J., Amelung, W., Renaud, F. G., Khoi, C. M., Duong, M. V., & Sebesvari, Z. (2019). Pesticides and antibiotics in permanent rice, alternating rice-shrimp and permanent shrimp systems of the coastal Mekong Delta. *Vietnam. Environment International, 127*(November 2018), 442–451. https://doi.org/10.1016/j.envint.2019.03.038

Gautam, M., et al. (2022). *Repurposing agricultural policies and support: Options to transform agriculture and food systems to better serve the health of people, economies, and the planet*. World Bank. https://openknowledge.worldbank.org/handle/10986/36875

General Statistic Office (GSO). (2023). Vietjam's statistics data National Account and State Budge

Haas, W., Krausmann, F., Wiedenhofer, D., Lauk, C., & Mayer, A. (2020). Spaceship earth's odyssey to a circular economy - a century long perspective. *Resources, Conservation and Recycling, 163*, 105076. https://doi.org/10.1016/j.resconrec.2020.105076

Hwangbo, W., & Park, Y. I. (2021). A study on the policy instrument for regional innovation system construction in the Mekong Delta region. *Asian Journal of Innovation and Policy, 10*(1), 039–067.

IUCN (2020). Ensuring effective nature-based solutions. Issue Brief (July 2020). International Union for Conservation of Nature, Gland, Switzerland.

Kompas, T., Che, T. N., Nguyen, H. T. M., & Nguyen, H. Q. (2012). Productivity, net returns, and efficiency: Land and market reform in Vietnamese Rice production. *Land Economics, 88*, 478–495. https://doi.org/10.3368/le.88.3.478

Laborde, D., Lallemant, T., McDougal, K., Smaller, C., & Traore, F. (2019). Transforming agriculture in Africa & Asia: What are the.

Long, L. Q. (2014). Cultivation and marketing of organic rice in Vietnam - an economic analysis. University of Agricultural Sciences, Bangalore.

MARD (Ministry of Agriculture and Rural Development). (2018). Report on 10 years of implementing Resolution 26 on Agriculture, Farmer, and Rural development.

MARD (Ministry of Agriculture and Rural Development). (2022). Data on agricultural trade. MARD database.

Navare, K., Muys, B., Vrancken, K. C., & Van Acker, K. (2021). Circular economy monitoring—How to make it apt for biological cycles? *Resources, Conservation and Recycling, 170*(March), 105563. https://doi.org/10.1016/j.resconrec.2021.105563

Nguyen, Q.H., Tran, D.D., Dang, K.K., Korbee, D., Pham, L.D., Vu, L.T., Luu, T.T., Ho, L.H., Nguyen, P.T., T. Ngo, T.T. and Nguyen, D.T., 2020. Land-use dynamics in the Mekong delta: From national policy to livelihood sustainability. Sustainable Development, 28(3), pp.448–467.

OECD 2015. *Agricultural Policies in Viet Nam 2015*, OECD Food and Agricultural Reviews, OECD Publishing, Paris, doi:https://doi.org/10.1787/9789264235151-en.

OECD, 2022. Agricultural Policy Monitoring and Evaluation 2022: Reforming Agricultural Policies for Climate Change Mitigation, OECD Publishing, Paris, doi:https://doi.org/10.1787/7f4542bf-en.

Rasiah, R. (2025). Building sustainability and food security in Southeast Asia, In R. Rasiah, R., V. Anter. & N.I. Annizah (eds), *Sustainable Agriculture and Food security in Southeast Asia: Towards Climate Resilience and Digitalization*, Singapore: Springer Nature.

Thanh, N. C., Van Manh, N., Van An, N., Thị, P., & Thao, P. (2016). Some initial results on research and Modeling of organic Rice production in the Mekong Delta. *Vietnam. International Journal of Energy and Environmental Science, 1*(1), 29–36. https://doi.org/10.11648/j.ijees.20160101.15

The Government (2017). Resolution 120 on sustainable and climate-resilient development of the Mekong Delta. No 120/NQ-CP, dated November 17, 2017; Ha Noi, Vietnam.

Tu, N.M., Nguyen, T.B., Dang, K.K., Luu, T., Thach, P.H., Nguyen, K.L.P., Nguyen, H.Q. Current and potential uses of agricultural by-products and waste in Main food sectors in Vietnam—A circular economy perspective. In Circular economy and waste valorisation; Springer: Cham, Switzerland, 2022; pp. 131–151.

TW, 2021. Resolution the 13th National Party Congress.

VCCI, 2020. Vietnamese Mekong Delta annual report.

VCCI, 2021. Vietnamese Mekong Delta annual report.

Hoang Thi Minh Vo, Gerardo van Halsema, Petra Hellegers, Andrew Wyatt, Hong Quan Nguyen (2021). The Emergence of Lotus Farming as an Innovation for Adapting to Climate Change in the Upper Vietnamese Mekong Delta. Land. 2021; 10(4):350. doi:https://doi.org/10.3390/land10040350.

Willer, H., & Lernoud, J. (2018). The world of organic agriculture. Statistics and emerging trends 2018. In *Research Institute of Organic Agriculture FiBL and* IFOAM—Organics International. doi:https://doi.org/10.4324/9781849775991.

World Bank 2016. Transforming Vietnamese Agriculture: Gaining More from Less. Vietnam Development Report. Washington, D.C. World Bank.

Agricultural Development and Food Security in Cambodia

9

Kum Kim

Abstract

This chapter presents the history, as well as the development of agriculture in Cambodia with a focus on its significance to the national economy, policies that the government has implemented and their impact on the livelihood of the people of Cambodia. Important issues addressed include food security and rice self-sufficiency, the effects of policies and climate change effects on vulnerability, digitalization, and strategies designed to alleviate agriculture-related poverty. The chapter raises concerns over the overuse of self-sufficiency rates as it does not measure those in need of food but without the capacity to purchase.

Keywords

Agriculture · Food security · Digitalization · Climate resilience · Cambodia

Introduction

According to the Food and Agriculture Organization of the United Nations, food security exists when people have physical and economic access to sufficient, safe, and nutritious food that meets their dietary needs and preferences for an active and healthy life (FAO, 2008). This definition consists of four components, namely availability, stability, accessibility, and utilization. A food system is said to be vulnerable when one or more of the four components of food security are uncertain and insecure. In other words, for food security objectives to be realized, all four components must be fulfilled simultaneously. The focus on food security in the context of food availability from crop production for rural households in Cambodia is the main objective of this chapter. Food availability addresses the "supply side" of food security and is determined by the level of food production, stock levels, and net trade.

Agriculture is a major sector of the Cambodian economy, which employs the vast majority of the country's workforce. The sector contributed to 22.2% of GDP in 2022 and accounted for 39% of

K. Kim (✉)
KOTRA Phnom Penh, Phnom Penh, Cambodia
e-mail: kumkimy@yahoo.com

R. Rasiah et al. (eds.), *Sustainable Agriculture and Food Security in Southeast Asia*, Sustainable Development Goals Series, https://doi.org/10.1007/978-981-95-0477-0_9

total employment in 2021.[1] Since 1990s, agriculture has experienced rapid growth as the government has made strong efforts with the objective to transform the sector from traditional subsistence to commercial plantation. Rice is the mainstay of agricultural sector growth (CDRI, CARD, IPRI, 2011). Rice production rose from 9.3 million tons in 2014 to 11.6 million tons in 2022 and its production surplus increased from 4.7 million tons to 6.8 million tons during the same period of time.[2] Other major crops include cassava, maize, mung bean, and soybean. Rice is also the major crop and staple food of Cambodia. It contributes to about 60% of agriculture GDP that is grown on over 3.2 million hectares of Cambodia's agricultural land, which is about one-third of the total agricultural land in the country (ADB, 2021a, 2021b). Furthermore, around 76% of 16.5 million people who live in rural areas depend on agriculture, mainly engaged in rice production for their livelihoods (World Bank, 2021a, 2021b). In this regard, the development of agriculture and with rice production surplus are crucial for promoting economic growth and ensuring food security in the country.

Despite remarkable growth and the significance of the sector, Cambodia's food security remains the critical issue for most of rural households. Despite increasing food production, the prevalence of moderate or severe food insecurity in the population reached 51.1% in 2021 (World Bank, 2022). Besides, about 2.5 million Cambodians experience severe food insecurity, which comes to roughly 15% of the country's estimated population of 16 million following price hikes from the outbreak of COVID-19 and climate change (2023). Yet, Cambodia reported a surplus of rice production since 1995, which raises the question of needs falling below demand. Among other things, the chapter seeks to answer this question, why food security is still the critical issue for rural households? The chapter will also address constraints faced by the sector in the context of agro-industry development, landholding size, labour productivity, climate change, and digitalization.

Significance of Agriculture

Employment

Agriculture absorbed the largest share of employment followed by the industry and service sectors. The share of agricultural, industrial, and services employment in 1991 was 79%, 5%, and 16%, respectively (ADB, 2021a, 2021b). Although agriculture has retained its important role as the chief absorber of Cambodia's labour force, its share fell from 79.0% in 1991 to around 39% in 2021 (Fig. 9.1).[3] There was a slight improvement from 2016. In contrast, the shares of industry and services grew from 5.1% and 15.9%, respectively, in 1991 to around 25.4 and 32.7%, respectively, in 2021.

The higher wages offered the industry and service sectors have stimulated urbanization. Agricultural employment based on largely monocrop rice cultivation is characterized by low incomes and is vulnerable to climate risks (Sodeth, 2020). According to Cambodia (2019a, 2019b), the average monthly wage of agricultural workers was USD 175 while that of industry and services was US$251 and US$316, respectively.

Value-Added

The share of agricultural value-added in total GDP averaged around 32% in the period 1993–2021 (World Bank, 2022). Within the sector, crop production (11.3%) enjoyed the largest output contribution to total national GDP, followed by livestock (11.3%), fisheries (24.7%), and forestry (6.9%) in 2021.

[1] Estimations by the International Labour Organization (ILO).

[2] Note: Rice production surplus is the amount of remaining rice of the country after deducting for domestic consumption.

[3] Based on estimations by the International Labour Organization (ILO)].

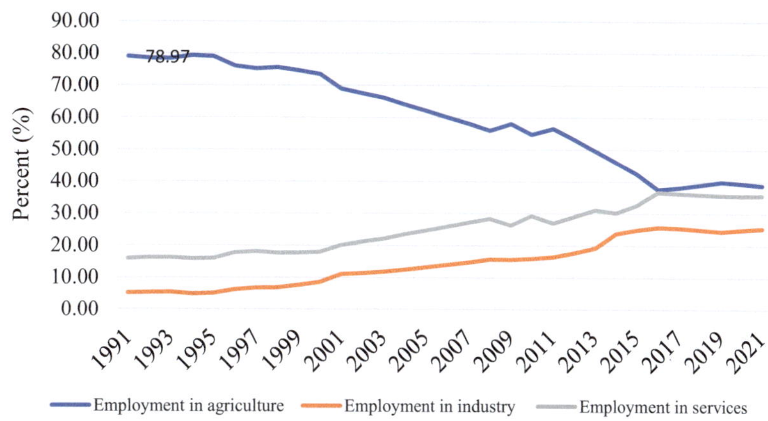

Fig. 9.1 Employment by sectors, Cambodia, 1991–2021. *Source* World Bank (2022)

The share of agriculture value-added in GDP declined in trend terms from 45.3% in 1993 to 21.9% in 2022 largely because of expansion in the industrial sector (Fig. 9.2), with its share in GDP rising from 5.1% in 1993 to 24.1% in 2022 (CDRI, CARD, IPRI, 2011).

Exports

Cambodia's share of agricultural raw material export in total merchandise export was 44.7% in 1962, which fluctuated sharply, though data was not available over the period 1972 till 2000 (Fig. 9.3). The share of agricultural exports in merchandise exports fell sharply from the late 1990s as garment exports rose sharply so that its share fluctuated between 2.4% in 2000 and 2.6% in 2021).[4]

Exports of agricultural goods from Cambodia are often underestimated as significant amounts of rice cross the border to Vietnam informally. For example, in the harvest season of 2020, about 1000 tons of rice grown in the Eastern provinces of Cambodia crossed into Vietnam daily.

Although the share of agricultural exports in merchandise exports has fallen, the export value of agriculture products to international markets increased considerably from USD 191.6 million in 2012 to USD 3.9 billion in 2020 and USD 5.0 billion in 2022 (Asia News Network, 2023; Piseth et al., 2021).[5]

Expansion in Agriculture

Agricultural Production

Agriculture production has notably increased in the past 2 decades, with a significant surplus of rice. Total paddy rice surplus increased from 4.6 million tons in 2015 to 5.7 million tons in 2020 and 6.3 million tons in 2022 (Fig. 9.4). The surplus of paddy rice production contributed to ensuring national food security and enabled Cambodia to become one of the main rice exporters in the region. In 2022, Cambodia exported 637,000 tons of milled rice and 3,477,886 tons of paddy rice, generating USD418 million and USSSD841 million, respectively.[6]

Cambodia produced 65.1 million heads of animals and 18.7 million tons of industrial crops in 2022. At the same time, the production of horticulture (corn, banana, mango, sweet potato, all kind of vegetables, chilies, and watermelon) rose strongly from 2.4 million tons in 2018 to 4.4 million tons in 2022, while fish production increased from 745,310 tons in 2018 to 862,200

[4] The share of garment export in merchandise export increased from 3.3% in 1995 to 70.3%in 2000 but fell gradually since to 43.8% in 2020 (World Bank, 2022).

[5] In current prices.

[6] Cambodianess's news article (January 6, 2023), Milled Rice Exports Top $400 Million. Accessed on September 8, 2023 from https://cambodianess.com/article/milled-rice-exports-top-400-million).

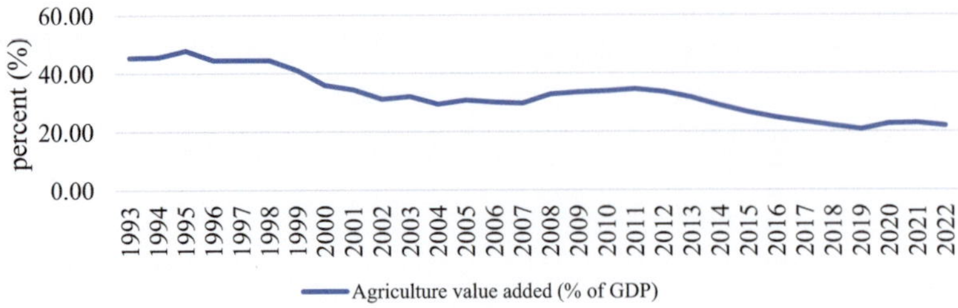

Fig. 9.2 Share of agricultural value added in GDP, Cambodia, 1991–2022. *Source* World Bank (2022)

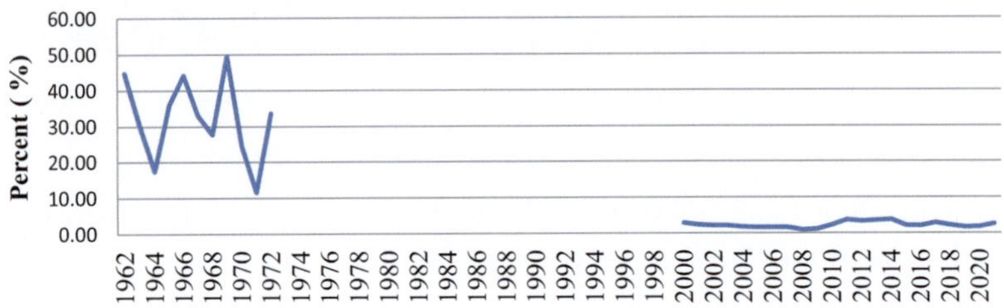

Fig. 9.3 Agricultural exports in merchandise exports, Cambodia, 1962–2022. *Source* World Bank (2022) [*Note* World Bank staff estimates through the WITS platform from the Comtrade database maintained by the United Nations Statistics Division]

Fig. 9.4 Agricultural production, Cambodia, 2018–2022. *Source* Cambodia (2023). Annual Report for Agriculture, Forestry and Fisheries 2022. Ministry of Agriculture, Forestry and Fisheries

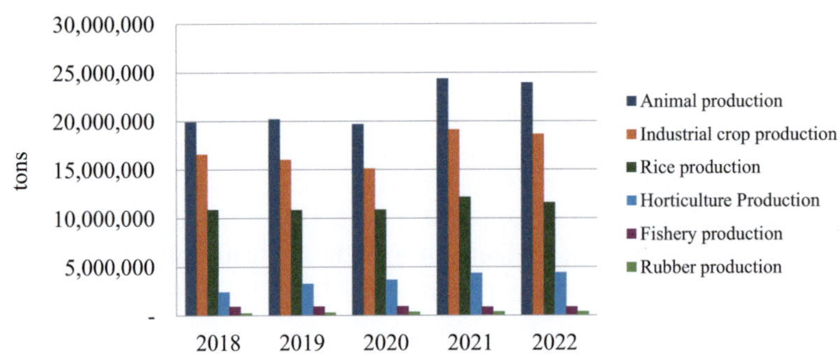

tons in 2022. Rice and rubber production reached 11.6 million tons and 0.4 million tons, respectively, in 2022 (Cambodia, 2023).

Expansion in Farmland

The share of agricultural land in total land in Cambodia increased from 19.9% in 1961 to 34.6%% in 2021 (Fig. 9.5)[7]; from 1971 to 1983, arable land decreased enormously due the costly civil war in the country. The Indochina and the civil war with the Khmer Rouge devastated rice production, with deleterious consequences to the livelihoods of the rural people as much of the rural infrastructures was destroyed. Between 1970 and 1974, rice-planted area declined by 77% while rice production fell by 84% of the 1970 level (Nesbitt, 1997). Things got worse in 1975–1979 when Cambodia came under brutal Khmer Rouge regime that accounted for 3 million lives from starvation and torture. Although the Khmer Rouge regime ended in 1979, the country's civil war lasted until 1998 (VOA, 2018). The share of agriculture land began to rise from 1985 particular with the growth of large commercial plantations

[7] World Bank (2020).

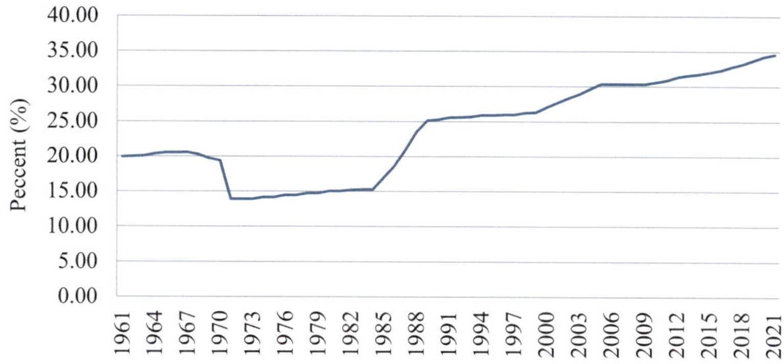

Fig. 9.5 Share of arable land, Cambodia, 1961–2021. *Source* World Bank (2022)

when economic land concessions (ELCs)[8] were introduced.

Official data from Cambodia's Ministry of Agriculture indicate that in 2018, rice plantation area accounted for 58.7% of total agricultural land area, which increased to 63.0% (3.4 million ha) in 2019 before falling gradually to 60.2% in 2022 (Fig. 9.6).

While agriculture has expanded, the size of farmland among farmers has continued to fall. Indeed, there is a positive relationship between farm size and food security among rural households in Cambodia. Around 76% of Cambodia's 16.5 million people in 2023 live in rural areas. They depend mainly on rice production for their livelihood, owning small farms that expose them to the risk of food insecurity.

One hectare of land is considered as the threshold to meet the milled rice requirements of a family of five in Cambodia (World Food Program, 2010). About 23% of rural households are landless, while 47.1% own farmland less than one hectare (10,000 square metres) (Table 9.1; see ADB, 2021a, 2021b). Therefore, only around 30% of rural households have enough land to produce rice to support their households' food consumption needs throughout the year. Consequently, improving access to agricultural land is key for rural households to enjoy food security. According to the World Bank, two-thirds of the country's rural households still face seasonal food shortages each year (World, 2019).

Table 9.1 Farm size of households, Cambodia, 2021

Agriculture land size	Household numbers ('000)	%
Less than 10,000 m^2	1080	47.1
10,000 m^2–19,999 m^2	557	24.3
20,000 m^2–29,999 m^2	256	11.2
30,000 m^2–39,999 m^2	137	6.0
40,000 m^2–49,999 m^2	76	3.3
50,000 m^2–99,999 m^2	137	6.0
100,000 m^2-and above	49	2.1
Total	2293	100

Source Cambodia (2022a)

A trend fall in farmland size is expected to persist as rural families continue to divide farmlands with their children (ADB, 2021a, 2021b). The division of farmland to the children is generally made after marriage.[9] The Cambodia Socio-Economic Survey 2021 found that only 2.1% owned more than 10 hectares in 2021 (Cambodia, 2022a).

Furthermore, the data from the World Bank indicate that arable land per person decreased from 0.5 ha in 1961 to 0.4 ha in 1984 and 0.2 ha

[8] Economic land concession refers to a mechanism to grant private state land through a specific economic land concession contract to a concessionaire to use for cultivation crops or industrial crops, including tree plantation, raising of animals and aquaculture, and construction, such as a plant or factory and facilities for the processing of domestic agricultural raw materials.

[9] It is a tradition that in Cambodia the division of the parents' land was often made when they were still alive, after the children are married. The land was equally divided between sons and daughters.

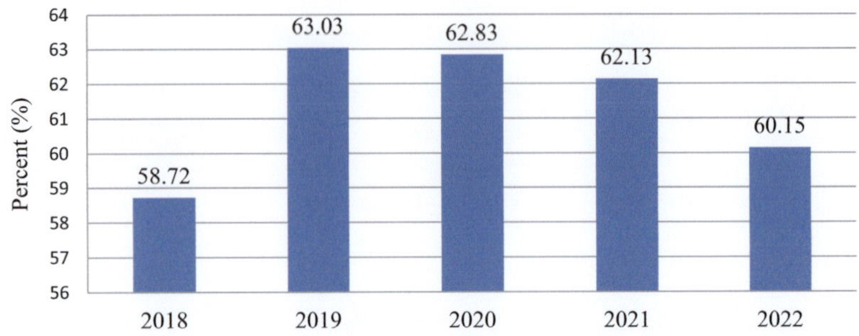

Fig. 9.6 Rice farming area in total agricultural land, Cambodia, 2018–2022. *Source* Cambodia (2023). Ministry of Agriculture, Forestry and Fishery's Annual Report 2022. Ministry of Agriculture, Forestry and Fisheries

Table 9.2 Organic farming, Cambodia, 2018 and 2019

2018-organic farm area (ha)	27,550
2019-organic farm area (ha)	25,757
2019-area of organic cereals (mainly rice and maize)	23,858
2009–2019 growth	218.60%
Organic farmland share in total farmland (2019)	0.50%
Number of producers (2019)	6360
2018-organic produce (ton)	8553
2019-organic produce (ton)	9480

Source GIZ (2022)

in 2021 World Bank (2022). Although there was a trend increase from 1971 till 1989, it has fallen gradually from then on (Fig. 9.7).

Organic Farming

Cambodia's organic farming started in the year 2000 with the implementation of the System for Rice Intensification (SRI). Between 2009 and 2019, the production area under certified organic agriculture increased by 218.6% (GIZ, 2022), though organic farming practice is still in a small scale. Cereals mainly rice production covered almost all of organic area (92%), with other crops consisting of pepper, cashews, fruits, herbs, and other vegetables. As such in 2018–2019, only 0.5% of Cambodian farmland (25,757 ha) was certified as organic involving 6,360 producers (GIZ, 2022; Table 9.2).

Organic farming is generally done by small-scale farmers and supported by international organizations with benchmark farming practices and certifications. Activities related to organic agriculture are mainly coordinated by the Department of Horticulture and Subsidiary Crops within the Ministry of Agriculture, Forestry and Fisheries (MAFF). The most recent development was the finalization of the organic Agriculture Regulation signed by the ministry on 16 April 2019. The regulation covers the scope of organic agricultural practices in the production and processing of food crops and commercial organic agricultural products, including fruits, vegetables, and other food crops. The regulation will adjust the implementation of Cambodia organic agriculture rules to ensure that all food products are of good quality and safe for human consumption. It also encourages the participation of farmers, agricultural communities, and the private sector in promoting agriculture practice.[10].

Agricultural Policies

As agriculture contributes to economic growth and ensuring food security, a number of policies have been formulated by the government to promote growth of the sector. Agricultural policies and administration are managed by the Ministry of Agriculture, Forestry and Fisheries (MAFF). The main agricultural policies introduced in Cambodia since 2015 are discussed here.

Agriculture extension policy 2015: The Cambodian government formulated the agricultural extension policy in 2015 to modernize

[10] Phnom Penh Post's articles on government make push for organic veggies (April, 2020). https://www.phnompenhpost.com/business/government-makes-push-organic-veggies

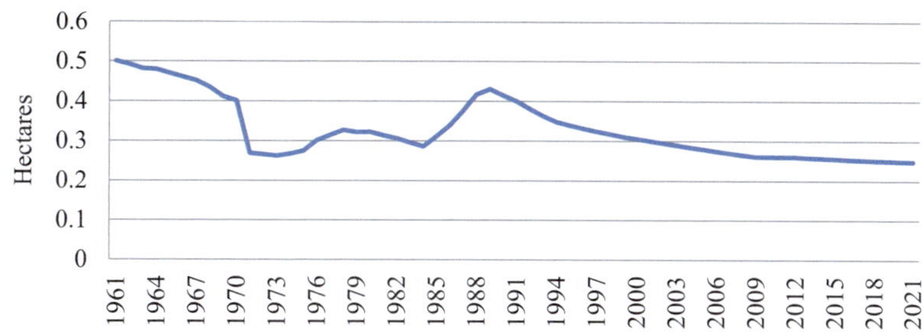

Fig. 9.7 Arable land, Cambodia, 1961–2021 (hectares per person). *Source* World Bank (2022)

country agriculture by increasing value added in the sector and promoting agricultural value chain responsiveness to market demand and regional and global competition. The policy presents the following goal: all Cambodian farmers and farming communities will access and adopt improved agricultural knowledge, information, and technologies to enhance agricultural productivity, diversification, commercialization, and sustainable natural resources management.

The new Rectangular Strategy, which was introduced in September 2018[11] through four phases, seeks to enhance the infrastructure for agricultural research, supporting technology improvement (including seeds), as well as pre- and postharvest technology development, to promote good agricultural practices, strengthen a range of agricultural services (including agricultural extension and information provision), mechanization, technological upgrading, improve land use management, and advocate resilient livestock and fisheries practices.

The policy document on the Promotion of Paddy Production and Export of Milled Rice of 2013 aimed to promote Cambodia as an internationally recognized exporter of milled rice by shifting from the informal export of paddy to the formal export of milled rice (Cambodia, 2013). The subsequent Agricultural Sector Strategic Development Plan 2019–2023 sought to: (i) enhance agricultural productivity, diversify, and commercialize; (ii) promote animal health and production; (iii) manage fisheries and development of aquaculture; (iv) manage and develop sustainable forestry and wildlife resources; and (v) improve efficiency of support services and human resources development (Cambodia, 2019a).

Meanwhile, the National Agricultural Development Policy (NADP) 2022–2030 provides broad guidance for the sector implementation to line ministries of the government, development partners, and the private sector. The policy outlines four main areas, viz. (1) modernization and commercialization of the agricultural value chain; (2) promote public and private investment in the agriculture sector; (3) stimulate sustainable growth and climate resilience; and (4) address institutional reforms and cross-cutting issues towards raising agricultural growth, competitiveness, and inclusivity.

The Ministry of Agriculture, Forestry and Fisheries also has sub-sector-specific policies such as Strategic Planning Framework for Fisheries 2015–2024, National Cassava Policy 2020–2025, and the Strategic Planning Framework for Livestock Development 2016–2025. Other overall policy frameworks to ensure socio-economic and environmental sustainability include the National Forest Programme 2010–2029, the Cambodia Climate Change Strategic Plan 2014–2023, the National Strategic Plan on Green Growth 2013–2030, and the Cambodia Sustainable Development Goals. In general, these policies were introduced to guide the future direction of the agricultural sector in the country.

[11] The Rectangular Strategy is the principal socioeconomic policy agenda of the Government of Cambodia which covers four growth factors, viz. (i) agricultural development, (ii) infrastructure rehabilitation and development, (iii) private sector development and employment creation, and (iv) capacity building and human resources development.

Land Development

Since the demise of the Democratic Kampuchea's regime in 1979, a number of land policies have been introduced to administer, manage, utilize, and distribute land in an equitable, transparent, and sustainable manner to contribute to achieving the national goals of poverty alleviation, ensuring food security, natural resources, and environmental protection, and socio-economic development. According to Land Law 1992, all land in Cambodia belongs to the state and shall be governed and protected by the state. The government stopped recognizing the land property rights that existed before 1979. Land Law 1992 and Land Law 2001 are the main laws on land ownership that are recognized in Cambodia.

Cambodia's Land Law 1992: The 1992 Land Law maintained the rights of possession of agricultural land with the state as the legal owner, while creating ownership rights for residential properties. Two types of state land were recognized, i.e. state public land and state private land, and that only the latter can be released for concessions or alienations. The 1992 Land Law did not provide a solid platform for full tenure security or for effective land management.

Land Law 2001: The 2001 Land Law incorporated a number of changes and enhancements to provide a better foundation for land administration, management, and distribution. Its reforms include extending private ownership rights to both residential and agricultural land, and officially certified ownership in a government document known as a title certificate, which enabled delegation of land administration from the central to the capital/provincial level.

White Paper on Land Policy 2012: In 2012, the government issued a new *White Paper on Land Policy* to promote land use, management, and natural resource management for sustainable and equitable socio-economic development. The goals of this policy are:

1. Strengthen security of land tenure to ensure confidence and efficiency of land markets.
2. Guarantee the management, protection, and use of land and natural resources in a transparent and effective manner to ensure equity of socio-economic development both in rural and urban areas, and to maintain environmental sustainability while preventing disputes over land use.
3. Provide a clear direction in transparent and equitable distribution and use of state land for public and private purposes.

Food Security

Cambodia was a significant food exporting country in the 1960s, particularly of rice but the civil wars in the 1970s started considerable shortfalls in food availability, though rising food production since 1981 almost brought self-sufficiency by the late 1980s (Tickner, 1996).

Based on the recent report of the Ministry of Agriculture, Forest and Fisheries, Cambodia has achieved food self-sufficiency and has produced quantities of rice that have exceeded domestic consumption since 1995 (Cambodia, 2022a, 2022b). Rice is the major crop and staple food of Cambodia. Rice production has notably increased since the past 2 decades with a significant surplus exported to several countries, including China. Total rice surplus[12] of Cambodia increased from 4.8 million tons in 2013 to 5.9 million tons in 2020 and 6.3 million tons in 2022 (Fig. 9.8).

The agriculture sector plays a central role for national food security in Cambodia as it produces food and provides jobs to the vast majority of the people, especially rural people. Given the surpluses recorded in rice production, one might be mistaken that Cambodia has seemingly achieved food security at the national level. This surplus, however, has been achieved in demand (measured by the capacity to pay) terms only as those who cannot afford are not registered in this demand. The benefits from rice production are not distributed evenly across the country, especially with rural households. The acute inequality

[12] Rice surplus is the amount of remaining paddy rice of the country after deducting for domestic consumption.

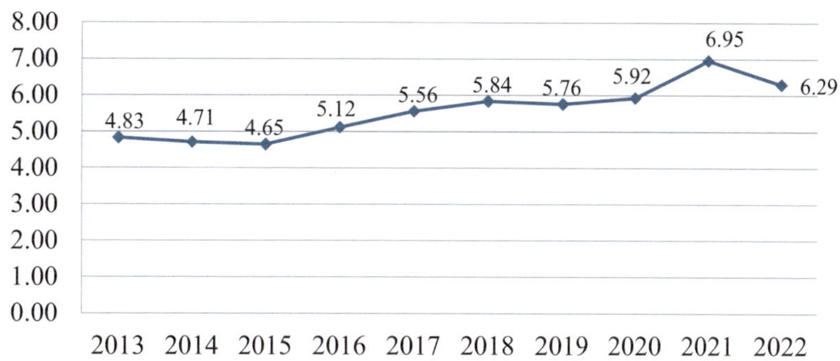

Fig. 9.8 Rice production surplus, Cambodia, 2003–2022 (million tons). *Source* Cambodia (2023). Ministry of Agriculture, Forestry and Fishery's Annual Report 2022. Ministry of Agriculture, Forestry and Fisheries

faced by rural dwellers, including the majority smallholder farmers, has little access to food.[13]

More than 10% of the rural population in Cambodia are landless while 47.1% of the households owned less than one hectare (World Bank, 2019). Due to small-scale nature of land holding, two-thirds of the country's rural households still face seasonal food shortages each year. In addition, the availability of food from own production in poor farm households is often not sufficient in quantity and quality to meet their food requirements. To make matters worse, floods and droughts, owing to climate change, frequently threaten food supply in the country.

Data from the World Bank (2022) indicate that the prevalence of moderate or severe food insecurity in the population is high with 51.1% of Cambodians living in moderate or severe food insecurity situation in 2021. Figure 9.9 indicates the prevalence of moderate and severe food insecurity in the population.

To tackle the issue of food security, a number of programmes and projects aimed at enhancing household food security and nutrition have been implemented in Cambodia through various government ministries, international and local non-governmental organizations (NGOs), and the United Nations. Social safety nets for food security and nutrition (school feeding, scholarships for the poor, nutrition programmes which include special nutritious food products for small children, targeted food distribution for disaster affected households, and Food for Work (FFW)/ Cash for Work (CFW) programmes) were implemented through the WFP in cooperation with the Ministry of Education, Youth and Sport, Ministry of Rural Development, and NGOs (WFP, 2010).

Following the rising food and agricultural input prices since 2008, large-scale food security programmes to assist the poor and food insecure to cope with the impact of the crisis were designed, which included the Emergency Food Assistance Project (EFAP), the Food Facility from the European Union (EU-FF), and the Global Food Crisis Response Program Grant from the World Bank. These programmes focused on immediate support to the poor and vulnerable through distribution of food and agricultural inputs, creation of temporary income opportunities, and strengthening emergency response capacities.

Furthermore, the government of Cambodia took a significant step to address issues faced by farmers by deploying Agricultural Technical Officers (ATOs) Cambodia's communes aimed at modernizing the agricultural community to ensure food security and rural development. The first phase of this ambitious plan involved recruiting 250 commune agriculture officers, which followed a meticulous and disciplined selection process that was completed on 17 December 2023 (Khmer Times, 2023).

The primary role of the ATOs is to collect information on local agricultural production and assist farmers to address their specific challenges. This information will be used to formulate timely solutions and requests made to the Ministry of Agriculture, Forestry, and Fisheries.

[13] OECD (2016), a territorial approach to food security and nutrition policy: The case of Cambodia (https://www.oecd-ilibrary.org/docserver/9789264257108-5-en.pdf?expires=1701679014&id=id&accname=guest&checksum=1E5AF8633239EA99CACC826F3B1C62A2).

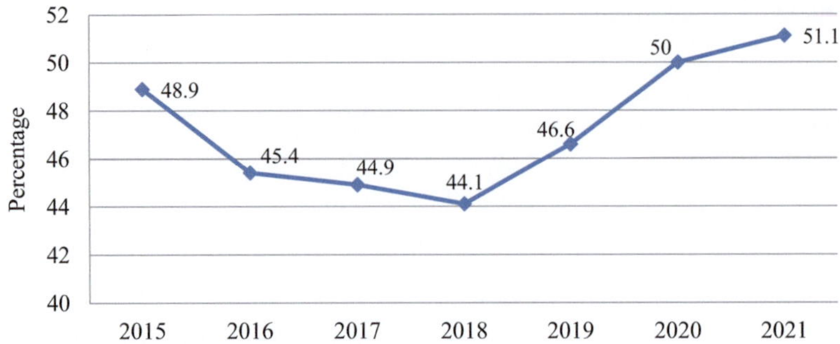

Fig. 9.9 Prevalence of moderate or severe food insecurity, Cambodia, 2015–2021. *Source* World Bank (2022) (https://data.worldbank.org/indicator/SN.ITK.MSFI.ZS?view=chart&locations=KH)

By bridging the gap between farmers and government authorities, the ATOs seek to address the concerns of the farming community effectively.

The government expected to deploy 250 ATOs by 2023, and an additional 800 in 2024, and 550 in 2025. This comprehensive plan aims to support 1550 potential communes and agricultural activities to serve approximately 2 million farming families (Khmer Times, 2023).

Underdeveloped Agro-Industry

Cambodia's agro-processing sector is underdeveloped due to limited investment from the public, as well as the private sector. The share of agro-industry investment in total investment fell from 3.1% in 2018 to 1.9% in 2022 (Fig. 9.10).[14]

The underdeveloped status of agro-industry has hindered upgrading to high value-added export of agricultural products thereby widening food trade deficit. Cambodia's agro-processing industry is predominantly made up of micro, small, and medium enterprises (MSMEs) producing primarily for the domestic market and the majority of these businesses do not follow food safety and internationally certified hygiene protocols (IDE, 2022).

Given the underdeveloped status of agro-industry in the country, the exports of agriculture products from Cambodia are mostly unprocessed (Piseth, Monyoudom & Tynarath, 2021). Only around 10% of Cambodia's harvested crops are processed within the country, while processed agriculture exports contributed only 5% of the country's agricultural exports in 2020 (Cambodia, 2022a, 2022b). Food imports increased significantly from US$111 million in 2000 to US$1.3 billion in 2020, while food exports only increased from US$2.8 million in 2000 to US$136.9 million in 2020, which has translated into a widening deficit amounting to US$1.1 billion in 2020 (Fig. 9.11).

Falling prices involving some crops have also undermined rural households' incomes. For example, pumpkins, mangoes, longans, and even rice have suffered drastic fall in prices due to the lack of local processing capacity. Some farmers are better off leaving the products to rot rather than harvesting owing to the financial burden.[15] For instance, in 2022, some farmers in Siem Reap were forced to throw away 30 tons of mangoes from their plantations because of a lack of demand, while farmers in Pailin Province abandoned mango and longan harvest. The lack of demand drove farmers to indebtedness, which resulted in around 90% of Cambodian farmers borrowing money from banks or microfinance to support them.[16]

Hence, Piseth et al. (2021) argue that promoting investments (both local and foreign) into domestic processing is a critical step to

[14] Data supplied by the Council for the Development of Cambodia (CDC).

[15] Cambodianess's news article (May 02, 2022), Opinion: Farmers At the Mercy of Crop Markets,
Accessed on October 15, 2023 from https://cambodianess.com/article/farmers-at-the-mercy-of-crop-markets.

[16] Cambodianess's article (June 13, 2022), Debt and Land Degradation: Lake Farmers' Plight, Accessed on October 15, 2023 from https://cambodianess.com/article/debt-and-land-degradation-lake-farmers-plight.

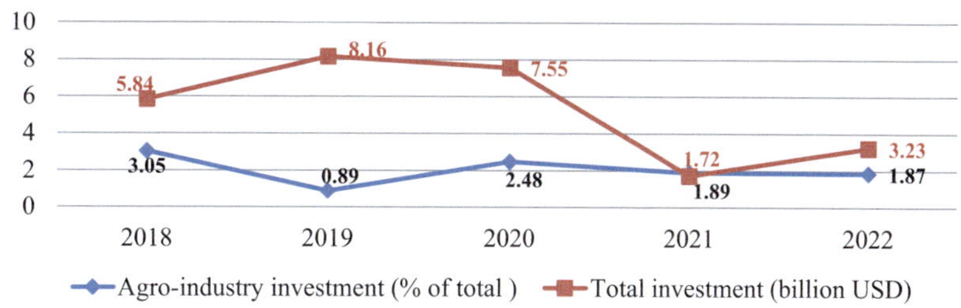

Fig. 9.10 Agro-industry investment in total investment, Cambodia, 2018–2022 (%). *Source* Plotted using data supplied by the Council for the Development of Cambodia

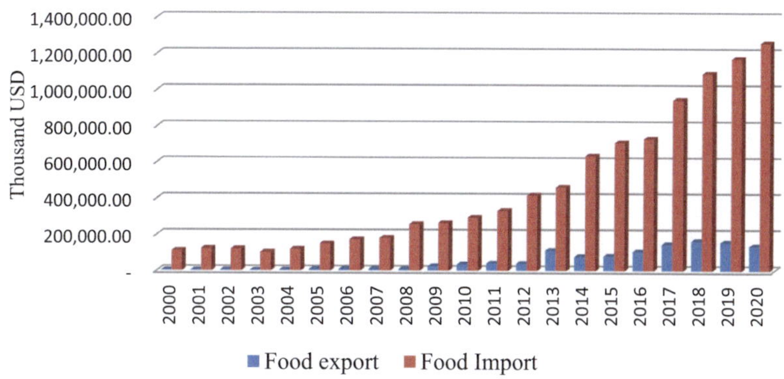

Fig. 9.11 Cambodia's food imports and exports, Cambodia, 2000–2022. *Source* World Bank (2022) on Cambodia Trade and Development Indicators. Accessed on October 15, 2023 from https://wits.worldbank.org/CountryProfile/en/Country/KHM/Year/2020/TradeFlow/Import/Partner/all/Product/16-24_FoodProd)

increase agro-food exports and value-added in the agricultural sector. The development of agro-industry will create more jobs, expand market, and provide better prices for farmers' crops.

Low Labour Productivity

Cambodia's labour productivity lags behind most regional countries. Low labour productivity derived from weak human capital and low total factor productivity. According to the World Bank (2021a, 2022), Cambodia's human capital index is low relative to its neighbouring countries. Using the impact of investments in children today on future productivity and long-term economic growth, Cambodia ranked 118th out of 174 countries, which was on par with India and Myanmar, but far behind the other countries in Table 9.3. Education is Cambodia's biggest obstacle to improving human capital (World Bank, 2021a).

Table 9.3 Human capital index, selected countries, 2020

Country	HCI score	HCI rank
Vietnam	0.69	38
Malaysia	0.61	62
Thailand	0.61	63
Philippines	0.52	103
India	0.49	116
Cambodia	0.49	118
Myanmar	0.48	120
Bangladesh	0.46	123

Source World bank (2021a)

Cambodia's education faces a number of challenges and a key challenge in the education system is low quality of education services. Public educational institutions lack the requisite administrative and financial autonomy to modernize their services. Low professional qualification standards for educators also lower the quality of education services, alongside low teacher capacity. Cambodia also faces challenges related to health, particularly child nutrition, which has weakened human capital and learning outcomes.[17]

In terms of agricultural labour productivity, though it gradually increased over the period 1995–2018, Cambodia ranked below most of ASEAN's member states. Agriculture labour

[17] Ibid.

productivity of Cambodia was at 5.4 (thousands of USD) in 2018 only above Lao PDR and Myanmar (Table 9.4).

Agricultural labour productivity is correlated with food insecurity because lower agricultural labour productivity results in lower food supply, higher food costs, lower farm income, and ultimately reduced buying power to meet the essential food needs of households (Abimbola et al., [2022]).

Climate Change Consequences

With rice being the staple food and principal income of rural households and its production highly vulnerable to climate change, rural households are severely exposed to food insecurity (Timothy et al., 2013).

Climate change affects food production and availability, access, quality, utilization, and stability of food systems. Extreme weather-related disasters are increasing that have reduced crop yields with ramifications for food security (Adisu et al., 2021). Drought and floods as a result of climate change have frequently happened in Cambodia, which has reduced the availability of water for irrigation. Figure 9.12 shows rice crop area damaged by droughts and floods in 1996–2021. As a result, rural households relying on rainfed agriculture face decreased food production, income loss, and heightened food insecurity.

Each year, Cambodia experiences crop losses to drought while floods destroy crops in other areas (SUMERNET [2021]). A severe drought in 2015–2016 affected 18 of 25 provinces and over 2.5 million people, while rice production losses caused by droughts associated with the 2019 *El Nino* were estimated at USD100 million.

The heaviest rainfall in three years caused severe flooding in 14 provinces in September 2022, with some 85,000 households affected (World Bank, 2023; Fig. 9.12). Climate projections suggest that future flood risk and heat stress will worsen in Cambodia. Without proper adaptation and mitigation measures, climate change could cost up to 9%of Cambodia's GDP by 2050 while increasing the poverty rate by up to 6%age points. Climate change is projected to lower staple crop and fisheries yields, thereby posing challenges to rural households' food security in Cambodia.

Consequently, the government of Cambodia has committed to address the challenge of climate both locally and internationally and to support conducive dialogue with the private sector to create an enabling environment for green investments and green finance. To formulate policies to support green growth development in Cambodia, the National Council for Sustainable Development (NCSD) was established as a governmental policy-making body to improve the coordination of climate change activities in Cambodia and to promote a strong, comprehensive, and effective climate change response. The NCSD has introduced the following legislations, policy documents, and plans to support green growth and sustainable development in Cambodia:

1. National Environment Strategy and Action Plan 2016–2023
2. National Biodiversity Strategy and Action Plan 2016–2020
3. National Green Growth Strategic Plan 2013–2030
4. National Policy on Green Growth 2013–2030
5. Cambodia Climate Change Strategic Plan 2014–2023
6. Sectoral Climate Changes Action Plans 2014/5–2018
7. Long-Term Strategy for Carbon Neutrality (LTS4CN)" 2021
8. Environment and National Resource Code 2023
9. Climate Change Financing Framework 2015

As shown in Table 9.5, the share of climate change expenditure in GDP rose from 1.0% in 2016 each year to 2.2% in 2020.

In the agriculture, forestry, and fishery sectors climate change expenditure continued to increase in 2020 strongly (31% increased from 2019) and climate change expenditure on agriculture in 2020 was 10.26% of total climate change expenditure (World Bank, 2023). The main source of climate

Table 9.4 Agricultural labour productivity, ASEAN, 1995–2018 (USD '000 per employed person)

Country	1995	2000	2005	2010	2015	2018
ASEAN	3.5	3.7	4.3	6	7.5	8.8
Brunei Darussalam	125.2	112.8	58.5	40.2	55.4	57.5
Cambodia	2.1	2	2.8	3.6	4.2	5.4
Indonesia	3.8	3.7	4.1	7	9.8	11.5
Lao PDR	4.7	5.8	5.4	5.7	4.7	5.1
Malaysia	23.5	19.5	26.2	38.8	37.9	43.7
Myanmar	2.9	3.6	3.3	2.6	2.2	3.5
The Philippines	4.8	4.8	4.7	5.6	6.9	7.9
Singapore	70.2	46.8	28	17.3	10.6	9.2
Thailand	3.7	3.6	5.4	7.2	8.6	8.7
Vietnam	2	2.5	3	4.1	5	6

Note in 2017 constant PPP prices
Source ASEAN (2021). Regional study on labour productivity in ASEAN. Jakarta: ASEAN Secretariat

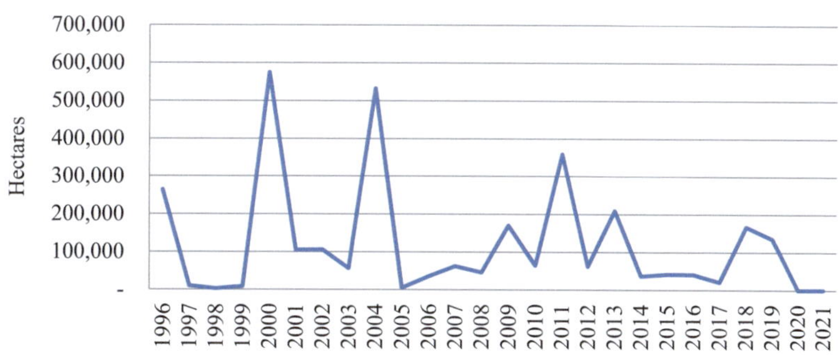

Fig. 9.12 Rice crop area damaged by droughts and floods, Cambodia, 1996–2021. *Source* National committee for disaster management (NCDM)

change financing on agriculture sector was mainly driven by the external finance, which represented 79% in 2020, reflecting high donor support for climate action in agriculture.

To mitigate the impact of climate change, Cambodia is drafting plans and strategies to make Cambodian agriculture climate resilient. The formulation of the Climate Change Priority Action Plan for the agriculture sector (CCPAP III) is currently underway, with the vision to make Cambodia's agriculture resilient by 2030 as it moves towards a sustainable agrifood system using climate-smart technology and innovations (World, 2023). Furthermore, the Ministry of Agriculture, Forestry, and Fisheries is drafting the Agricultural Sector Master Plan, which represents the long-term strategic framework for a comprehensive roadmap to lead Cambodian agriculture. Its additional mission is to provide high-quality agricultural support services with scientific methods, regulatory compliance, and clearly defined policies to support sustainable agricultural development in the country (World Bank, 2023). Since the mid-2000s, over USD1.3 billion (80% by development partners, 20% by the government) has been spent on the rehabilitation of existing irrigation schemes and the construction of several large irrigation schemes and reservoirs.

In addition, with funding from the Climate Investments Funds-Pilot Program for Climate Resilience (PPCR), with support from the Asian Development Bank and the World Bank, the government has prepared the Strategic Program for Climate Resilience (SPCR), comprising seven investment projects in four sectors (agriculture,

Table 9.5 Climate change expenditure, Cambodia, 2015–2020

	2015	2016	2017	2018	2019	2020
Climate expenditure (% of GDP)	1.4	1.0	1.2	1.8	2.0	2.2
Climate change expenditure (USD million)	252.9	200.2	262.9	433.9	546.4	591.3

Source Cambodia (2021a). Cambodia climate public expenditure review 2020. Ministry of Economy and Finance

water resources, urban development, and transport). The total budget for this programme is USD 561.8 million over the implementation period from 2011 to 2023 (Cambodia, 2024).[18] A number of agricultural projects were implemented within the framework of this programme (see Table 9.6).

Digitalization

Cambodia is currently undergoing digital transformation with the government and private sector investing in the development of digital infrastructure and services, which includes the expansion of broadband internet access and the creation of digital platforms and services, such as e-commerce and online government services. As digitalization is crucial for government and business, numerous policies have been formulated to promote digitalization. The government began to expand e-government by establishing the National Information Communication Technology Development Authority (NiDA) at the Office of the Council of Ministers in 2020.

The government established the National Institute of Posts, Telecommunications, Information and Communication Technology (NIPTICT) in 2014 as an agency of the Ministry of Posts and Telecommunications to support training of human capital, skills, and digital literacy to civil servants and to promote digital research and innovation. The institute was renamed as Cambodia Academy of Digital Technology (CADT) in 2021. T*he Law on Telecommunications* was promulgated in 2015 for the purpose of ensuring (1) the use of infrastructure and network and the provision of telecommunication services with efficiency, safety, quality, reliability, and affordability; (2) to promote and encourage the private sector to participate in the development of this sector; and (3) to generate fair competition and consumer protection.

The government approved in 2016 *the Development Policyfor Telecommunications, Information and Communication Technology 2020*, which set out measures and goals to strengthen the development of the telecommunications, information, and communication technology sectors, and digital security to promote industrial development and the use of information and communication technology. Subsequently, the Cambodia Digital Economy and Society Policy Framework 2021–2035 and Cambodia Digital Government Policy 2022–2035 were formulated in 2021 and 2022, respectively, to promote digital adoption and transformation. The government established the National Digital Economy and Society Council in 2021 to lead, facilitate, and foster the development of a vibrant economy and society by laying the digital foundation, adoption, and transformation of all social actors, including states, citizens, and businesses.

Several government ministries and institute are launching digital projects, such as digital registration for land ownership, vehicle registration, and taxation. The revenue management of state revenue collection units is gradually being digitalized, especially the introduction of electronic filing system (E-Filing System) incorporating the electronic value added tax system (E-VAT System), consisting of value-added tax (VAT) credit, VAT payable, and VAT refund; the implementation of a National Single Window system;

[18] National Council for Sustainable Development's program on climate change. Accessed at https://ncsd.moe.gov.kh/dcc/program/strategic-program-climate-resilience-spcr.

Table 9.6 Projects of SPCR for climate resilience related to agriculture

Project name	Budget and period	Adaptation activities
Rainwater harvesting and drip irrigation for high-value crop production in Cambodia (Spice Value Chain Development Project)	USD 29 M (2014–2018)	• Introduction of rainwater harvesting and drip irrigation technologies Promotion of climate-resilient agricultural inputs, such as high-quality seeds, plants, fertilizers, and organic pesticides
Enhancement of flood and drought management in Pursat Province	USD 48 M (2012–2019)	• Upgrading of water management infrastructure for flood and drought mitigation: irrigation scheme to provide wet-season supplemental irrigation for 16,100 hectares and full irrigation to a smaller command area during the dry season, and a new headworks structure to facilitate peak-flood diversion • Improvement of national flood and drought forecasting and establishment of flood and drought warning system • Improvement of hydraulic design standards and guidelines for the Mekong Delta • Improvement of capacity of farmer water user committees to manage drought and flood risk, and of communities and local governments to improve the disaster preparedness of communities
Promoting climate-resilient agriculture in Koh Kong and Mondulkiri Provinces	USD28.9 M (2015–2019)	• Construction of rainwater harvesting ponds and drip irrigation for high-value crop production in Koh Kong and Mondulkiri • Adoption of drought-resistant crops • Climate proofing of irrigation infrastructure and adoption rice intensification system in Mondulkiri • Construction of bioengineered sea barriers to reduce saltwater intrusion, and adoption of salt-tolerant crops in Koh Kong • Ecosystem-based adaptation to improve forest cover and soil and water management through catchment development planning in Mondulkiri

Source Cambodia (2021b). National Council for Sustainable Development's Program on Climate Change. National Council for Sustainable Development. Accessed on May 20, 2024 from https://ncsd.moe.gov.kh/dcc/program/strategic-program-climate-resilience-spcr

and the online business registration platform. In addition, the digital adoption for businesses, especially in the service sector, has gained momentum through the operations of e-commerce and the demand for e-products and services, particularly driven by the growing number of young people who can use technology in line with the growth of the use of social media, other digital platforms, and the growing use of FinTech.

Cambodia experienced fluctuating changes in mobile phone and internet subscriptions (Fig. 9.13). Mobile phone subscribers in Cambodia fell from 2016 till 2017 before rising in trend terms till 2019. It has since fallen till September 2022. Almost all Cambodians own smartphones, which has translated to rising surfing on social networks.

Furthermore, internet subscriptions rose from 8 million subscribers in 2016 to 18 million subscribers in 2022 (September figures) (Fig. 9.14). The high numbers of internet users show how fast Cambodians are embracing technology.

The government has strategized to strengthen and expand mobile infrastructure and high-speed internet through the introduction of 4th generation mobile services (4G) and to lay the foundation for the 5th generation mobile services (5G). The 4th generation mobile network (4G) is providing services to approximately 80% of the population (Telecommunication Regulator of Cambodia, 2021). The use of digital devices, e-commerce transactions during the pandemic, and financial

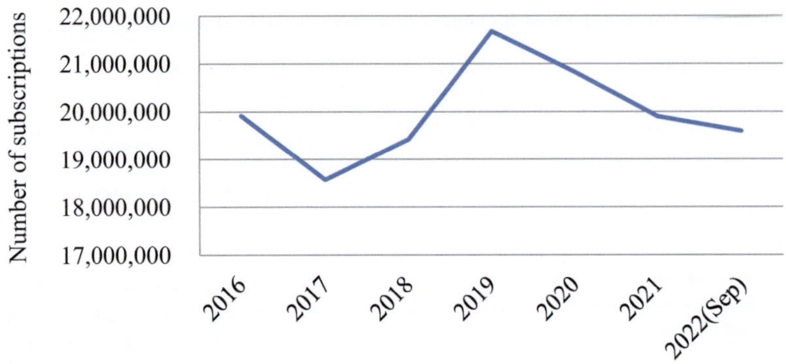

Fig. 9.13 Mobile phone subscriptions, Cambodia, 2016–2022. *Source* Telecommunication Regulator of Cambodia (TRC)

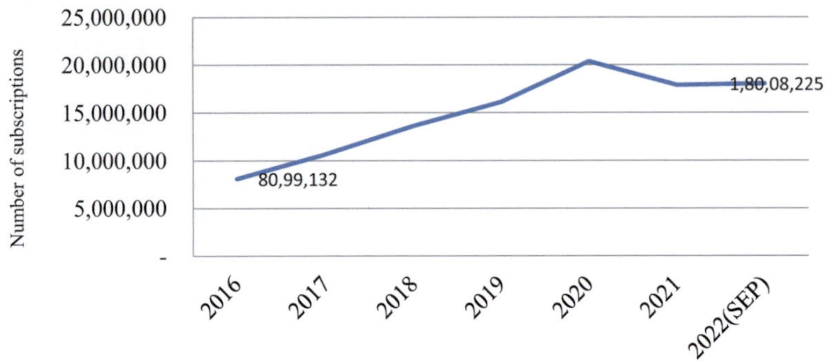

Fig. 9.14 Internet subscriptions, Cambodia, 2016–2022. *Source* Telecommunication Regulator of Cambodia (TRC)

technology boosted the use of internet and mobile services in Cambodia.

The proliferation of digitalization into agriculture will contribute to the promotion of farmers' income by improving production efficiency, broadening sales channels, and promoting the upgrading of the agricultural structure help farmers make more precise decisions, and improve access to information, knowledge, and markets (Zhang & Fan, 2023).

The digitalization of agriculture in Cambodia is still in its infancy stage. Although some key strategies and policies, such as the Rectangular Strategy Phase IV, the National Strategic Development Plan (NSDP 2019–2023), the e-Government Master Plan (2017–2020), the Strategic Plan for Agricultural Development (2019–2023) highlighted the need to further develop the digital economy and increase investment in agriculture modernization, the adoption of digitalization in the agricultural sector is still slow. There is no sizable project on the digitalization in agriculture being implemented by the government. The majority of the key project implementation initiatives are from NGOs, private sector, and development partners. However, the projects implemented by them are generally small in scale (Table 9.7).

Table 9.7 Initiatives related to digitalization in agriculture, Cambodia

Private Sector/NGOs/Donors	
AMK Microfinance Institution Plc	Implements E-Farmer Support App Activity through Tonlesap App. The Tonlesap App objective is to provide agricultural technical knowledge and agricultural market information, key techniques in farming, livestock, and information in agricultural inputs as well as loan access for business expansion
Spien	Connects agricultural products to consumers and wholesales' shops through product display on the website. Focus on product promotion through website and Facebook page. It targets B2B (business to business), and the next step is to target individual consumers
Oxfam	• Launched BlockChain technology for the project, Livelihoods for Organic Cambodian Rice (BlocRice) that introduced cashless payments to farmers
USAID	Implemented and funded different projects related to modernization and digitalization of agriculture
IFAD	• Provided funds through the government to modernize the agricultural value chain in Cambodia
Netherlands Space Office and the International Cocoa Organization (ICCO)	• Began the Angkor Salad Project to use geodata to help at least 100,000 Cambodian vegetable farmers increase production and save money. The project has five information services farmers can access via SMS, voice and mobile app. Farmers can access irrigation advice to figure out how much water they need to use for their crops based on geospatial data and their crop cycle

Source Chan et al. (2020). Digital challenges and opportunities for agribusiness enterprises in Cambodia: Implications for the Cambodian economies. Phnom Penh: Centre for policy studies

Conclusion

Cambodia managed to generate market surplus in rice production but still suffers from food security problems because a significant share of the rural population cannot afford it, which reminds us of the problem of distribution in food allocation (Rasiah, 2025). Rural households are strongly affected by climate change as well as lack of access to sufficient land and a fall in crop yield.

Small farmland holding and landlessness are also found to be critical issues that make rural households unable to produce sufficient food crops for whole-year consumption. Landlessness and inefficient farm size have increasingly become acute because of the fragmentation of farms in rural areas through marriage. Furthermore, the lack of processing not only impedes high value-added export of agricultural commodities, it also impacts negatively on farmers' incomes. With limited processing facilities, the price of harvested crops is generally low, especially of mongos, pumpkins, and longan, which farmers often dump or leave unharvested because of a lack of demand. The government has recognized that low labour productivity and low level of digitalization have negatively affected incomes of rural households, while exposing them to food insecurity.

References

ADB. (2021a). Cambodia Agriculture, Natural Resources, and Rural Development Sector Assessment, Strategy, and Roadmap, Manila: Asian Development Bank (ADB). Accessed on August 19, 2023 from https://www.adb.org/sites/default/files/publication/718806/cambodia-agriculture-rural-development-road-map.pdf

ADB. (2021b). *Additional Financing Report on Climate Resilient Rice Commercialization Sector Development Program.* Manila: ADB, Accessed on August 25, 2023 from (https://www.gafspfund.org/sites/default/files/inline-files/44321-014-afr-en_1.pdf)

Abimbola O. Adepoju, Akubuike C. Obialo (2022), Agricultural Labor Productivity Transitions Among Maize Farming Households in Rural Nigeria, University of Ibadan: Nigeria. Downloaded on November 12, 2023 from https://scindeks-clanci.ceon.rs/data/pdf/0352-3462/2022/0352-34622204093A.pdf

Adisu, M., Ayele, T., Zerhun, G., & Ashenafi, H. (2021). Climate Change Impacts on Household Food Security and Farmers Adaptation Strategies. *Journal of Agriculture and Food Research, 6*(12), Article 100197.

ASEAN Secretariat. (2021). *Regional Study on Labour Productivity in ASEAN*. The ASEAN Secretariat.

Asia News Network. (2023). *Cambodia's agriculture exports down value climbs.* Accessed on October 18, 2023 from https://asianews.network/cambodias-agricultural-exports-down-value-climbs/#:~:text=Last%20year%2C%20Agriculture%20Minister%20Dith,7.9%20million%20tonnes%20in%202021

CDRI, CARD, IPRI. (2011). *Review of Agricultural Policy and Policy Research: A Policy Discussion Paper.* CDRI.

Cambodia. (2012). *White Paper on Land Policy 2012.* Downloaded on March 2024. https://data.opendevelopmentcambodia.net/laws_record/land-policy--white-paper/resource/f030466c-90e4-4352-b13f-7df434d7371e

Cambodia. (2013). *Policy Document on the Promotion of Paddy Rice Production and Export of Milled Rice.* Royal Government of Cambodia.

Cambodia. (2019a). *Agricultural Sector Strategic Development Plan 2019–2023.* Ministry of Agriculture, Forestry and Fisheries.

Cambodia. (2019b). *Cambodia Labor Force Survey 2019.* Ministry of Planning: National Institute of Statistic.

Cambodia. (2021). *Cambodia Socio-Economic Survey 2021.* Ministry of Planning: National Institute of Statistics.

Cambodia. (2021a). *Cambodia Climate Public Expenditure Review 2020.* Ministry of Economy and Finance.

Cambodia. (2021b). *National Council for Sustainable Development's Program on Climate Change.* Accessed on May 20, 2024 from https://ncsd.moe.gov.kh/dcc/program/strategic-program-climate-resilience-spcr

Cambodia. (2022a). *National Institute of Statistics (2022).* Report of Cambodia Socio-Economic Survey 2021.

Cambodia. (2022b). *National Agriculture Development Policy 2022–2030.*

Cambodia. (2023). *Annual Report for Agriculture, Forestry and Fisheries 2022.* Ministry of Agriculture, Forestry and Fisheries.

Cambodia. (2024). *Programs on climate change.* Accessed on March 12, 2024 from https://ncsd.moe.gov.kh/dcc/program/strategic-program-climate-resilience-spcr?page=1

Chhun, C., Bora, B., & Sothy, E. (2015). *Effect of Labour Movement on Agricultural Mechanisation in Cambodia.* Cambodia Development Research Institute (CDRI).

Construction Property Magazine. (2022). *Government receives more than US$ 2 million as ECL rent income.* (https://construction-property.com/govt-receives-more-than-us-2-million-as-ecl-rent-income-as-the-total-land-area-is-1-16m-hectares/)

FAO. (2008). *An Introduction to the Basic Concepts of Food Security.* Food and Agricultural Organization of the United Nations. Accessed on September 23, 2023 from https://www.fao.org/3/al936e/al936e00.pdf

GIZ. (2022). *Sector Brief Cambodia: Organic Market.* Downloaded on October 25, 2023 from https://www.giz.de/en/downloads/giz2022-en-sector-brief-cambodia.pdf

IDE. (2022). *SMEs and food processing markets in Cambodia.*

Khmer Times. (2023). *Big Push: PM's New Strategic Policy to Enhance Farmers Income.* November 21, Accessed on May 20, 2024 from https://www.khmertimeskh.com/501395304/big-push-pms-new-strategic-policy-to-enhance-farmers-income/

Nesbitt, H. J. (Ed.). (1997). *Rice production in Cambodia.* International Rice Research Institute.

OECD. (2016). I Library, a territorial approach to food security and nutrition policy: The case of Cambodia (https://www.oecd-ilibrary.org/docserver/9789264257108-5-en.pdf?expires=1701679014&id=id&accname=guest&checksum=1E5AF8633239EA99CACC826F3B1C62A2)

Piseth, S., Monyoudom, Y., & Tynarath, H. (2021). *Cambodia's Agri-Food Trade: Structure, New Emerging Potentials, Challenges & Impacts of Covid-19.* CDRI.

Rasiah, R. (2025). Strengthening Agricultural Sustainability and Food Security in Southeast Asia, In R. Rasiah, V. Anter & N.I Annizah (Eds), *Sustainable Agriculture and Food Security in Southeast Asia: Towards Climate Resilience and Digitalization*, Singapore: Springer Nature.

Sodeth L. (2020). *Job and Labor Productivity in Cambodia's Agriculture Sector*. World Bank Group. https://www.nbc.gov.kh/download_files/research_papers/english/3.1.Agriculture-labor-productivity.pdf

Sophal, C., Chansovy, N., & Metta, H. (2020). *Digital Challenges and Opportunities for Agribusiness Enterprises in Cambodia: Implications for the Cambodian Economies.* Centre for Policy Studies.

SUMERNET. (2021). *Climate Change Threats on Food Security in Asia: A look at Cambodia and Myanmar.* September 10, Accessed on May 20, 2024 from https://www.sumernet.org/story/climate-change-threats-on-food-security-in-asia-a-look-at-cambodia-and-myanmar

Tickner, V. (1996). *Food Security in Cambodia: A Preliminary Assessment*. United Nations Research Institute for Social Development. https://www.files.ethz.ch/isn/28980/dp80.pdf

Timothy S. et al (2013). *Cambodian Agriculture: Adaptation to Climate Change Impact*. International Food Security Research Institute. file:///C:/Users/Kum%20Kim/Downloads/Cambodian%20agriculture%20adaptation%20to%20climate%20change%20impact.pdf

VOA. (2008). *Cambodia Ends Long-Running Debate as it Celebrates 20 Years of Peace*. (https://www.voanews.com/a/cambodia-ends-long-running-debate-as-it-celebrates-twenty-years-of-peace/4716904.html)

World Food Program. (2010). *Cambodia Food Market Analysis and Survey Report*. World Food Program. https://documents.wfp.org/stellent/groups/public/documents/ena/wfp244779.pdf

World Bank. (2019). *Cambodia—Providing Land and Opportunity for Landless and Land-poor Families*. https://www.worldbank.org/en/results/2019/10/22/cambodia---providing-land-and-opportunity-for-landless-and-land-poor-families

World Bank. (2020). *Agricultural land (% of land area)—Cambodia*. Accessed on March 15, 2024 from https://data.worldbank.org/indicator/AG.LND.AGRI.ZS?locations=KH

World Bank. (2021a). *Resilient Development: A Strategy to Diversify Cambodia's Growth Model*. The World Bank Group.

World Bank. (2021b). *Climate Risk Country Profile*. https://climateknowledgeportal.worldbank.org/sites/default/files/2021-08/15849-WB_Cambodia%20Country%20Profile-WEB.pdf

World Bank. (2022). *Main Findings of the Cambodia Country Economic Memorandum (CEM)*. https://www.worldbank.org/en/country/cambodia/publication/main-findings-of-the-cambodia-country-economic-memorandum-cem)

World Bank. (2023). *Country Climate and Development Report*. The World Bank Group. https://documents1.worldbank.org/curated/en/099092823045083987/pdf/P17887106c6c2d0e909aa1090f3e10505c1.pdf

Zhang, X., & Fan, D. (2023). *Can Agricultural Digital Transformation Help Farmers Increase Income? An Empirical Study Based on Thousands of Farmers in Hubei Province*. (https://link.springer.com/article/https://doi.org/10.1007/s10668-023-03200-5)

Kum Kim is a senior specialist at KOTRA Phnom Penh, a Korean governmental organization engaged in promoting trade and investment between Korea and Cambodia. He holds a bachelor's degree in business administration from Norton University in Cambodia, and an international masters in regional integration (IMRI) from the Asia-Europe Institute, Universiti Malaya. He has undertaken numerous research projects on social development topics in Cambodia.

Agriculture Development in Lao PDR: Toward Raising Productivity, Resilience, and Sustainability

10

Latdavanh Songvilay and Khammerng Bannalath

Abstract

The agricultural sector in Lao PDR has been targeted by the government for development and to drive socio-economic growth since a majority of the country's population rely on agriculture for their livelihood and income. In recent decades, the government of Lao PDR has made significant efforts to enhance and advance the agricultural sector. Nevertheless, the agricultural sector has yet to fully utilize its potential, despite the abundant availability of natural resources. In addition, with improved infrastructural conditions for access to large markets, especially with the completion of the Lao PDR–China railway recently, it opens up a window of opportunity for the development of agricultural production in Lao PDR. However, the productivity of agriculture remains low due to various factors, including traditional farming practices that heavily depend on natural conditions, the adverse effects of climate change, the ongoing degradation of soil quality with limited adaptive capacity, and the limited utilization of technology. Additionally, existing government policies may not effectively stimulate or improve agricultural productivity. Therefore, doing research on agricultural development is crucial in order to investigate its future capacity for improving production while also ensuring adaptability to the prevailing circumstances of climate change, with the ultimate objective of attaining sustainable agriculture.

Keywords

Productivity · Resilience · Green and sustainable agriculture

Introduction

The government of Lao PDR has always given priority to the agriculture sector due to the significant participation of its population that till its abundant land. Lao PDR encompasses a total area of 236,800 square kilometers. Within this territory, agricultural land spans approximately 4.5 million hectares, constituting approximately 19 percent of the country's total land area. Agricultural land can be classified into three distinct categories in Lao PDR. Firstly, there are approximately two million hectares of flat land that is well

L. Songvilay (✉)
Lao Academy of Social and Economic Sciences, Macroeconomic Research Institute, Vientiane, Laos
e-mail: s.latdavanh@gmail.com

K. Bannalath
Ministry of Finance, Lao Statistics Bureau, Vientiane, Laos

R. Rasiah et al. (eds.), *Sustainable Agriculture and Food Security in Southeast Asia*, Sustainable Development Goals Series, https://doi.org/10.1007/978-981-95-0477-0_10

suited for the cultivation of rice and short-span plants and crops. Secondly, there are 1.8 million hectares of moderately sloped land that is suitable for the cultivation of cash/commercial crops. Lastly, there are 0.65 hectares of natural grassland areas that are conducive to animal husbandry. In addition, the country also has other flat and forested areas of three million hectares that have yet to be included in the above classifications (MPI, 2021).

Throughout the country's development, notably after Lao PDR implemented the New Market Mechanism in 1986 and transitioned from a centrally planned economy to a market-oriented one, the agriculture sector has been identified as the primary sector for socio-economic development. Therefore, the government has made great efforts to implement a series of policy guidelines and measures aimed at fostering the growth and development of this sector. Consequently, the agricultural sector has experienced significant growth. A key outcome of the development of agricultural infrastructure is a progressive reduction in dependence on traditional farming methods that rely heavily on nature. As a result, Lao PDR has achieved self-sufficiency in food production, as well as generating exports, notably since the year 2000 (MAF, 2015).[1]

Although the agricultural sector enjoyed an average annual growth in value added of 3 percent, its contribution to GDP fell from 26 percent in 2000–2010 to 16 percent in 2010–2020. The persistence of low productivity means that the sector has yet to realize its potential despite the abundance of natural resources in the country. This problem is further compounded by the utilization of low technology, which demonstrates the weak resilience to face global climate change problems. Improving agricultural productivity is essential for Lao PDR to achieve a more sustainable future.

Consequently, this chapter aims to examine the progress of the agricultural sector in Lao PDR and its capacity to enhance productivity and sustainability. The rest of the chapter is organized as follows: based on secondary data the chapter assesses the dynamics of agriculture in Lao PDR followed by government policies implemented to strengthen the sector. In doing so, the chapter uses data from administrative records and surveys gathered by organizations, such as the statistics yearbook and agricultural census.

Changing Dynamics of Agriculture

As with most economies, the contribution of agriculture to the Lao PDR economy has shown a trend fall as industry and services began to expand. This section presents the contribution of agriculture to the Lao PDR economy by value added and employment.

Value-Added

The agricultural sector has contributed significantly to the economic growth of Lao PDR, though its contribution to GDP fell from 41.6 percent in 2002 to 23.3 percent in 2010 and 17.8 percent in 2022 (Fig. 10.1). The fall in the relative share is largely a consequence of faster expansion in industry and services as the share of industry increased from 18.1 percent in 2002 to 27.2 percent in 2010 and to 34.0 percent in 2022. Meanwhile, the services share expanded in trend terms from 40.3 percent in 2002 to 49 percent in 2010, before falling slightly to 48.1 percent in 2022.

Within agriculture, agricultural cropping has remained the strongest contributor over the period 2002–2022, contributing 62.2 percent of agricultural value added in 2022 (Fig. 10.2). The fisheries sector contributed 16.1 percent followed by livestock at almost 16.0 percent, and the forestry and logging sector at 6.3 percent in 2022 fell from 17.7 percent in 2002. The significant drop in forestry and logging is due to the government actions in order to achieve its target of preserving forest coverage at 70 percent forest coverage (MPI, 2016) by issuing Prime Minister's Decree No. 15, which focuses on increasing the inspection of

[1] We acknowledge that self-sufficiency rates do not equate to the absence of starvation and poor nutrition in Lao PDR (see Kum, 2025; Rasiah, 2025).

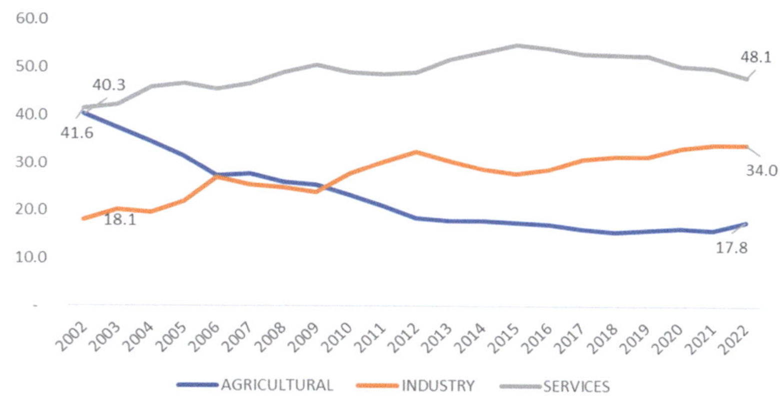

Fig. 10.1 Share of agriculture in GDP, Lao PDR, 2002–2022. *Sources* Lao Statistics Bureau, 2022

timber exploration, timber movement, and timber business activities, and the export of timber to other countries with the aim to manage the exploration of forestry and logging in a sustainable manner.

Although the agricultural sector plays an important role in the Lao economy, its growth remains slow when compared to the service and industrial sectors. Whereas the annual average growth of the agriculture sector only reaches 3 percent per annum, the service and industry sectors grew by 7 percent and 10 percent per annum over the period 2022–**2022** (LSB, 2022).

Employment and Labour Productivity

The share of agricultural employment in total employment in Lao PDR has also shown a decline in the period 2015–2020: from 72.0 percent in 2015 to 70.8 percent in 2020 (Fig. 10.3). Also, according to the population and household census report, the share of employed workers in agriculture in 1985 was 89 percent, which fell to 85 percent in 1995, 78.5 percent in 2005, 77 percent in 2015, and 70.8 percent in 2020 (Bannalath & Bounthideth, 2020).

The agriculture sector is also characterized by lower productivity and wealth distribution than the industry and service sectors due to its limited value-added and higher employment share (Table 10.1). Agricultural productivity in Lao PDR in 2005 was USD 508 per person, which increased to USD 824 per person in 2015, before falling to USD 785 in 2020. The service industry experienced a notable increase in its value over the years, recording USD 4148 per person in 2005, USD 7154 per person in 2015, and USD 7247 per person in 2020. Another sector that exhibits the highest level of productivity is the industry sector, which recorded USD 6427 per person in 2005, USD 15,855 per person in 2015, and USD 19,814 per person in 2020. All figures are in constant prices.

One major reason for the low labor productivity recorded in agriculture is because of inadequate capacity utilization. Also, the level of education attained by individuals in the farming community can serve as an indicator of their proficiency in various areas, including crop selection, cultivation techniques, farming practices, and marketing abilities. The 2019/2020 agriculture census reveals that out of a total farm population of 3.5 million in Lao PDR, approximately 27.8 percent had no formal education, while 30.8 percent have only completed primary school. Another 13.8 percent of the population has attained lower secondary education, while 6.2 percent have obtained secondary education. While 3.8 percent have achieved intermediate level of education, 1.2 percent have obtained a bachelor's degree or above. A negligible proportion of individuals, comprising less than 1 percent, received proper training (Fig. 10.4). These shortcomings have constrained the use of cutting-edge technology, which has inhibited productivity growth in the agriculture sector. According

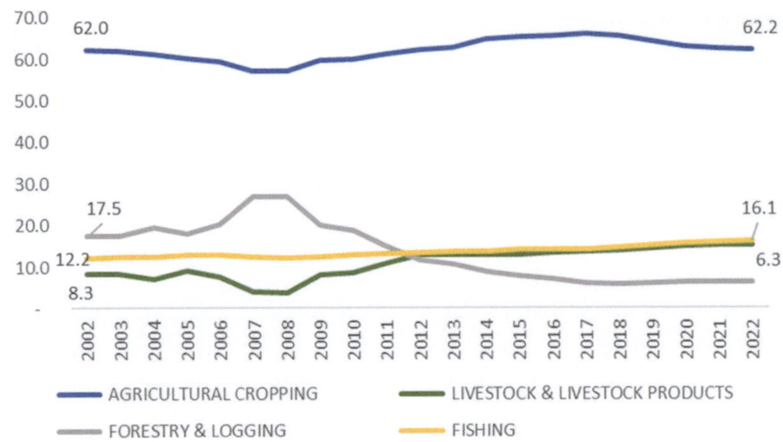

Fig. 10.2 Share of agriculture by sub-sectors, Lao PDR, 2002–2022. *Source* Lao Statistics Bureau

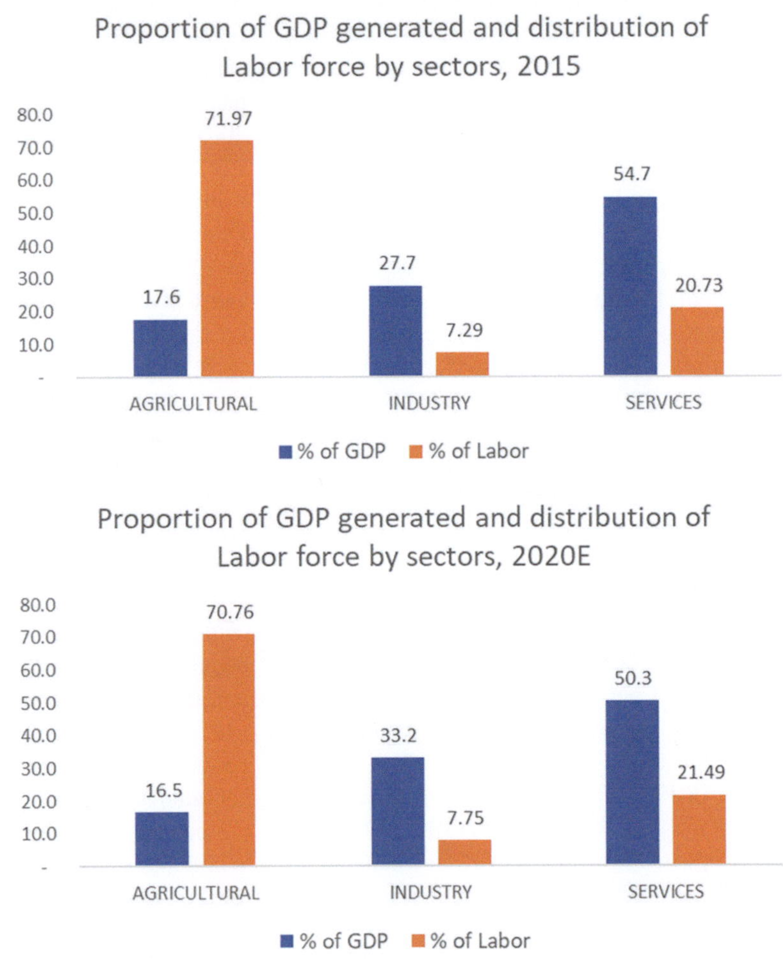

Fig. 10.3 GDP and employment by sectors, Lao PDR, 2015 and 2020. *Source* Macroeconomic Research Institute, Lao PDR

to economic census data from 2016, only 2 percent of agricultural enterprises used information technology (IT) tools like computers, internet access, and websites. While these figures have improved to reach a significant 37 percent in 2019, 63 percent of agricultural enterprises have not yet included IT in their operations (Fig. 10.5).

Table 10.1 Share of labor and labor productivity by economic sectors, Lao PDR, 2005–2020

Sector	Share of labor_Unit: percent			Labor productivity_Unit: USD		
	2005	**2015**	**2020E**	**2005**	**2015**	**2020E**
Agriculture	78.50	71.97	70.76	508.00	824.00	785.00
Industry	6.25	7.29	7.75	6427.00	15,855.00	19,814.00
Service	15.25	20.73	21.49	4148.00	7154.00	7247.00
Total	**100.00**	**100.00**	**100.00**	**1438.00**	**3233.00**	**3654.00**

Sources Macroeconomic Research Institute, Lao PDR. Note: Productivity is estimated by the GDP value added divided by the number of labors. E is estimated by the Macroeconomic Research Institute, Lao PDR

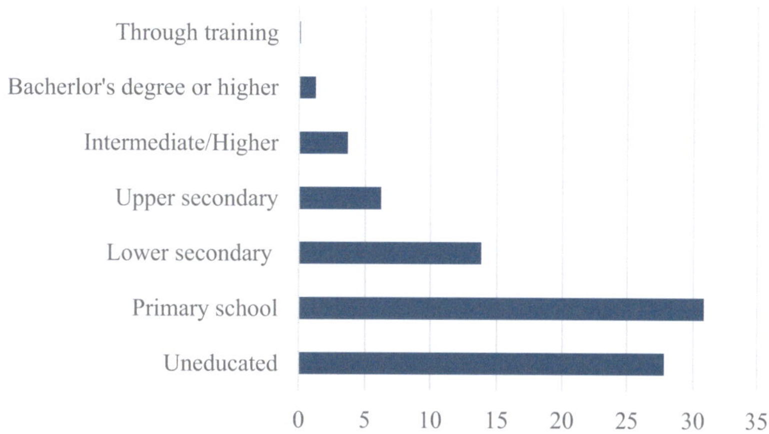

Fig. 10.4 Level of education of farm population, Lao PDR, 2019/2020. *Source* Lao Agriculture Census 2019/2020, Lao Statistics Bureau

Table 10.2 2 Labour productivity, Lao PDR and Thailand, 2020

Sector	Lao PDR (USD)	Thai (USD)	Thai/Lao PDR
Agriculture	785.00	3810.00	4.85
Industry	19,814.00	18,450.00	0.93
Natural resource sector	83,947.00	89,954.00	1.07
Manufacturing	8984.00	19,496.00	2.17
Service	7247.00	14,380.00	1.98
Total	**1438.00**	**3233.00**	**2.25**

Sources Macroeconomic Research Institute, Lao PDR

Furthermore, while comparing the productivity of the agricultural sector between Lao PDR and its neighboring nation, Thailand, it is evident that Lao PDR's agricultural labor productivity is significantly lower, which is approximately 4.9 times lower than that of Thailand (Table 10.2). Additionally, the productivity of other sectors in Lao PDR is also comparatively lower when compared to Thailand.

Tariff structures, trade balances, and agricultural labor productivity

In the period 2001 until 2018, Lao PDR has experienced trade deficits (Fig. 10.6). However, it has enjoyed a trade surplus in agriculture throughout the period 2001–2022 (Fig. 10.7). The exports' share of agricultural goods in GDP (which fell in 2010–2012) has shown a trend rise to 11 percent in 2022.

Lao PDR' trading structure has undergone limited changes during the course of its development. Based on the ASEAN Tariff Harmonized System (which categorizes trade into 22 sectors), it can be observed that the export of mining and power has constituted the highest share of Lao PDR exports, accounting for around 35 percent of the total export value over the previous decade. In contrast, agricultural products comprise approximately 20 percent of the export value. However,

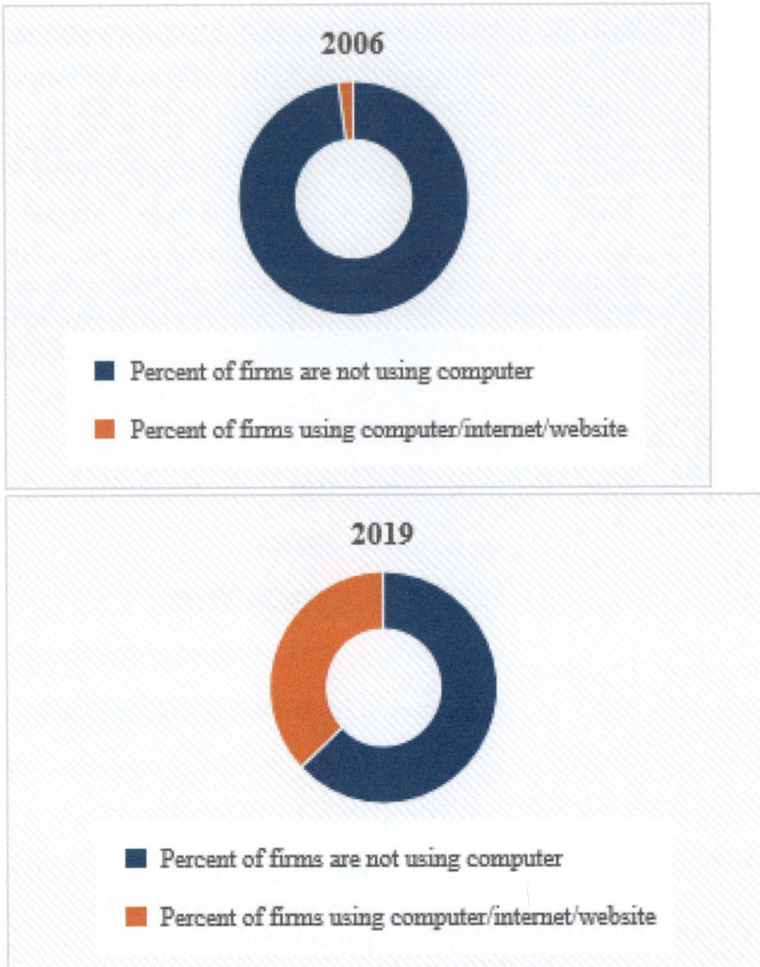

Fig. 10.5 IT utilization in business, Lao PDR, 2006 and 2009. *Sources* Lao Statistics Bureau

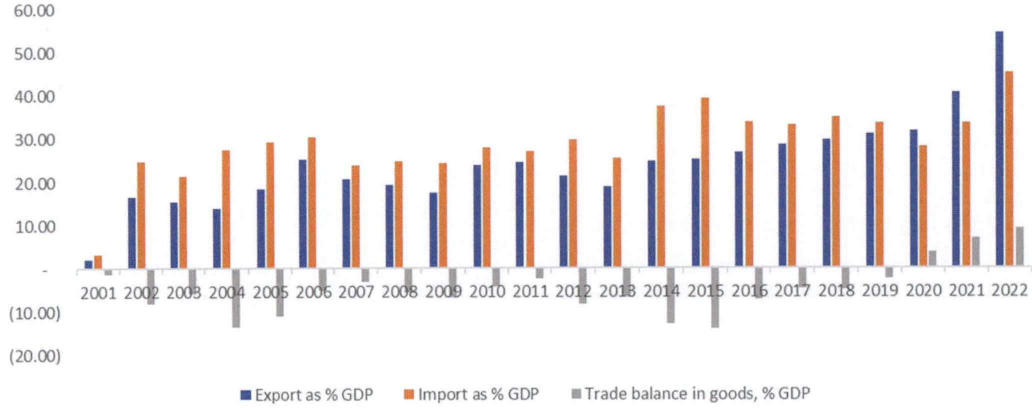

Fig. 10.6 Overall trade balance, Lao PDR, 2001–2022. *Source* Lao Statistics Bureau

Fig. 10.7 Trade balance in agriculture, Lao PDR, 2001–2022. *Source* Bank of Lao PDR; Lao Statistics Bureau

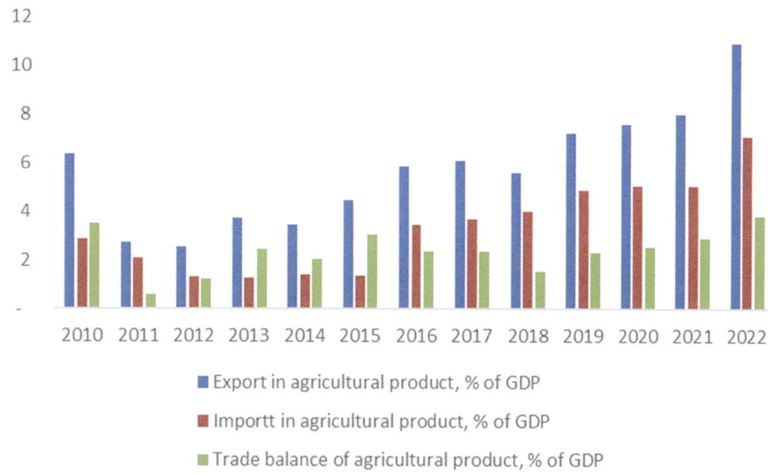

there has been a trend improvement in agricultural exports as a result of acquiring a greater export quota from China. The primary agricultural exports consist of cash crops, such as cassava, sugarcane, rice, bananas, and watermelon. Additionally, food goods, such as sugar, cassava flour, and coffee, as well as live animals, including cows and buffaloes, are also exported (see Fig. 10.8).

Despite Lao PDR experiencing a trade surplus in agricultural products, there has been a noticeable increase in the import of some products, which has risen from less than 10 percent in 2010–2015 to over 16 percent in 2016–2020 (Fig. 10.9). The primary reason for this trend is Lao PDR' limited capacity in agro-processing, which is partially attributed to its constrained economy of scale. Consequently, Lao PDR predominantly exports raw materials that are further processed into semi- and final products or food in neighboring countries, which is then sent back to Lao PDR. The primary import of agricultural products includes food products, live animals, and vegetables.

Government Policies to Promote Agriculture

The ninth Five-Year National Socio-Economic Development Plan (2021–2025) sought to enhance the agricultural sector's potential to optimize the trade advantages that Lao PDR currently enjoys, with the ultimate goal of increasing income and employment opportunities for its population. The plan has identified development guidelines, including a focus on crops and livestock with a market with links to processing production chains to meet domestic and international demand, shifting from single-crop to diversified farming multi-crop farming, promoting organic farming practices for domestic consumption, and to link them with tourism. Furthermore, in order to enhance the productivity of agriculture, the plan promotes the use of modern equipment, new techniques, and innovative approaches, which targets transitioning from traditional and fragmented farming practices to collective and new cooperative production models somewhat similar to Malaysia's FELDA (Rasiah, 2025). The plan also seeks to establish reliable agricultural infrastructure that is resilient to natural disasters (MPI, 2021).

The Lao PDR government supports and encourages both domestic and foreign investment in the agriculture sector with the goal of increasing productivity, creating job opportunities for locals, and eventually eliminating poverty in Lao PDR. The agricultural law enacted in 1998 states that the government actively supports and facilitates multi-ethnic people and various economic entities, both domestic and foreign, to undertake extensive and contemporary investments in agricultural production. The state protects the rights and interests of individuals engaged in agricultural activities through the establishment of rules, laws, provisions, and policies. These measures

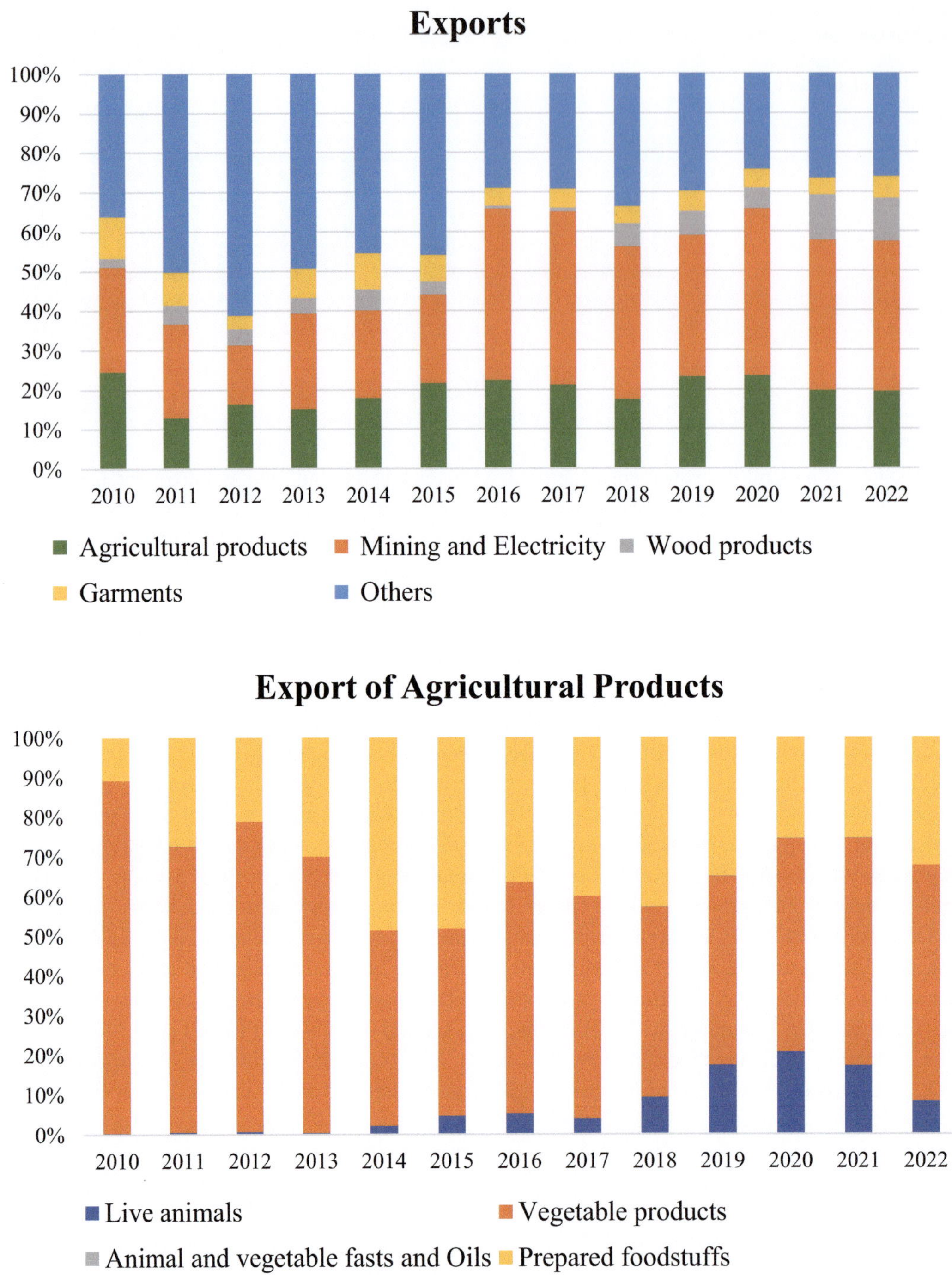

Fig. 10.8 Overall and agricultural exports, Lao PDR, 2010–2022 (%). *Source* Lao Statistics Bureau

include credit policies, policies for tariff and tax exemptions, and restrictions on the importation of non-essential agricultural products. The primary objective of these measures is to foster agricultural production in Lao PDR. According to the revised investment promotion law implemented

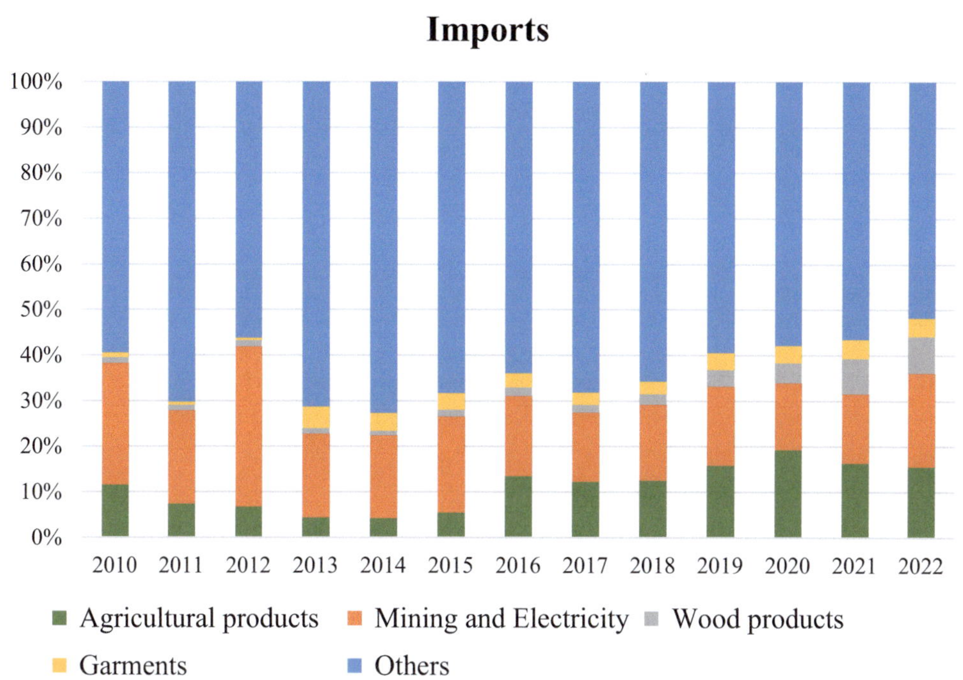

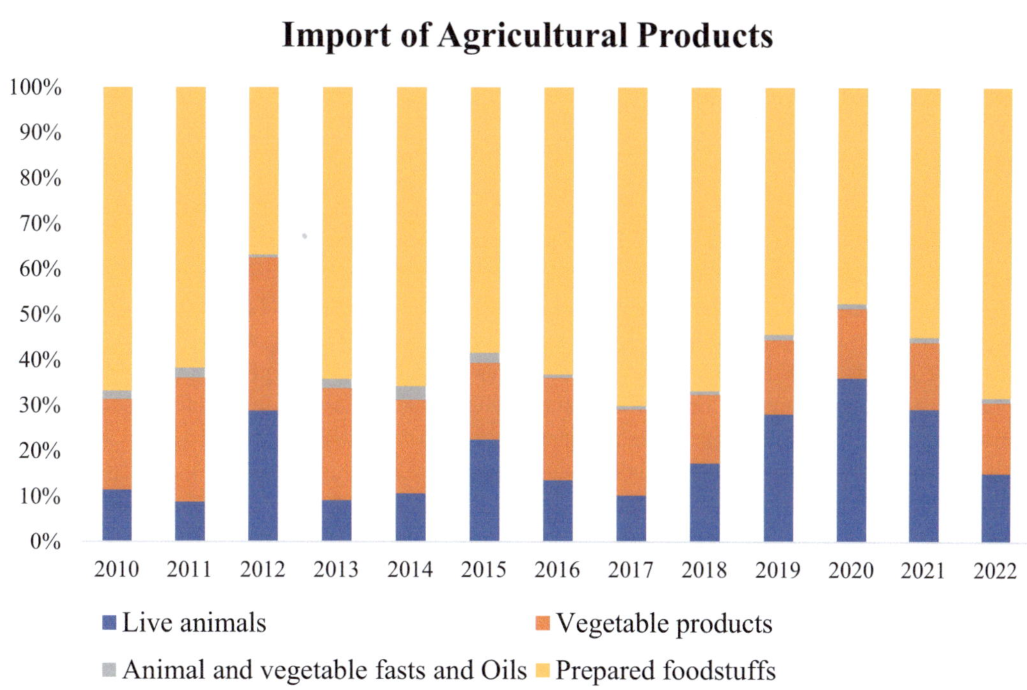

Fig. 10.9 Major imports by sectors, overall and agricultural, Lao PDR, 2010–2020 (%). *Source* Lao Statistics Bureau

Table 10.3 Investment incentives enjoyed by each zone, Lao PDR, 2016

Incentive	General Exemption		Additional Exemption (**)	
	Profit Tax	Royalty/Concession Fee	Profit Tax	Royalty/Concession Fee
Zone 1: Rural and remote areas with inadequate socio-economic infrastructure that are favorable to facilitating investments	10 years	10 years	5 years	5 years
Zone 2: Zones with socio-economic infrastructures that are partially facilitating investments	4 years	5 years	3 years	3 years
Zone 3: Special economic zones	Subject to the specific regulation of special economic zone			

(**) if invest in the clean agriculture, toxic-free agriculture, agro-processing, education and health care.

Source Investment Promotion Law (2016)

in 2016, the agriculture sector is among the nine sectors that are eligible for investment incentives from the government. Specifically, these incentives are provided for investments related to clean agriculture, toxic-free agriculture, seed production, animal breeding, industrial crops and plantations, forestry development, environment and biodiversity protection, activities promoting rural development, and poverty reduction. Furthermore, the government offers investment incentives through the implementation of zoning policies (Table 10.3). The following table presents the incentive policies for each respective zone.

Meanwhile, to simplify and streamline the investment process, the government has made joint attempts to enhance the operational mechanisms of the relevant agencies involved to start businesses. The establishment of the Investment Promotion and Management Committee at the central and local levels through the enactment of Decree no. 05/PM dated in 2018 has enabled the committee to function as a think tank for the government tasked with identifying investment applications and evaluating to approve qualified investments. Furthermore, the One Stop Service Office functions as a committee's secretary with the objective of streamlining the investment process by minimizing and eliminating unnecessary approval procedures required by several government sectors. Furthermore, the government has demonstrated its commitment to enhance the business climate by implementing Order No. 02/PM and No. 03/PM, which is specifically aimed at improving regulations and mechanisms that enable company operations in Lao PDR, with the ultimate goal of creating a more business-friendly environment in the country.

Being classified as a least developed country, Lao PDR has been granted the advantage of enjoying special tariff preferences within bilateral and multilateral trading systems. In order to promote these endeavors, the government of Lao PDR has adopted a set of guidelines and instructions aimed at facilitating investment and trade activity. These recommendations provide a vital point of reference for investors and traders to assist them with their commercial operations. For example, the Lao PDR government has achieved success in negotiating with trading partners, resulting in the export of a total of 48 plants, with 16 plants and 1 livestock product gaining access to the Chinese market, 16 plants in the Vietnamese market, and 15 plants in the Thai market. The issuance of clear guidelines pertaining to sanitary and phytosanitary (SPS) procedures has helped the successful export of these plants and livestock.

The agriculture sector in Lao PDR received promotional incentives from the government. Farmers and individuals who invest in the agriculture sector are able to access finance, including the fund for small and medium producers, and enjoy tariff and tax exemptions on the import of inputs, equipment, and technology for agriculture

Table 10.4 Key policy measures to support livestock and aquatic farming, Lao PDR, 2022

1. The government has implemented a policy to cut the cost and service charge for issuing land titles for livestock and aquatic farming by 50 percent. Additionally, a five-year exemption for land use fees has been granted specifically for livestock farming.
2. The government provides a reduced electricity tariff rate to individuals or entities that invest in animal feed production, animal medicine, animal slaughter, and farm processing operations that comply with technical standards from the Department of Livestock and Fishery.
3. Import tariff exemptions are provided for the purpose of facilitating the importation of machinery, equipment, and technology that are necessary for production and treatment services.
4. Exemption three-year land tax. According to article 29 of income Tax law no. 67/NA, dated June 18, 2019, small and medium-sized enterprises (SMEs) with an annual revenue of less than 50 million kip are exempt from income tax.

Source From Decree No. 196, date 6 July 2022 on National Policy on Promotion of livestock and aquatic farming, Lao PDR

Table 10.5 Key policy measures to support crop plantation, Lao PDR, 2023

1. Plantation area that has consistently met the good agricultural Practices (GAP) and organic agriculture (OA) standards for three consecutive years will be eligible for a land tax reduction program.
2. Offers exemptions on fees associated with obtaining GAP and OA certifications, grants a five-year exemption on property taxes, and provides access to low-interest rates.
3. Import tariff exemptions are granted for seeds, fertilizers, and insecticides that cannot be domestically produced, with the aim of promoting production, research and development, and mitigating the impact from natural disasters.
4. The government provides support and offers a lower electricity tariff for investors who engage in large-scale agricultural and smart farming practices, good agricultural practices.
5. The government identify and introduce a credit policy for farmers and investors with access to funding at a subsidized interest rate.

Source From Decree No. 226, date 10 July 2023 on National Policy on Promotion Crop Plantation, Lao PDR

production. They can also apply to seek low electricity tariffs. Two national policies, one focused on the promotion of livestock and aquatic farming and the other on the promotion of crop plantation, were introduced in 2022 and 2023, respectively. These policies have explicitly identified certain supporting measures, including selected, specific, and explicit policies from two national policies. The key measures implemented among these policies in 2022 and 2023 are shown in Tables 10.4 and Table 10.5.

The management and development of agricultural land is currently inadequate due to many factors, which include the absence of extensive surveys and consensus among regions regarding land classification and allocation; the legal framework governing agricultural land management is incomplete. Hence, the Ministry of Agriculture and Forestry is formulating the Strategy and Vision 2030–2040 on Agricultural Land Management and Development, which is expected to be launched anytime soon. The primary objective of this strategy is to strengthen the management and development of agricultural land in the new era. This is done with the intention of preserving agricultural land areas, which amounted to 4.5 million hectares in 2023, or approximately 19 percent of the total land area. Additionally, the strategy aims to optimize the utilization of agricultural land in a sustainable manner, which will contribute to the growth of the national economy and enhance the quality of life for the people of Lao PDR. Based on the draft strategy, the aggregate agricultural land area is allocated as follows: approximately two million hectares (44%) is designated for rice cultivation; one million hectares (22%) for short-life crop plantation land; 0.8 million hectares (18%) for fruit tree or long-life crop plantation land; and 0.7 million hectares (16%) for grass field intended for animal husbandry.

Changes in Commercial Crop and Extensive Farming Acreages

Given the lack of data for analyzing commercial cultivation and the breakdown of extensive and intensive farming, this chapter uses crop farming data for both consumption and commercial purposes. According to the agricultural census

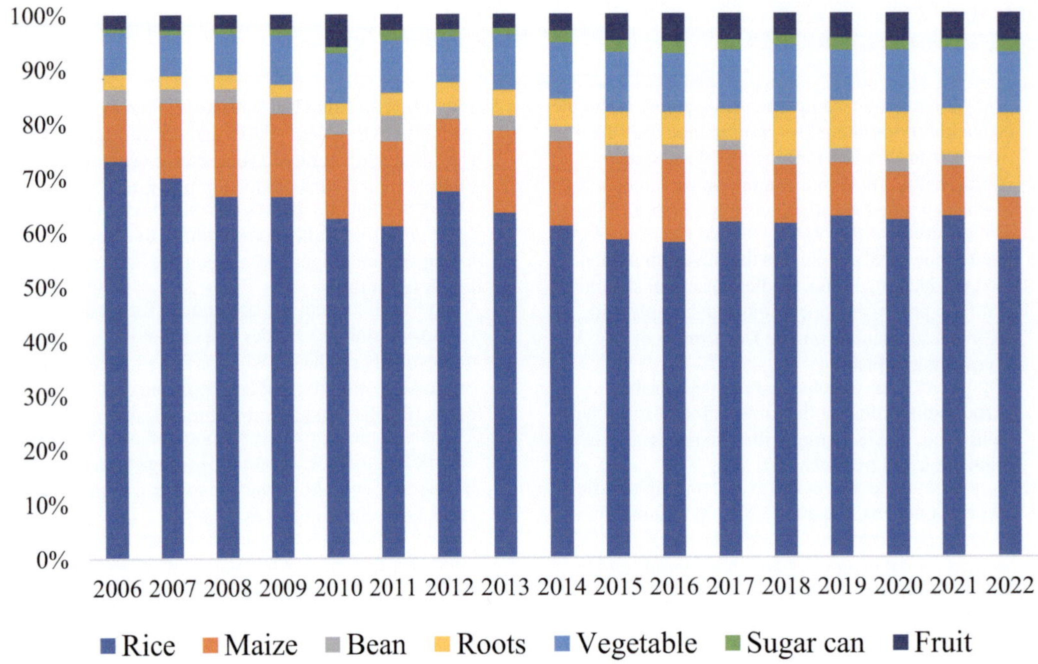

Fig. 10.10 Cultivated land, under primary cash crop, Lao PDR, 2006–2022 (%). *Source* Ministry of Agriculture and Forestry

by the Lao Statistics Bureau, there was a 4 percent decrease in the total area dedicated to agricultural purposes between 2011 and 2019 as a consequence of a shift in focus toward other sectors, such as manufacturing and services, as indicated in the LSB's Economic Census Report of 2019. Furthermore, there has been a shift in cropping patterns among household farmers and enterprises with a transition toward planting crops that are in higher demand. According to the statistics report released by the Ministry of Agriculture and Forestry in 2022, there has been a notable growth in the cultivation of key commercial crops, such as rice, maize, beans, roots, vegetables, sugarcane, coffee, and fruits. The total land area for such cultivation increased from approximately 1.0 million hectares in 2006 to 1.3 million hectares in 2010, and further to 1.5 million hectares in 2022 driven largely by increased access to the Chinese market through the export of raw materials.

Rice is the main crop, followed by maize, as well as a variety of root crops and vegetables. The primary crops, including rice, maize, and vegetables, make up an average of 63 percent, 13 percent, and 10 percent of the cultivated land, respectively (Fig. 10.10). However, there has been a decrease in the ratio of cultivated rice and maize over other crops, which can be attributed to two factors, namely one, the demand for cash crops like sugarcane, cassava, beans, and fruits from neighboring countries has grown strongly; and two, the prolonged mono-crop cultivation without sufficient soil nourishment has led to a reduction in soil moisture. In 2006, the proportion of planted rice and maize was 73 percent and 14 percent, respectively, which by 2022 fell to 58 percent and 8 percent, respectively. In contrast, the cultivation of cassava and fruits had a significant rise from 3 percent each in 2006 to 14 percent and 15 percent, respectively, in 2022 (Fig. 10.11).

Food Farming, Intensive Farming, and Food Security

According to the findings of the Lao Economic Census, it is evident that the agriculture sector has not attracted strong interest from investors

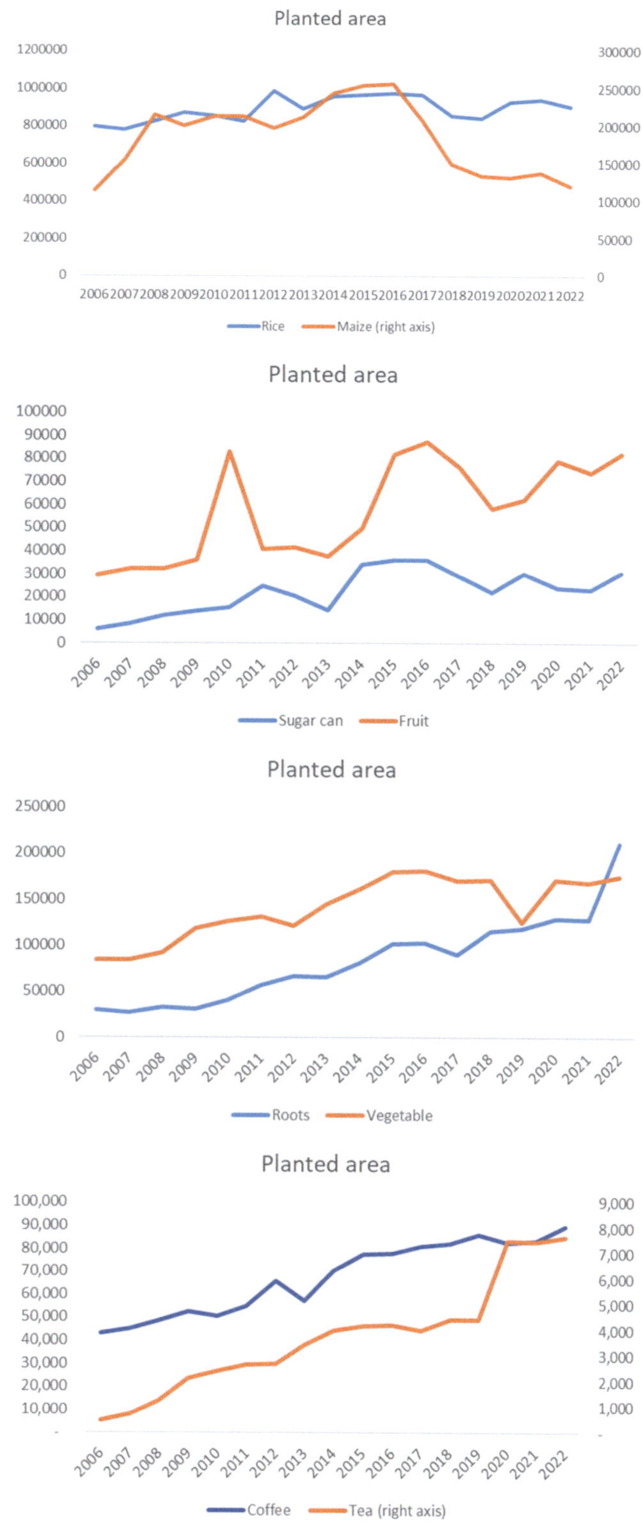

Fig. 10.11 Planted area of main crash crops, Lao PDR, 2006–2022 (ha). *Source* Ministry of Agriculture and Forestry

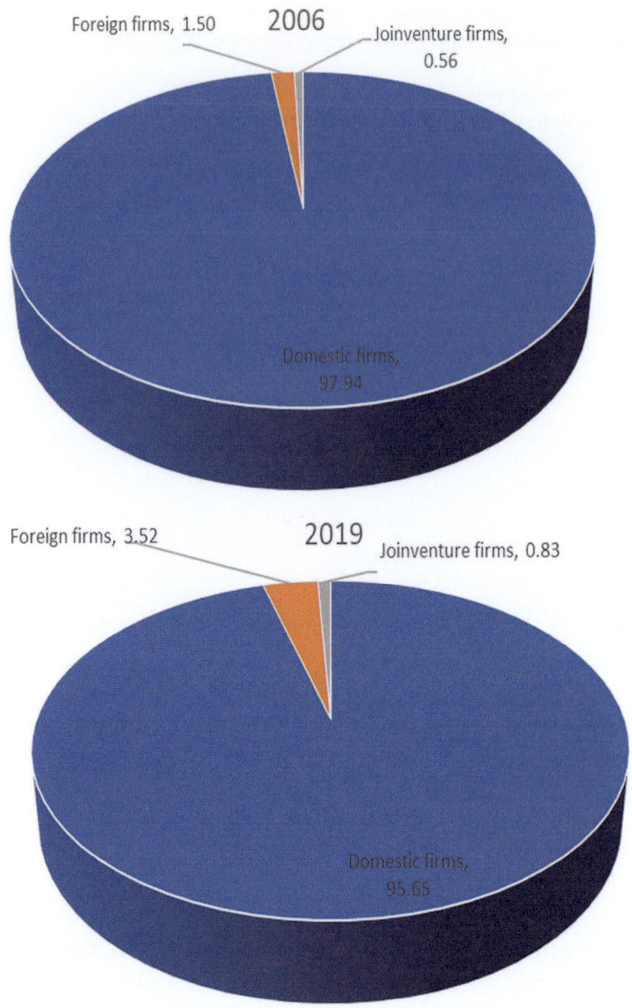

Fig. 10.12 Agricultural firms by ownership, Lao PDR, 2006 and 2019. *Source* Lao Statistics Bureau

in comparison with other sectors like services and natural resources. The number of agricultural enterprises in Lao PDR fell from 4319 firms in 2006 to 2188 firms in 2013, further decreasing to 1817 firms in 2019 (Table 10.6). Consequently, there was a decline in the proportion of agricultural enterprises, which accounted for 3.4 percent of the total number of firms in Lao PDR in 2006, decreasing to 1.8 percent in 2013 and further declining to 1.4 percent in 2019 (LSB, 2019). The observed decline in numerical figures between 2006 and 2013 can be attributed to the closure of sawmills and wood processing facilities. This outcome was a direct consequence of government measures implemented to combat illegal logging and illicit trafficking of timber in the early 2010s.

Although the number of agricultural enterprises has stagnated over the years, there is currently a growing interest among international businesses to invest in the agriculture industry. According to the findings of the economic census, there has been a notable rise in the proportion of foreign enterprises engaged in agricultural activities, with an increase from 1.5 percent in 2006 to 3.5 percent in 2019 (Fig. 10.12). Investors from China, Vietnam, and Thailand are the main foreign investors to the agricultural sector in Lao PDR.

Despite most of the labor being involved in agricultural activities, the supply of agricultural products, with the exception of vegetables, is not sufficient to meet domestic demand. The World Bank (2022) found that the supply of

Table 10.6 Number of firm and its share by economics sectors, Lao PDR, 2006–2019. *Source* Lao Statistics Bureau

Sector	No. of firms			Share (%)		
	2006	2013	2019	2006	2013	2019
Wholesale & retail trade	81,780	78,407	81,998	64.44	62.79	61.19
Manufacturing	24,331	15,573	19,714	19.17	12.47	14.71
Accommodation & food service activities	3439	14,549	14,248	2.71	11.65	10.63
Transportation & storage	3779	3509	3419	2.98	2.81	2.55
Construction	628	642	2300	0.49	0.51	1.72
Agriculture	**4319**	**2188**	**1817**	**3.40**	**1.75**	**1.36**
Electricity, gas, water supply	378	289	450	0.30	0.23	0.34
Human health & social work activities	375	421	1056	0.30	0.34	0.79
Public administration & support service	755	854	997	0.59	0.68	0.74
Arts, entertainment and recreation	1013	866	711	0.80	0.69	0.53
Mining & quarrying	297	257	545	0.23	0.21	0.41
Other services	5819	7318	6742	4.59	5.86	5.03
Total	**126,913**	**124,873**	**133,997**	**100**	**100**	**100**

rice, beef, and fish products could only meet 90 percent, 50 percent, and 59 percent, respectively, of domestic demand. Therefore, Lao PDR continues to depend on imports of these items from its neighboring countries. Furthermore, the census of 1998, 2011, and 2019 suggests that the cultivation practices of households in Lao PDR primarily serve their own needs rather than commercial demand. According to the most recent census report of 2019, 72 percent of households were engaged in the cultivation of crops in regions characterized by rainfall. Notably, the primary focus of this agricultural activity was self-consumption, as families mostly grew crops for their own consumption. Conversely, a comparatively smaller proportion of agricultural households, i.e., 18 percent focused on crop production for commercial purposes. Indeed, there has been no significant change in the share of households engaged in commercial production over the three censuses. The primary reasons for agricultural households engaging in subsistence production are attributed to their limited adoption of new technology, such as new crop varieties and high-tech equipment in their farming practices. Approximately 70 percent of agricultural households use traditional varieties that are low yielding, whereas only 30 percent utilize improved varieties. Furthermore, most households in Lao PDR lack capital assets, specifically equipment. This is evident in the agricultural sector, where a significant number of households lack the necessary equipment to support farming. For example, only 61 percent and 11 percent of agricultural households utilize two-wheel and four-wheel trucks, respectively, in their farming activities.

Climate Resilience and Sustainable Agricultural Practices

Due to its heavy reliance on natural resources and low capacity for adaptation, Lao PDR is very vulnerable to climate change effects. The World Bank (2021) identified Lao PDR as a nation highly vulnerable to projected climate change patterns, ranked at 142 out of 181 countries in the 2020 ND-GAIN Index. The government has recognized that agricultural development in Lao PDR continues to encounter difficulties owing to climate change and natural disasters, including but not limited to floods, droughts, outbreaks of pests and plant diseases, outbreaks of animal diseases, soil erosion, soil compaction, and soil degradation. Consequently, the government is taking efforts to address these challenges to mitigate their negative impact (MAF, 2015).

Following the ratification of the United Nations Framework Convention on Climate Change (UNFCCC) in 1995 and the Kyoto Protocol in 2003, the government has undertaken several measures to meet its obligations. The National Adaptation Program of Action to Climate Change (NAPA) was established in 2009 with the purpose of identifying adaptation objectives and priorities for various sectors, including agriculture, forestry, water resources, and health. These objectives and priorities are meant to direct actions and strategies in these sectors from 2010 to 2020. The Plan of Action for Disaster Risk Reduction and Management in Agriculture (2014–2016) was implemented from 2014, followed by the enactment of the Decree on Climate Change in 2019. These legislative measures serve as the foundational guidelines for sector-specific adaptation and the enhancement of resilience, particularly in the agriculture sector.

The government has acknowledged the importance of prioritizing climate change adaptation and resilience as a crucial national goal, which is aimed at mitigating the risks and negative consequences associated with climate change, including the agricultural sector. Those strategies include conducting research to study and develop crop varieties that can withstand floods and droughts, implementing climate-smart agriculture techniques, such as direct seeding and greenhouse usage, adopting water management practices that incorporate water harvesting techniques, establishing agro-climate information services, enhancing the resilience of agricultural infrastructure to support the agribusiness value chain, promoting crop diversification, and facilitating commercialization.

The Ministry of Agriculture and Forestry in 2021 has outlined its mission to strengthen its ability to address natural disaster risks, including implementing risk management and mitigation measures at the district and village levels. These measures involve ensuring an adequate reserve of agricultural inputs, machinery, medicine, and designated areas for livestock evacuation. Enhancing the study and forecasting of climate change‘s effects on agricultural farming, which is highly vulnerable to floods, droughts, outbreaks of pests, and plant diseases throughout the year, assists farmers to develop technical skills and to demonstrate how to use technology and technique to adapt agriculture to climate change. Attempts are also being made to develop and enhance ICT communication tools (mobile phone apps) with the goal of providing farmers with information on weather-related forecasts, as well as emergency action plans to treat livestock and plant diseases. The ministry in addition conducts comprehensive research and development on rice and crop seeds that are durable, resilient, and adaptable to climate change. The national policy on climate change of the Lao PDR Vision 2050 has put out the following strategies, and programs of action for the year 2030 (MTC, 2021). The following priority projects on climate change adaptation and resilience for agricultural sector in 2021–2030 have been identified (Table 10.7).

Lao PDR has favorable conditions to implement organic farming for domestic consumption and export, which include the utilization of low external input systems to facilitate a smooth transition to an organic system; Lao PDR goods are widely recognized for their minimal use of pesticides, and hilly terrain that provide favorable conditions for the cultivation of fruits and vegetables during "out of season" periods (Khamxay & Walter, 2016). The launching of the potash fertilizer project, Asia-potash International Investment (Guangzhou) Co., Ltd. in Central Lao PDR, is promising one of the world's leading potash production sites but also one likely to support both domestic consumption and exports (Belt and Road Portal, 2023). In fact, Huawei's support of smart mining technologies since 2022 may be the trigger Lao PDR needs to quicken digitization and the diffusion of industry 4.0 technologies (Huawei, 2022). According to the World Bank (2022), Lao PDR exhibits the lowest utilization of synthetic fertilizers and agrochemicals in Southeast Asia, with a recorded average of 0.03 kg/Ha over the period of 2010–2015. This notable characteristic presents a potential path for enhancing profitability through the promotion and marketing of environmentally sustainable agricultural goods, including organically farmed crops. Besides, the government in general, the Ministry

Table 10.7 Priority projects, agriculture, Lao PDR, 2021–2030

1. Conduct studies and assessments of the impacts of climate change on agriculture, and the vulnerability and risks of the agriculture sectors, sub-sectors, and value chains. Develop and expand sector-specific early warning systems, including agro-met.
2. Enhance information dissemination, knowledge transfer, advisory services, and support to increase the preparedness and adaptation capacity of farmers and agro-businesses.
3. Enhance technology transfer for adaptation and resilience to climate change impacts and specific hazards.
4. Promote research, development, and deployment of climate resilience crops, focusing on cash crops, crops for food security, commercialization, conservation, and reduction of risks related to climate change.
5. Develop and enhance the resilience of agricultural infrastructure and facilities to climate change and disasters.
6. Enhance prevention, monitoring, and early warming of livestock disease outbreaks related to climate change.
7. Develop and enhance capacity for the management of water sources, supply systems, and flood recovery system.
8. Develop and enhance capacity for the implementation of adaptation plans in the agriculture sector.
9. Promote study agribusiness related to climate change.

Source: National Strategy on Climate Change of the Lao PDR, Vision for the Year 2050, Strategy and Programmes of Action for the Year 2030 (MONRE, 2021)

of Agriculture and Forests particularly, is strongly committed to promoting organic farming.

Efforts to draft regulations and standards for organic farming, as well as to implement a regional certification program, began in the 1990s starting with the 1993 end of plant protection through chemical subsidies, the development of integrated pest management (IPM), the establishment of biofertilizer factories, the promotion of biopesticides, the creation of pesticide-free zones, and NGO initiatives emphasizing organic agriculture. Green and Sustainable Agriculture (GSA) has been given priority by the government to encourage the development of a healthy society and environment in Lao PDR. The green and sustainable agriculture framework for Lao PDR 2021–2030 was launched in 2021, which aims to provide guidance in the selection of priorities and strategic directions for the development of programs and activities in green and sustainable agricultural production. Five policies for GSA have been identified in the framework, which correspond to the seven directions for agriculture indicated in the National Green Growth Strategy in conformity with international sustainable agriculture standards: which include: (i) improving the resilience of the agricultural system; (ii) sustainable input management; (iii) value generation; (iv) multi-functional agro-ecological land use planning; and (v) capacity building and green extension (MAF, 2021).

Diffusion of Digitalization and Digitization in Agriculture

The government of Lao PDR is aware that the digitalization of infrastructure can help with modernizing agriculture to support sustainability and resilience. While the entry of Huawei to smartify potash mining in Central Lao PDR is likely to quicken digitalization in the country (Huawei, 2022), digital transformation in Lao PDR otherwise is relatively new and is still in its early stages of development, particularly in the agriculture sector, and hence, there is not much data on its effects and adaptation in Lao PDR. The ICT Strategy of the Ministry of Agriculture and Forestry, which was endorsed in 2016, has identified various areas of intervention for facilitating the use of ICT in the agriculture sector. These areas include sector-wide data collection and analysis, data management for research and innovation, support services for production and value chains, and the management of agricultural knowledge (MAF, 2016). In addition, according to the National Digital Economy Development Vision and Strategy for 2021–2040, which was introduced in 2021, the digitalization of agriculture is one of fourteen programs that are in the pipeline for development, with the hope of better management, as well as enhancing productivity, minimizing labor usage, and transforming farmers into entrepreneurs in smart agriculture.

The core mission of the government on digitalization is to focus on (i) encouraging the use of digital technology in the management and production of agricultural products from planting or feeding till marketing management with the goal of strengthening the value chains within each supply chain; (ii) advocating the transition from conventional agricultural practices to digital entrepreneurship in the agricultural sector, with a particular focus on leveraging digital technologies for trading, marketing, packaging, delivery, and customer services; (iii) enhancing and developing the learning platforms pertaining digitalization of agriculture by guaranteeing its user-friendliness and practical applicability; (iv) promoting research and innovation to shift from traditional to smart agriculture to boost productivity; and (v) promoting investment in the digitalization of agriculture. The aforementioned initiatives encompass several projects and activities, which include projects focused on the enhancement and development of smart farming through the utilization of digital technologies, such as 5G, Internet of Things (IoT), drones, and robots in agriculture. Additionally, a project has been launched centered on e-commerce and the distribution of agricultural products, specifically the One District One Product (ODOP) initiative, which aims to connect local producers directly with domestic and international consumers and another dedicated to establishing an e-marketplace for agricultural products (MTC, 2021).

Conclusion: Synthesis and Implications for Theory and Policy

The government of Lao PDR has focused its efforts on agriculture development, recognizing that the major share of its population relies on the sector, and hence, much has to be done to reduce poverty in the country. Indeed, significant efforts have been taken to enhance and promote this sector through investments into infrastructure and support for local farmers to enhance their production. Furthermore, the government has been actively promoting and facilitating local and foreign private sector investment in the agricultural sector through the implementation of various incentive schemes. Despite the various efforts made, the current level of agricultural production is not achieving its maximum growth potential. However, there is an opportunity to improve productivity and increase value added in the agricultural sector of Lao PDR to align with market demands for high-quality and safe products that are environmentally sustainable.

The agricultural sector is dominated by subsistence farming where farmers struggle to meet their own food consumption rather than having surpluses for commercial sale. However, large-scale farming remains limited and its performance is still not very productive as it is facing supply-side problems that hinder productivity and market engagement. Its problems include: (a) limited availability of climate-smart technologies, specifically improved seeds that are resilient to drought and heat; (b) inadequate utilization and ineffective use of fertilizers; (c) diminishing soil fertility in locations characterized by inappropriate farming practices; (d) constrained coverage and efficacy of agriculture extension and animal health services; (e) limited access to irrigation and drainage services. Furthermore, the absence of collective farming that is based on cooperation among farmers and inadequate rural infrastructure presents challenges to enhancing production and integrating agricultural activities for generating exports. Also, the absence of well-established commercial relationships among participants throughout the value chain presents challenges in establishing efficient linkages and accessing markets, which leads to missed opportunities for rapid expansion into larger markets. The launching of potash farming with smart technology from Huawei in Central Lao PDR offers an opportunity not only for raising agricultural productivity but also generating exports.

While there is a serious need to smartify rural small farm agriculture, Lao PDR is also in a position to synergies from agricultural trading in the medium to long term, particularly through an increase in export quotas from China under the Lao-China Economic Corridor (LCEC) program. Therefore, there should incentives given to technopreneurs seek to smartify the production of particularly food crops, including its diversification to raise farm incomes. The government on its part should directly improve the infrastructure to support smart farming and climate resilience, as well as institute the legal and health procedures to seek the compliance of sanitary and phytosanitary (SPS) standards in the farming and preparation of animals and plants. To fulfill these requirements, it will be essential to strengthen the workforce with professionals exposed to advanced training who can then man the upgraded laboratory infrastructure. In addition, it is crucial to build a comprehensive database that encompasses information on illnesses and pests, which among others things should enable efficient the monitoring of avian influenza, foot and mouth disease, and organic pesticide usage.

Fortunately, Lao PDR demonstrates considerable prospects for agricultural expansion that is focused on organic and non-chemical agricultural production. The highlands have a significant comparative advantage because of their enormous reserves of fertile soils and excellent climatic conditions. The large-scale mining of potash in central Lao PDR will only support this further. In addition to essential crops, such as rice, maize, and cassava, hilly terrain and tropical climate is also excellent for the production of fruits and vegetables that offers higher prices in the export market. Such initiatives should take advantage of the strong power supply projects being developed in the country. The existing specialization in non-processed farm items should be transformed into processing within Lao PDR to support higher value added to be recorded in the country.

References

Bannalath, K., & Bounthideth, S. (2020). *Labor projection till 2030*. Macroeconomic Research Institute.

Belt and Road Portal. (2023). Potash fertilizer project of Asia-potash International listed in Belt and Road case study. Retrieved from https://eng.yidaiyilu.gov.cn/p/0H1SO4D8.html

Huawei. (2022). Laos' First Smart Mine. Retrieved from https://www.huawei.com/en/media-center/multimedia/photos/laos-mine-call-300m-down

Khamxay, S., & Walter, R. (2016). Organic agriculture production and Marketing in Lao PDR.

LSB. (2019). *Economic Census Report*. Ministry of Planning and Investment.

LSB. (2022). Statistics Yearbook. Ministry of Planning and Investment.

MAF. (2015). *Agriculture Development Strategy to 2025 and Vision to the Year 2030*. Ministry of Agriculture and Forestry. Retrieved September 10, 2023, from https://maf.gov.la/wp-content/uploads/2016/01/MDS-2025-and-Vision-to-2030-Eng.pdf

MAF. (2016). *ICT Strategy for Ministry of Agriculture and Forestry 2025*. Ministry of Agriculture and Forestry.

MONRE. (2021). *National Strategy on Climate Change of the Lao PDR, Vision for the Year 2050, Strategy and Programs of Action for the Year 2030*. Ministry of Natural Resource and Environment.

MPI. (2016). *The 8th Socio-Economic Development Plan*. Ministry of Planning and Investment.

MPI. (2021). *9th Five Year National Socio-Economic Development Plan (2021–2025)*. Ministry of Planning and Investment. Retrieved September 14, 2023, from https://data.laos.opendevelopmentmekong.net/en/library_record/9th-five-year-national-socio-economic-development-plan-2021-2025

MTC. (2021). *National Digital Economy Development Vision and Strategy for 2021–2040*. Ministry of Technology and Communications.

Rasiah, R. (2025). Building sustainability and food security in Southeast Asia, Rasiah, R., Venus, A. & Annizah, N.I. (eds), *Sustainable Agriculture in Southeast Asia and Food Security in Southeast Asia: Towards Climate Resilience and Digitalization*, Singapore: Springer Nature.

Sustainable Agriculture in Myanmar: Towards Climate Resilience and Food Security

11

Naw Shareen

Abstract

Myanmar's agricultural industry, which contributes to 21% of exports, employs 56% of the labour force and generates 32% of the national GDP, has been negatively impacted by the COVID-19 pandemic and the 2021 military takeover, resulting in numerous challenges for its agricultural sector. This study investigates the correlation between food security and climate-resilient agriculture, as well as sustainable agriculture within the Myanmar context. The research delves into the current status of food security and the policy framework surrounding the implementation of Climate-Smart Agriculture (CSA), which aims to address climate change issues in agriculture. This chapter presents an analytical review of Myanmar's sustainable agriculture, studying problem trees and objective trees of the current agriculture contexts. In order to enhance agricultural productivity and mitigate the impacts of climate change, this chapter examines the underlying causes of food insecurity in the country. Through literature reviews, and content analysis on policies, it explores various approaches aimed at alleviating food insecurity, poverty and climate resilience agriculture in Myanmar. It evaluates the effectiveness of these approaches and provides recommendations for the future implementation of sustainable agriculture initiatives involving relevant stakeholders, organizations and research institutions.

Keywords

Sustainable agriculture · Food security · Climate resilience · IR4.0 technologies · Poverty · Myanmar

Introduction: Agriculture and Climate Change Vulnerabilities in Myanmar

Agriculture is a major economic activity of Myanmar; which contributed to 20% of the National GDP, employing 46% of the labour force, and 3% of the country's agricultural raw material exports in 2022 (World Bank, 2023e). The nation's food security depends on agriculture, which is also a key source of national and household income. Overall, 12.8 out of 67.6 million hectares of land is cultivated in Myanmar where 70% of the population lives in rural areas and agriculture is their main source of income (Eurocham,, 2020c).

N. Shareen (✉)
Kasetsart University, Bangkok, Thailand
e-mail: naw.s@ku.th

R. Rasiah et al. (eds.), *Sustainable Agriculture and Food Security in Southeast Asia*, Sustainable Development Goals Series, https://doi.org/10.1007/978-981-95-0477-0_11

Myanmar ranked as the second most affected country by extreme weather events over the last 20 years (Eckstein et al., 2021). The negative consequences of climate change contribute to food insecurity and flood danger in the country (Boori et al., 2017); the socio-economic status of farmers has been affected by climate change impacts and unpredictable natural disasters, such as droughts, floods, land degradation, desertification, deforestation and major landslides. Extreme temperature variations have become common as natural disasters are occurring more frequently each year with human activities worsening climate change in the country. This grim situation has had a negative effect on food production.

This chapter seeks to examine the changing contribution of agriculture in Myanmar's economy, the agricultural policies that the government has pursued, and their implications for climate resilience. Also discussed are the problems arising from food security issues and climate change in Myanmar. Consequently, this chapter also Myanmar's agriculture in the context of productivity, food insecurity and sustainability. Following Rasiah (2025), this chapter takes on the arguments on sustainable agriculture with a focus on Myanmar towards building food security that is resilient to climate change.

Contribution of Agriculture to the National Economy

Myanmar is generally ideal for agricultural activities due to the availability of fertile lands and freshwater supply. Agriculture is the third largest contributor to the GDP of the country; the first and second are the industry (41.1%) and services sector (38.6%), respectively (Fig. 11.1). Its contribution to GDP amounted to around US$12.1 billion US$ GDP in 2022 (World Bank, 2023d)), which has contracted significantly since 2001, falling from 37.4% in 2010 to 20.3% in 2020 (Fig. 11.2).

About 25 million tonnes of over 40 different types of crops are produced from the cultivation of 29 million acres annually, which amounts to 18% of the total cultivatable land in Myanmar.

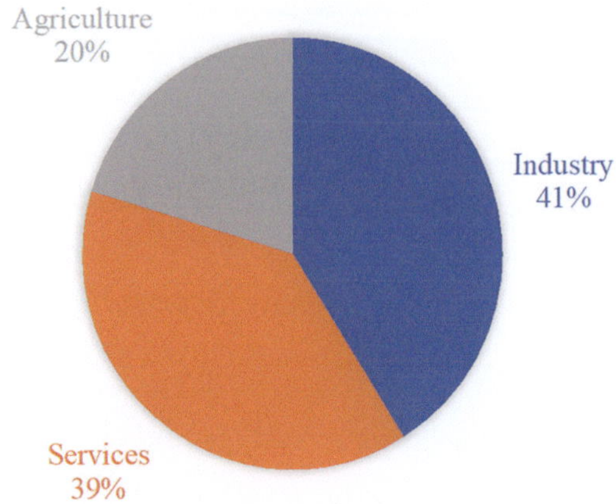

Fig. 11.1 GDP composition by sectors, Myanmar, 2022. *Source* World Bank (2023)

About 72% of agricultural output comes from crop production, which is derived from rice, bean pulses, and oil seed.

There are four agroecological zones in Myanmar: the hilly and mountainous area, the central dry zone, the delta area and the coastal area (see Fig. 11.3). In the eastern and western parts of the country, which are hilly and mountainous, farmers cultivate a variety of rain-fed tree crops and horticultural products in addition to rice, maize and pulses. In the central dry zone, rice is cultivated from monsoon rain or through the irrigation system, as well as pulses, sesame and oilseed crops. The delta areas are primarily devoted to lowland rice cultivation; major cash crops such as oil palm, rubber and fruits, and aquaculture production characterize the coastal areas (Eurocham, 2020a, 2020b, 2020c; Flanders, 2020; Lambrecht et al., 2022).

Rice (which is also known as paddy before it is husked) is the leading crop of Myanmar and farmers typically grow rice and maize during the monsoon season and beans and pulses during the dry season. However, in the temperate highlands, they frequently strive for a second harvest of rice and maize if there is enough water left over after the rains. Other types of crops grown are pulses; sugarcane, beans, sesame, groundnuts, and timber are among the other principal crops. Chickpeas, black gram, and green gram are the most often cultivated pulses and beans in Myanmar (Eurocham, 2020a, 2020b, 2020c).

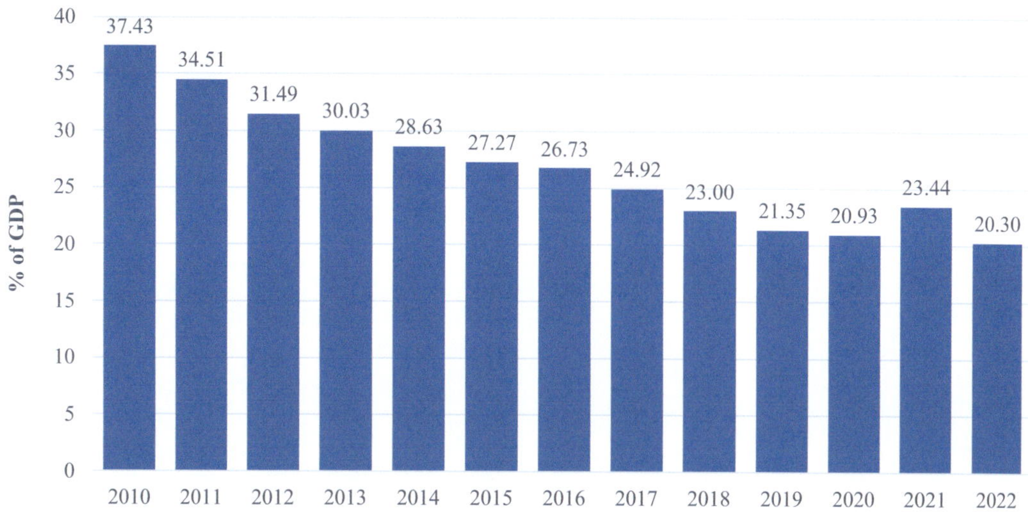

Fig. 11.2 Value-added share of agriculture in GDP, Myanmar, 2010–2022. *Source* The Global Economy (2023)

Rice production value rose to 24.9 million tons in 2t, followed by sugar cane at 11.7 million tons, both of which rose significantly from 2012 (Statista, 2023c). However, Myanmar's dry season rice production decreased by 7.5% from 2021 to 2022, with the largest plot reporting an average yield of 1,163 kgs per acre or 3.1 tons per hectare in 2022 (MAPSA, 2023e). The production volume of main crops in Myanmar is illustrated in Fig. 11.4. Rice is grown primarily for home consumption, with a marketed surplus of 21% (Fang & Belton, 2020a, 2020b, 2020c)

Myanmar has a comparative advantage in pulse production, derived from its extensive cultivatable land, diverse agroecological zones, abundant water and labour resources, which led to a boom in the pulse sector upon transitioning to a market economy (Boughton et al., 2015). More than 20 different kinds of pulses are produced in Myanmar, such as lima beans, kidney beans, pigeon peas, chickpeas, mung beans (green grams), black grams and cowpeas (Maung, 2020). Myanmar is the world's second-largest exporter of beans and pulses, behind Canada, and is the biggest exporter in Asia. However, the Indian Government imposed import restrictions on black beans, pigeon peas and mung beans from Myanmar, which affected exports as India is the biggest market for Myanmar, which has been importing beans and pulses from Myanmar for over 30 years (Eurocham, 2020a, 2020b, 2020c). Figure 11.5 shows the annual cultivation of black gram, green gram, pigeon pea, chickpea and other pulses.

Maize is the most important cash crop of the Shan state, accounting for 54% of the value of crops sold by households in that state, which is located in the Eastern part of Myanmar (Fang & Belton, 2020a, 2020b, 2020c). In Myanmar, land-limited households are more likely to produce paddy than maize due to limited land. Small landholdings lack suitable land for cultivation, and land-scarce households prioritize subsistence paddy production. Large landowners tend to allocate land to cultivate maize, but most households endowed with adequate land cultivate both crops. Most corn in Myanmar is cultivated from hybrid seeds imported from Thailand (Maung, 2020).

Many factors contribute to low agricultural productivity, such as a lack of high-quality public goods, growing wages in non-agricultural economic sectors, variations in working capital costs and shifting land prices as a result of demands from the industrial and urban development sectors, which affect agricultural productivity and lead to a decline in the agriculture workforce (Fang & Belton, 2020a, 2020b, 2020c).

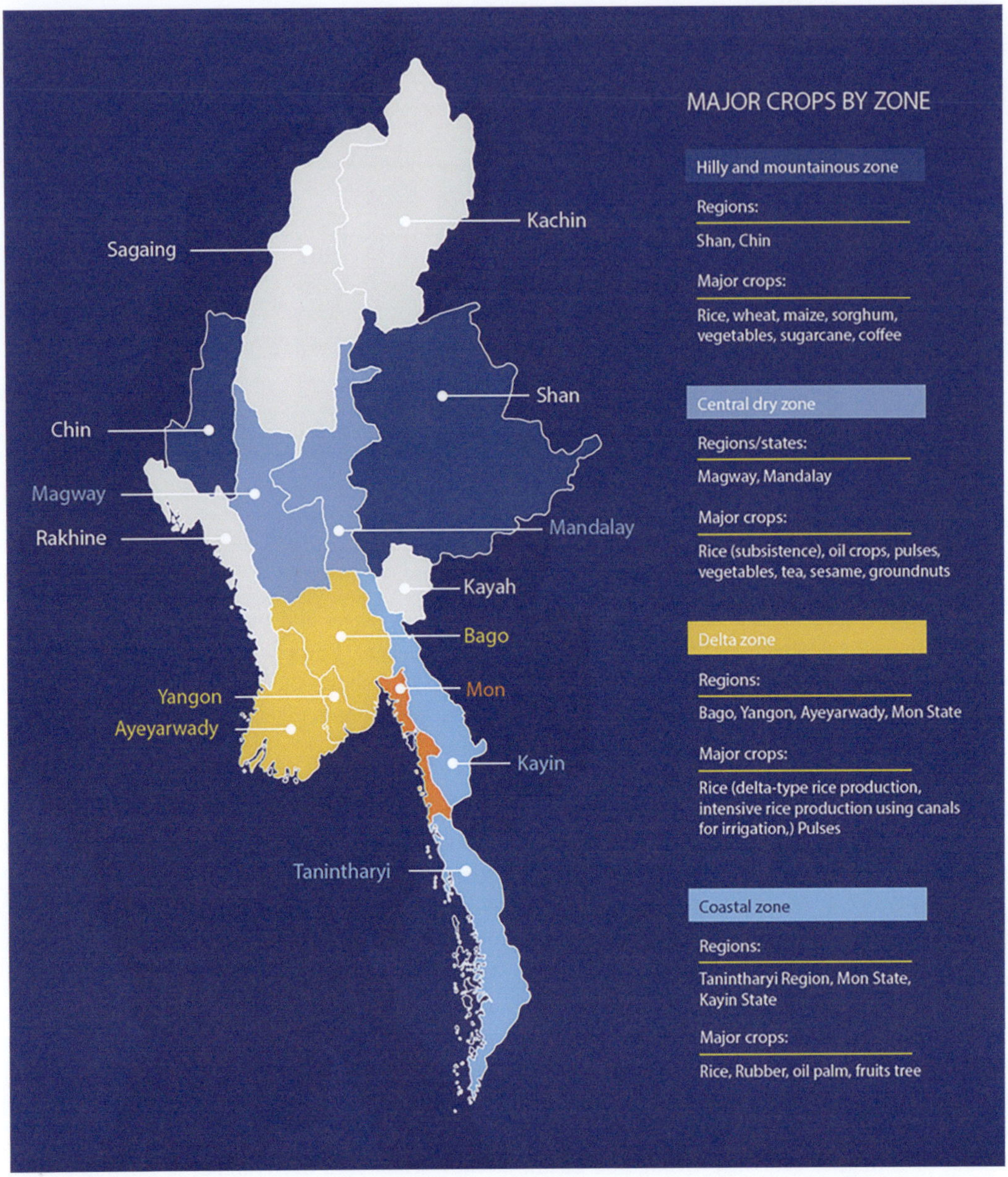

Fig. 11.3 Agroecological zones and major crops, Myanmar. *Source*: Eurocham (2020a, 2020b, 2020c)

Myanmar has established trade partnerships with many countries globally, including China, Thailand, India, Japan and Singapore. China has been Myanmar's main trading partner with raw materials being the prime export, including minerals, lumber and agricultural items. Singapore and Thailand are the next important markets for Myanmar (see Fig. 11.6). Due to its advantageous position, Myanmar is a major hub for regional commerce, and its economic connections reach beyond Asia to the Middle East and Europe. The main export items of Myanmar are gas, oil, wood, rice, beans, seafood, copper, clothing, and precious and semi-precious stones. Myanmar imports fuels and lubricants, consumer goods, semi-finished products, machines and equipment, metal, oil, cement, fertilizers and medicines (MOC, 2023).

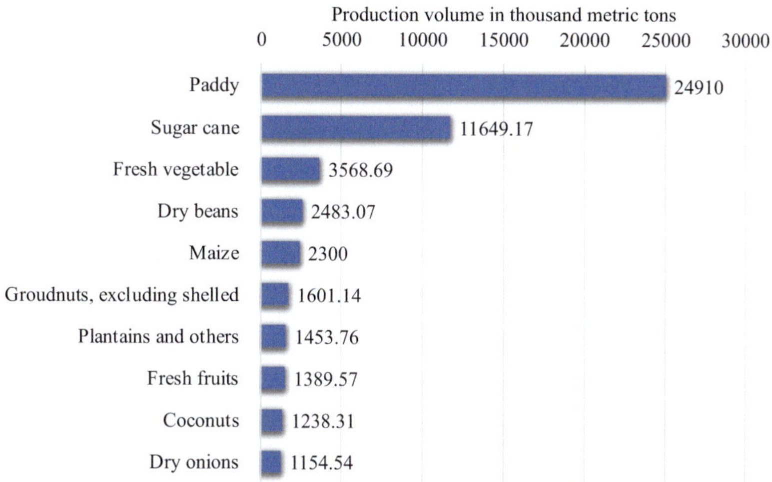

Fig. 11.4 Production of main crops, Myanmar, 2021 ('000 metric tons). *Source* Statista (2023c)

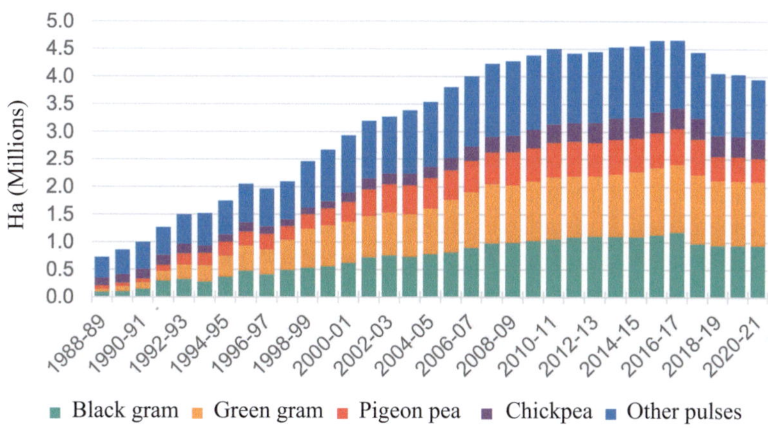

Fig. 11.5 Acreage of major pulses, Myanmar, 1988–2021. *Source* Department of Planning, Ministry of Agriculture, Livestock and Irrigation (2021)

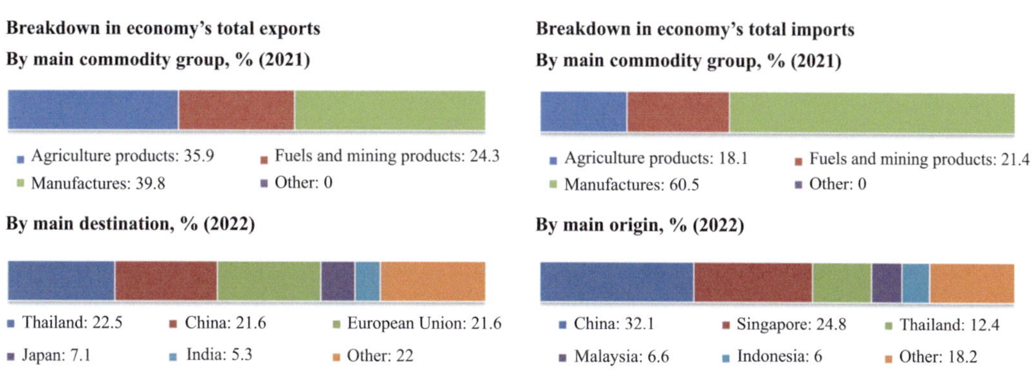

Fig. 11.6 Top five export and import partners of Myanmar, 2021–2022. 2023 *Source*: World Trade Organization ()

Food Security

Although Myanmar has rich and productive cultivatable land and natural resources, barriers threaten to restrict the achievement of food self-sufficiency targets because of political instability. With 13.2 million people already insecure, the country faces challenges, such as violence, economic crisis, climate-induced shocks, population displacement and COVID-19 (FAO, 2022a, 2022b, 2022c). Agriculture, often highlighted as a crucial sector for resilience and food security in the country, is now in danger owing to the political crisis in Myanmar. In 2022, 4% of the households experienced moderate to severe hunger, with the highest levels encountered by Kayah (9.8%), Chin (9.9%) and Tanintharyi (12.6%) (MAPSA, 2023a, 2023b, 2023c, 2023d, 2023e, 2023f, 2023g, 2023h, 2023i).

The political chaos in the country has left key aspects of policymaking removed from emphasis. Five months after the coup, farmers in Myanmar face significant challenges in agriculture, including increased fertilizer and fuel prices, and limited access to credit. A civil war can have direct and indirect impacts on food, economic insecurity and environmental destruction. In some areas, arable land areas have been seriously affected, leading to a decline in food production and rising unemployment rates, resulting in starvation to threatening the regional food supply system.

Almost one in four people in Myanmar has already experienced food insecurity as a result of the country's fast-worsening food security problems (FAO, 2022a, 2022b, 2022c). The percentage of people facing severe food insecurity has been increasing from 1.9% in 2019 to 5% in 2021 (Fig. 11.7). The persistence of the moderate to severe food insecurity population was reported at 22.2% in 2019, 25.5% in 2020 and 29.3% in 2021. Consequently, food security in Myanmar faces challenges in the provision of adequate, safe, diversified and nutrient-rich foods for healthy diets (Kyi, 2018). Addressing these challenges requires revisions in the food system and the country needs to become more nutrition sensitive. Climate change, extreme weather events and food security are interconnected, thereby highlighting the need for a more sustainable food system.

Government Policy

Prioritising to promote sustainable agriculture, food security and reducing rural poverty, the Ministry of Agriculture and Irrigation (MOAI) and the Ministry of Livestock, Fisheries and Rural Development (MLFRD) have established policies, goals and strategic plans for the development of the agriculture sector, which directly and indirectly, objectives and strategies encompassing mitigation and adaptation; actions on climate change in the country closely linked to national food security and nutrition policies (Hom et al., 2015).

Myanmar's Agriculture Policy (2016), which is also known as the second five-year short-term plan, was developed in 2016 to increase access to food, nutrition and safety while enabling smallholder farmers to increase their incomes through higher productivity and diversified production in response to market demand (Myanmar National Portal, 2023). The National Strategy on Poverty Alleviation and Rural Development (NSPARD) was established in 2011 to focus on the growth of agro-based industries, the equal distribution of resources across states and regions, the encouragement of domestic and international investment, the successful implementation of people-centred development and the alleviation of poverty. Land reforms, promoting financial access and creating work possibilities were some of the crucial reform initiatives of NSPARD (Hom et al., 2015). The Agriculture Development Strategy (ADS) aims to increase 50% of land and labour productivity by 2023 (MOALI, 2018a). These policies both explicitly and implicitly acknowledge the significance of combating climate change, contributing as the key elements of the Myanmar Sustainable Development Plan (2018–2030) (Barbon et al., 2022).

The climate change action plans have been incorporated into the pertinent policies, planning and budgeting processes of the agriculture, fishery and livestock sectors to promote

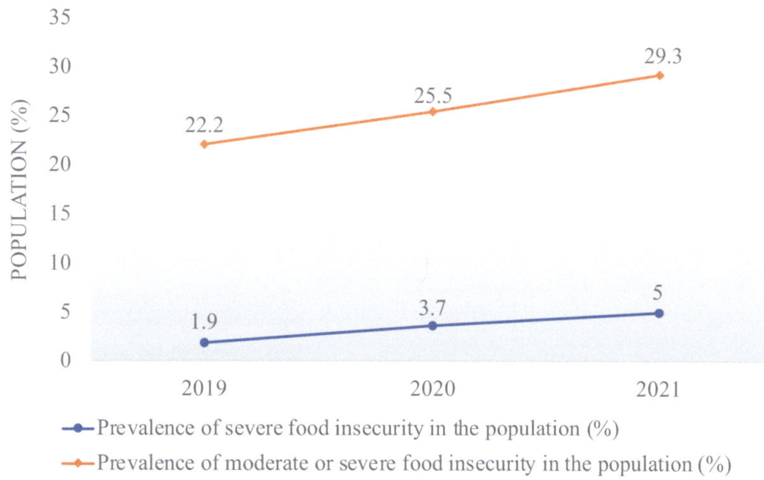

Fig. 11.7 Prevalence of severe and moderate food insecurity, Myanmar, 2019–2021. *Source* World Bank Open Data (2023)

climate-resilient productivity and climate-smart actions. Climate-Smart Agriculture (CSA) has been implemented for climate change adaptation and resilience building, greenhouse gas emission reduction and/or elimination, and sustainable agricultural output and income (FAO, 2020). Through livelihood diversification, the CSA, climate and weather monitoring, and knowledge-sharing interventions, the Ministry of Agriculture, Livestock, and Irrigation (MOALI) has collaborated with multiple stakeholders and organizations to advance climate-smart adaptation practices in the agriculture, fisheries and livestock sectors (MONREC, 2021). With support from local and international funders, many sectors have embraced adaptation techniques and technology that are both environmentally responsible and climate resilient. However, Myanmar's sustainable food system remains challenged by insufficient adaptation of agriculture innovations along the food supply chain and in the non-modernized agro-industry. Policymakers should prioritize the diversification of crops, promotion of quality management systems, improvement of value-adding procedures and enhancing logistics for efficient distribution (Kyi, 2022).

MOALI is the main agency that foresees the different aspects of agricultural development and is comprised of different departments to provide agricultural extensions, services and research and to promote sustainable agriculture in Myanmar. State-owned firms, such as the Myanmar Agriculture Development Bank (MADB), Myanmar Economic Bank (MEB) and the Cooperative Bank (CB), provide financial assistance to farmers, including smallholders. Research centres like the Yezin Agricultural University, under the management of MOALI, also contribute to advancing agricultural knowledge, research and technology. Private sector entities, such as farmer organizations and cooperatives, are important for building productive partnerships, commercialization and sharing farming techniques with the private sector and agribusinesses (MOALI, 2018b). Non-governmental organizations (NGOs) and foreign development agencies play crucial roles in implementing initiatives centred on rural development, capacity building and technology transfer to farmers. Agricultural development activities, which have been implemented in 8,313 village tracts and towns in Myanmar, account for 58% of the total village tracts by 39 organizations, such as International Non-governmental Organizations (INGO), National Non-governmental Organizations (NNGO), UN and International Organizations (IO), Community-Based Organizations (CBO) and Community Society Organizations (CSO) (MIMU, 2023). Figure 11.8 shows the types of agriculture development activities conducted by these organizations.

Food security depends on a person's capacity to obtain a stable supply of appropriate that is sufficiently nutritious to ensure (World Bank,

Org. Type	INGO	NNGO	UN & IO	Red Cross	CBO/ CSO/ Other	Total
All Sectors	91	64	21	9	8	193
Agriculture	27	7	4	0	1	39
CBO – Community-based Organization, CSO – Civil Society Organization						

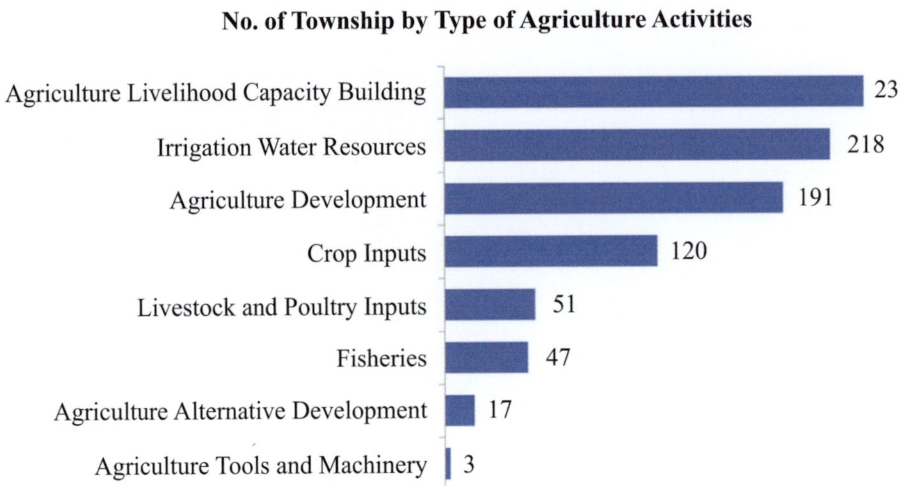

Fig. 11.8 Organizations and types of agriculture activities by townships, Myanmar. *Note*: August 2023 figures. *Source* Myanmar Information Management Unit (2023)

2007). Food availability, utilization, accessibility and stability are the four fundamental pillars that support food security in a nation. Food availability requires a sufficient amount of a country's food production and supplies, including imported agricultural commodities. If domestic output stagnates or drops, food insecurity becomes a greater issue. Because of rising food prices and declining consumer purchasing power, food and nutrition insecurity is increasingly becoming a serious problem in Myanmar. The devaluation of the Myanmar kyat and rising global fertilizer and chemical input prices owing to supply chain disruptions caused by the trade sanctions have affected agricultural input costs. Furthermore, poor infrastructure and rising transportation costs make food even more expensive in remote locations, while foreign exchange availability often hinders imports (World Bank, 2007).

Food utilization to achieve a state of nutritional well-being is important when all physiological demands are satisfied by a sufficient diet, clean water, sanitation and medical attention (Kyi, 2018). Malnutrition negatively impacts economic growth and productivity; food needs to be easily accessible, of the right kind and variety, prepared correctly, and consumed by healthy people. Lack of dietary diversity and poor food quality leads to micronutrient deficiencies or hidden hunger. Agriculture that combines conventional crop-breeding methods with biotechnology can improve food quality to combat hidden hunger. Biofortification (which refers to a combination of conventional crop-breeding methods and biotechnology) can enhance staple crop varieties and add more vitamins and minerals. Using new crop types can improve diet diversity to reinforce the relationship between food security and nutrition security (World Bank, 2007).

Access to food is crucial for individuals to sufficient allocation of food for a healthy diet. Access to food is influenced by the legal, political, economic and social structures of the society; entitlements are the collection of all commodity

Fig. 11.9 Average annual natural hazard occurrence, Myanmar, 1980–2020. *Source*: World Bank Climate Change Knowledge Portal (World Bank, 2023a)

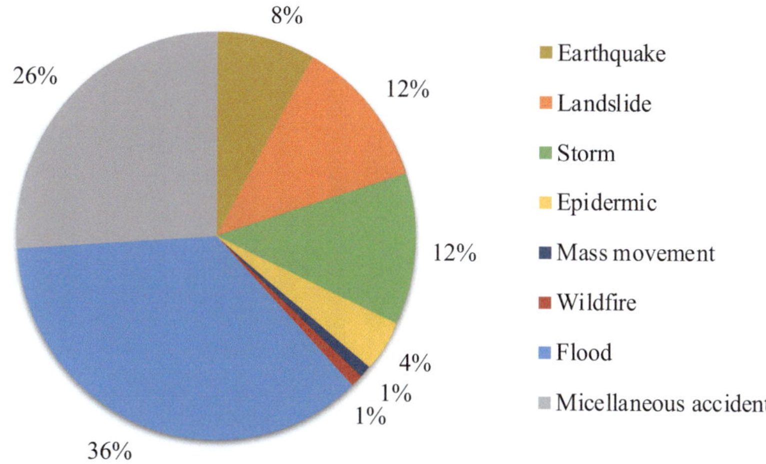

bundles over which individuals can access them (including traditional rights, such as access to shared resources) (Kyi, 2018). However, food scarcity is a greater issue for malnourished individuals, who are often net buyers of food rather than producers. Agriculture's ability to provide income for the underprivileged, particularly women, is more important than increasing local food supply availability (World Bank, 2007).

Access and availability components of food security, also known as stability, are essential for maintaining food security (Kyi, 2018). While equitable distribution rather than actual production is widely considered the cause of starvation (Rasiah, 2025), governments have the responsibility to ensure that food production generally meets the nutritional needs of the population.

According to the World Bank climate change study, as illustrated in Fig. 11.9, the average annual natural hazards that have occurred in Myanmar are quite significant: e.g. floods (36%), storms (12%), landslides (12%), earthquakes (8%), epidemics (4%), wildfire (1%), mass soil movement (1%) and miscellaneous accidents (26%) in the period of 1980–2020 (World Bank, 2023a).

According to another study on the dry zone (Swe et al., 2015), which contributes 35% of the country's cultivable lands, there have already been climate change impacts on agricultural production with about 95% of the respondents reporting that they have suffered from decreased yield, increased pest infestation and water scarcity. The indirect impacts have also been documented, including food insecurity, reduced incomes and changes in crop varieties. Other negative impacts include crop damage from heavy rain, low seed germination due to high temperatures, shortages in animal feed, increased water demand and alterations in the growing season due to irregular rainfall. As agriculture also considerably contributes to greenhouse gas (GHG) emissions, the long-term effects of climate change will have considerable impacts on agriculture and food security, requiring adequate adaptation of agricultural systems over time (Eitzinger et al., 2018; Hom et al., 2015). Rising temperatures and altered precipitation patterns lead to the risk of crop failures in the short term and falling production in the long term.

Methodology and Data

This study assesses Myanmar's food security, which focuses on climate change adaptation and sustainable agriculture. The relationship between agricultural food production and the current progress of food security is studied by using the content analysis method and online data sources to get a clear picture of the situation.

Secondary data sources and open data from the World Bank, Food and Agriculture Organization (FAO), Myanmar Information Management Unit (MIMU), International Food Policy Research

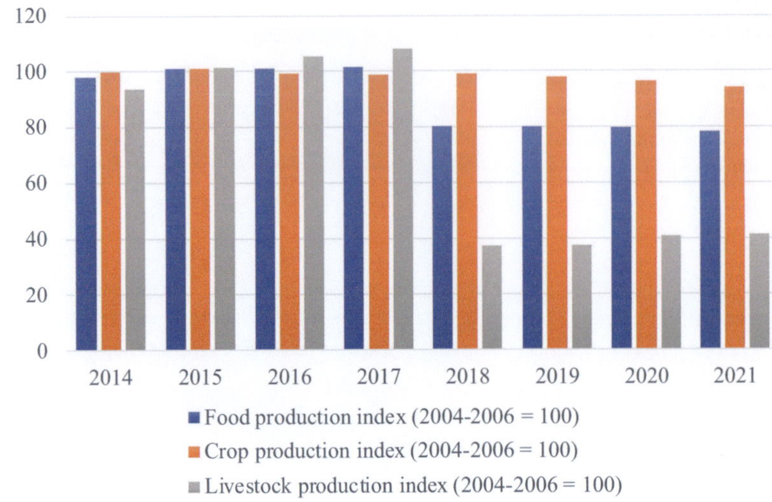

Fig. 11.10 Production index, food, crop and livestock, Myanmar, 2014–2021. *Source* The Global Economy (2023)

Institute (IFPRI) the Global Economy, etc., were used to analyse and interpret the current and future food insecurity status. Using content analysis methods and literature reviews, a problem tree is developed to locate the underlying causes of food security problems in Myanmar. The objective tree is constructed to find out the factors and sub-factors contributing to food security problems and towards improving food security. With consideration for the local context and climate change adaptation methodologies, the study offers insights into the food security situation in the country.

Results and Discussion

To examine the relationship between the country's food security status and food productivity, food production indexes are studied to get a picture of agriculture production trends. Myanmar's food production index fell in trend terms from nearly 98% in 2014 to 78.4% in 2021. Its crop production index fell in trend terms from 99.8% in 2014 to 93.9% in 2021, while its livestock production index fell from 93.6% in 2014 to 41.5% in 2021 (Fig. 11.10). The food production index covers the overall production of agricultural commodities, including crops and livestock production in the region. The indexes are calculated relative to the base period of 2004–2006, showing an increase or decrease in percentage of the annual production.

Underlying Problems of Food Insecurity: Towards a Conceptual Framework

The underlying problems of food security are examined here using a problem tree approach (Fig. 11.11). Through literature reviews, content analysis and desktop study, the problems are explored from the point of view of farmers and the vulnerable people with a focus on agricultural production and climate resilience.

Agricultural Employment

Myanmar's largest percentage of the labour force is in the agricultural sector. However, the employment rates dropped by more than 15% between 2017 and 2022, which saw a relative increase in employment in mining, construction and retail sectors. According to the World Bank (2023b; Fig. 11.12), which is why agricultural employment shows a declining trend since 2010 (52.3%), followed by a significant fall in 2018 (49.0%) and 2020 (46.5%). Meanwhile, experienced older workers who prefer self-employment have moved out of agricultural activities, while younger workers have less interest to enrolling in agriculture employment. Small farmers who have fewer acreage of land want to depend on family

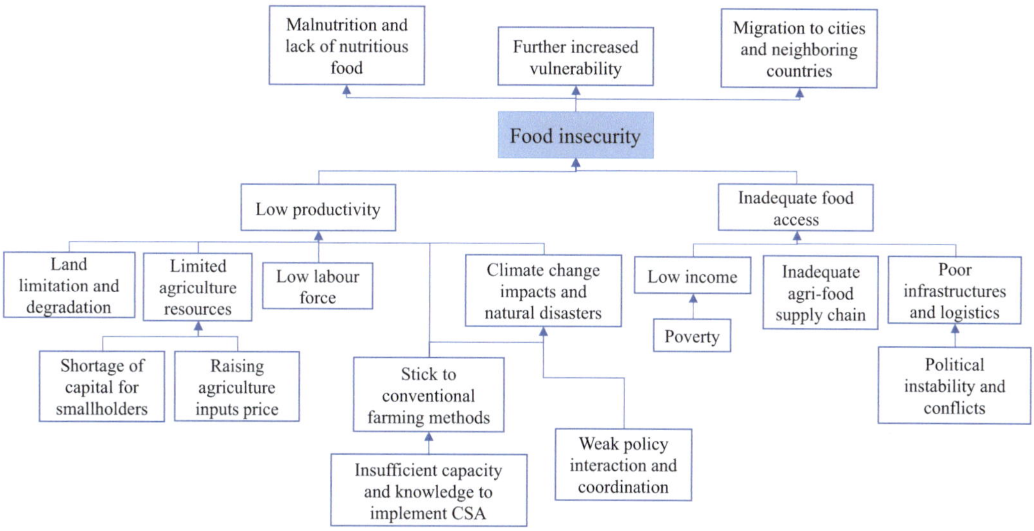

Fig. 11.11 Problem tree of food insecurity. *Source* Adapted by the author from (Kyi, 2019; MAPSA, 2022c; World Bank, 2023b; Tun Oo et al., 2023)

Fig. 11.12 Share of agricultural employment in total employment, Myanmar, 2010–2020. *Source* World Bankl ILO modelled estimates database" ILOSTAT (2023)

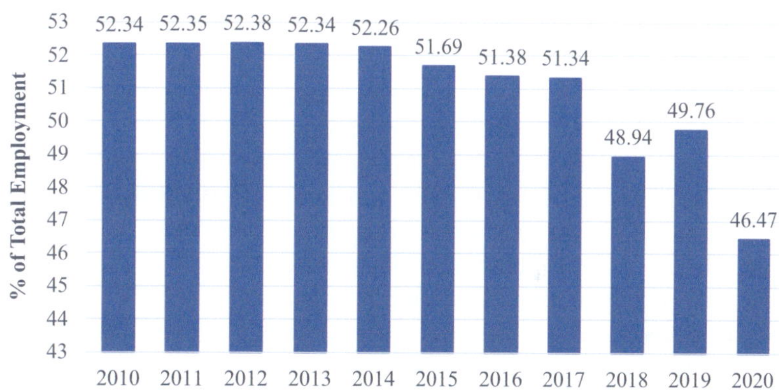

labourers as casual labourers, due to high agriculture costs, labour costs and transportation fees (World Bank, 2023c).

In addition, Myanmar has faced a severe pandemic after COVID-19 struck in 2020 and subsequently the coup in 2021; electricity shortage, rising costs and conflicts, crippled household income, which fell by around 20% in 2022. Consequently, in 30% of the households that used migration as a coping mechanism, a member moved to another rural area inside Myanmar, whereas in 56% of these households, a member moved to an urban location (World Bank, 2023b).

Monsoon paddy production generated an average net profit of $114 per hectare in 2013–2014, which was 10 times lower than that of China. In comparison, Myanmar produces just 23 kg of paddy per day of labour, compared to 62 kg in Cambodia, 429 kg in Vietnam and 547 kg in Thailand during the monsoon season. According to the World Bank (2016), farm revenues as a whole are insufficient to increase household per capita income over the local rural poverty level, which shows that the agriculture sector's promise of raising incomes has not been achieved (Zorya, 2016; World Bank, 2016).

Agricultural Productivity

All agricultural land has been owned by the state since 1962, and farmers have only been given cultivation (user) rights. Since those rights became formally transferable and eligible for mortgages in 2011 and cropping choice was liberalized, the scope of those rights has expanded, especially in recent years. However, there was still a great deal of uncertainty related to tenure because no record of rights has yet to be granted. As a result, the government has the power to cancel rights at any moment, and farmers only receive minimal compensation (Oberndorf, 2012; Raitzer et al., 2015). Moreover, smallholders have to compete with large agricultural companies to get land use permits. Many smallholder farmers were forced to give up access to their customarily farmed fields under the military government as a result of land purchases made to meet the military's development goals (Schneider et al., 2020).

The government is aware that the inability to address issues related to secure land tenure and land conflict could have detrimental effects on foreign direct investment, sustainable economic growth, environmental protection, social harmony and the development of a stable democratic system that upholds the rights of all citizens (Oberndorf, 2012). To address the issue of tenure insecurity, there was a need for the government to implement comprehensive agricultural land policies and laws with an agricultural land management system, to protect the specific land tenure regimes of smallholder farmers. Smallholder farmers, families and communities should have social protections, secure land tenure and specific protection of women's rights in policy and legal frameworks. To guarantee the rights of the ethnic minority, efforts to preserve people's rights and their involvement in land decision-making, policy should also explicitly recognize customary rules of land resources and offer mechanisms for community ownership of property.

Myanmar's labour and land productivity is significantly lower than that of other rice bowl countries in Asia. According to the International Labour Organization (ILO), Myanmar's GDP per hour worked per person productivity was US$5.1 in 2020 (ILOSTAT, 2023). In Myanmar, an agricultural labourer makes just US$1.8–US$2.5 per day during the rainy season and $3.0–$3.5 per day during the dry season. In contrast, an agricultural worker in Thailand makes $8.5 per day, while a farmer in the Philippines makes an average of $7 per day (Zorya, 2016 May).

Land degradation is one of the main reasons for low agricultural productivity. For instance, desertification in the dry zone has led to a decline in productivity, resulting in deforestation, soil erosion and salinization. Soil erosion is intensifying in the upland areas due to the high intensity of rainfall and rapid surface runoff. The main factor contributing to the dry zone's increased soil alkalinity is irrigation with saline groundwater. Overcropping and improper use of pesticides and fertilizers are also factors that have lowered soil productivity (Kyi, 2019). Approximately 50% of Myanmar's geographical area is either deforested or wooded, including several hilly regions and deforested hill slopes. Upland agricultural productivity is affected by significant soil deterioration, soil erosion and nutrient depletion brought about by deforestation and unsustainable practices (Kyi, 2019; LIFT, 2015).

Fragmented and Underfunded Farms

About 80% of the farmers in Myanmar possess land use rights in less than 10 acres of farmland. Access to revenue has remained constrained without verification of land use rights, which is a frequently acknowledged barrier to farmers' increased productivity. Other expenditures in increased productivity, such as tube well boreholes, improved land levelling and other upgrades, are disincentivized in the absence of assurance over future ownership. Myanmar has advanced significantly as a result of the recent increase in agricultural credit facilities provided by the MADB and the Myanmar Livestock and Fisheries Development Bank (MLFDB), which provide farmers with reduced interest rates (Raitzer et al., 2015). However, due to a financial crunch brought on by the banking crisis, agricultural credit, the primary government subsidy to the industry, has been essentially discontinued. The majority of farmers have had trouble repaying their prior seasonal loans, and the financial structure of the subsidy provides no incentives for farmers to comply (Tun, 2022).

A decline in yield was reported by 18% of farmers in 2021, while 24% of farmers experienced a fall in planted acreage. While the cost of production inputs rose, the earnings of smallholder farmers fell. International inorganic fertilizer prices and shipping costs significantly increased since 2021, leading to higher border prices for fertilizers in Myanmar. The political crisis and depreciation of the Myanmar currency have further increased domestic fertilizer prices. A survey of agro retailers in June 2021 estimated fertilizer prices to be 52% and 29% higher for urea and compound fertilizer, respectively (MAPSA, 2021b). Consequently, 89% of households experience food insecurity, thereby turning to unhealthy coping strategies due to falling yield and skyrocketing costs (FAO, 2022a, 2022b, 2022c).

Transportation challenges and a decline in demand are faced by agricultural producers, particularly in the value chains for vegetables and seafood (USDA, 2021). The cost of inputs has escalated; conflict and COVID-19 have hampered border trade; and shipping is no longer dependable, whether for the import of inputs or the export of goods (Tun Oo et al., 2023).

Climate ChangeConsequences

Due to agroecological changes brought about by climate change, Myanmar's agriculture is seriously threatened, with implications for food security and productivity. Around 46% of Myanmar's population works in agriculture, which mostly depends on small and medium-sized enterprises engaged in fishing, aquaculture and agriculture. The delta area is essential for rice production because of its fertile alluvial soils and temperate climate. Perennial crops like rubber, palm oil and tropical fruit trees predominate in the farming landscapes of the southern coastal districts of Myanmar, posing a further danger to the country's agricultural viability (NAPA, 2012; Tun Oo et al, 2023; Union of Myanmar et al., 2009).

Producing rice, a primary food crop and significant export from Myanmar, is mostly dependent on precipitation and is concentrated in places that are vulnerable to natural disasters, especially the delta region and central dry zone. Because of this sensitivity, rice production is more vulnerable to the effects of climate change, including fluctuations in temperature, rainfall, the timing of the monsoon season, delayed monsoon rains and more severe extreme weather events, including heat waves, floods and droughts (MONREC, 2019, 2021; Tun Oo et al., 2023).

In Myanmar, floods rank as the most frequent average yearly calamity, followed by accidents and storms. With a shortened monsoon season and predicted increases in precipitation of 11% by 2040 and 23% by 2070, Myanmar may see unexpectedly high levels of precipitation that might cause pluvial, riverine and flash floods (Horton et al., 2016). It is predicted that temperatures will rise by 1.1 °C by 2040 and 2.7 °C by 2070. The central dry zone is predicted to have temperature rises of up to 3 °C beyond 2040 (Horton et al., 2016). More saltwater intrusion is expected as a result of sea levels rising by 5 to 13 cm in 2020, 20 to 41 cm in 2050, and 37 to 83 cm in 2080 (Horton et al., 2016; MONREC, 2019; Nortajuddin, 2020; Tun Oo et al., 2023).

Inadequate Access to Food

According to the study by MAPSA, two out of every three persons in Myanmar were predicted to be impoverished by the end of 2022 (MAPSA, 2023e). High rates of inflation have had a substantial influence on poverty rates due to stagnating nominal earnings. In July 2022, for example, there were high inflation rates of 19.5% year over year, which saw the food price index increase by 49.7%, severely aggravating poverty rates (MAPSA, 2021b, 2023b).

Owing to their socio-economic traits, households in Kayah, Chin and Sagaing (which have been negatively impacted by previous military conflicts) became the most vulnerable. More than four out of five households in these areas employed a coping technique to get by daily; the most popular ones were conserving money and cutting back on expenses related to food and non-food items (MAPSA, 2023b). In Rakhine, around 80% of households had low incomes and used asset sales or mortgages as a coping mechanism.

Food and nutrition security declined dramatically, as poor food consumption rations increased from 9.4% in January 2022 to 15.7% in August 2022 (MAPSA, 2022a). There was a higher likelihood of poor food consumption scores in rural households as food insecurity and poor diet quality are associated with low income and lack of assets with the daily wage labourers being the most vulnerable (MAPSA, 2022b). Inadequate nutrition is negatively associated with remittances. The areas which are most hit by violence, such as Chin (48%), Kayin (23%) and Magway (23%), had the lowest rates of food consumption.

Underdeveloped Agri-Food Supply Chain

Constituting 84% of the agricultural GDP, five value chains of fish, rice, oilseeds, livestock and other agriculture in 2019, these sectors exert a major influence on the agrifood system (MOALI, 2018b). Enhancing these value chains can boost economic growth and lower poverty levels in Myanmar's main agroecological zones, especially through increased agricultural production and value-added processing, which is essential for mitigating migration and resettlement, as well as satisfying consumer demand for a more varied diet at lower prices. Indeed, the government's efforts to diversify agriculture from simply rice cultivation (MAPSA, 2023c).

The International Food Policy Research Institute (IFPRI) reported that transportation has become a serious problem since the coup took place in 2021, which has been made worse by restrictions and curfews impeding the trans-boundary rice market and curfews enacted at various administrative levels at erratic times and locations. Almost 30% of food sellers in September 2021 reported having trouble obtaining food supplies from outside their villages. In addition, gasoline prices increased dramatically after the military takeover, which is partly a consequence of changes in global oil prices. Still, they were made worse by the devaluation of the Myanmar Kyat, logistics issues, foreign commerce and problems with local gasoline delivery (MAPSA, 2022c).

Over only 9 million people live in communities with roads or tracks in Myanmar (ADB, 2017), where 40% of the rural population lacks access to all-season roadways (World Bank, 2020). In terms of logistics performance, Myanmar is ranked 137th out of 160 nations, whereas its peer countries in Southeast Asia (SEA) are ranked 26–44 (Arvis et al., 2018). Under these circumstances, farm commercialization is limited, transportation costs for agricultural products and inputs become higher, and diversification into higher-value perishable crops is frequently unfeasible.

To increase production and diversify into higher-value crops, the irrigation infrastructure that is currently in place has to be upgraded. Large areas of rice are flooded by public utilities, leaving farmers with little control over water management and no motivation to save it. Private irrigation systems overuse and underutilize groundwater supplies because they take recharge capacity into account when they extract it. To promote water conservation and enable agricultural system diversification into more productive ones, a thorough irrigation water management policy and investment plan are required (MAPSA, 2022c). To expand into high-value businesses like horticulture, investments in irrigation, mechanization services, and upgrade private finance services are also required, including grading, packaging, cold chain facilities and processing facilities.

Towards Fortifying Future Food Security

After analysing the underlying causes of Myanmar's food insecurity problems, a few approaches are examined here towards improving food security in the country (Fig. 11.13).

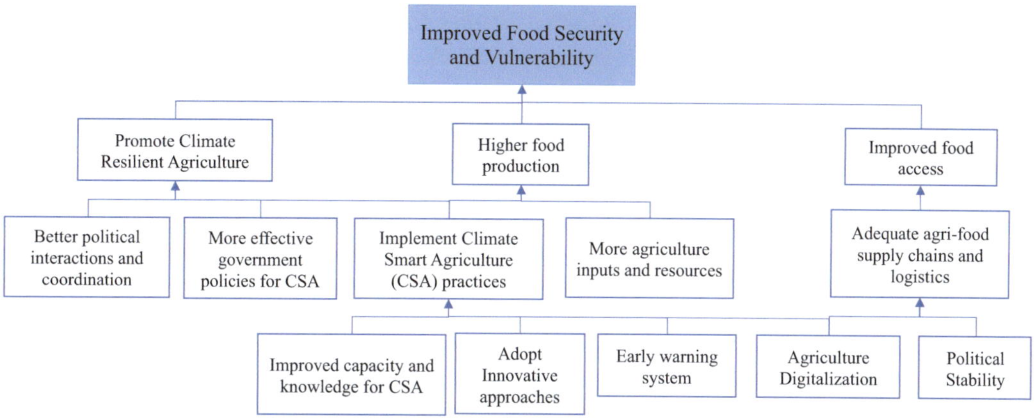

Fig. 11.13 Objective tree for addressing food insecurity and vulnerability. *Source* Adapted by the author from (Barbon et al., 2022; Thant et al., 2022; Tun Oo et al., 2023; Sundram, 2023)

Climate Resiliencein Agriculture

At the institutional level, major constraints to climate change adaptation include limited institutional capacity and capability, limited resources (financial and technical), weak policy interaction and implementation, and political instability (Tun Oo et al., 2023). Lacking efficient interministerial coordination makes policies incoherent and achieving climate targets more difficult. Furthermore, inadequate financial resources mean that it is mostly dependent on outside assistance, which hinders its ability to successfully adopt and execute climate measures.

Poor policy interaction and execution is a major obstacle preventing Myanmar from integrating climate policies. To make matters worse, political instability has undermined continuity and consistency in the execution of policies. Hence, the country's capacity to fulfil its climate change targets may be jeopardized by this instability, which might lead to a lack of political will to prioritize climate policy.

Policies like the Intended Nationally Determinant Contributions (INDCs) and the National Adaptation Programme of Action (NAPA) that address climate change adaptation are primarily top-down initiatives that get very little input from the public and other stakeholders. It surely will help to introduce participatory approaches to promote CSA implementation to create a policy environment to absorb CSA interventions. There is still a need to strengthen research in policymaking and reform—including market-based incentives, land tenure and government-driven agricultural extension (Barbon et al., 2022).

High-Technology Climate-ResilientFarming

Innovative sustainable agriculture practices have been practised in Myanmar for some time. Shan state, for example, which is crucial for food production and exports from the country, is facing economic instability and soaring seed and fertilizer prices. UNDP is providing support to farmers in the southern part of the state with affordable, climate-resilient technological solutions, solar pumping systems, fertigation and demonstration farms. Solar pumps help farmers reduce fuel costs to improve crop yields by displacing fertilizer in irrigation systems. Fertigation reduces fertilizer use from 25.0 kg to 1.5 kg, saving farmers time and money. A demonstration farm introduces farmers to new, more productive rice varieties and teaches sustainable practices and pest management. Basmati seeds, a better-quality grain with higher market demand and price, can be stored and used for multiple years to provide long-term benefits (Win, 2023).

The study found that in the past, farmers used to apply locally available indigenous climate change adaptation strategies. However, most farmers seem to have shifted to introduce adaptation strategies, the most popular strategy being to adjust the planting dates and sowing methods (56.1%) (Tun Oo et al., 2023). Furthermore, farmers identified several barriers that limit the adoption of adaptation methods, such as lack of awareness and capacity, weak technical knowledge and farm management, lack of access to credit, limited resources, sticking to traditional farming practices and less normative tendency to adopt new techniques (Tun Oo et al., 2023).

By encouraging the use of natural pesticides and fertilizers, the deployment of organic farming reduces the harmful effects of chemicals on the health of the soil and the purity of the water. Agroecology places a strong emphasis on incorporating ecological concepts into agricultural systems to promote biodiversity, nutrient cycling and organic pest management. By optimizing resource allocation via the use of cutting-edge technologies and data analytics, precision agriculture empowers farmers to make well-informed decisions about crop protection, fertilization and irrigation (FAO, 2023). These environmentally friendly methods reduce soil erosion, increase soil fertility and improve water usage efficiency to maximize agricultural yields and provide a steady supply of nutritious food (Çakmakçı et al., 2023). Furthermore, these innovative approaches tackle global issues as well as local dynamics, highlighting the significance of locally appropriate solutions, active engagement and the application of local expertise (FAO, 2019).

Improved Access to Food

Strengthen Agri-Food Supply Chains

Myanmar has implemented various initiatives to promote food safety in its food chains, but there are still gaps in the supply chain. Some approaches are being implemented by MOALI (2021) to solve problems related to increasing access to nutritious food, including scaling up efforts to enhance nutrition and raise awareness of malnutrition, expanding the supply of nutritious foods, supporting the agribusiness and agro-food sectors, and enhancing social protection programmes. Priority areas for implementation include enhancing food production in agroecological zones and promoting incomes on and off farms to support livelihoods by expanding cold chain technology to address current problems and challenges associated with nutritious food. Despite national inspection and certification systems, poor coordination and low investment in safe food production and distribution remain. Solutions include strict enforcement of protocols, strengthening legal and regulatory frameworks, capacity development of relevant agencies and promoting consumer awareness (MOALI, 2021).

Before production, a demand-driven strategy is required to obtain a more feasible and efficient system that connects production to demand, which requires market evaluation to determine high-value, high-demand items as well as the kinds of crops that have to be promoted and planted. Farmers may lower their production costs to boost their profits by using CSA techniques (Thant et al., 2022). Also, since Myanmar's health system is still in its infancy, and food hygiene requirements are deficient at every level of processing, it will help to raise food safety and quality awareness among farmers and consumers. Because of the underdeveloped market infrastructure, information sharing, and coordination, supply chain management, capacity development in food supply chain management skills and financial technologies is still necessary for farmers (Thant et al., 2022).

Digitalization

Myanmar is embracing agri-tech solutions to enhance its agriculture sector; the country's digitalization is facilitated by digital applications, such as "Tun Yat", "Green Way", "Golden Paddy" and "Htwat Toe". These apps connect farmers with agronomists to offer solutions for crop growth rates, pest infestations, fertilizers, weather forecasts, flood warnings and farming advice. Mobile applications create online platforms to help farmers find information about crop pricing, best practices and techniques. The digital agriculture revolution is expected to contribute to Myanmar's agriculture sector development, upgrading and transformation.

Utilizing digital technology like data analytics, remote sensing and precision agriculture may help farmers control crop diseases, maximize resource use and reduce climate risk. Digital technology may strengthen market integration, guarantee the robustness and reactivity of its food systems, monitor and manage illnesses and pests, and improve supply chain management (Sundram, 2023; Teng et al., 2021) to support evidence-based decision-making for farmers and agronomists through supporting precision agriculture and strengthening agricultural digitalization. Farmers can raise productivity through well-informed decisions thanks to real-time market pricing, weather predictions and best practices communicated via digital channels. To help farmers invest and increase production, mobile banking and digital payment systems may promote financial inclusion and give them stronger access to financing (OCED, 2021).

Conclusion: Towards Food Security and Climate Resilience

Even if Myanmar's agricultural sector can be ramped up to provide local food sufficiency, it has significantly lagged behind regional and international rivals, such as Thailand and Vietnam. To reach its full potential, Myanmar must address issues associated with food safety, market inefficiency, lack of accessibility and lack of awareness and farmers' capabilities. The inspection and certification agencies in charge of food safety and agricultural inputs should be strengthened at the border, while there should be stronger control on the quality of fertilizers. Awareness-raising programmes for the relevant stakeholders, consumers and traders are necessary with strict enforcement, strengthening of legislative and regulatory frameworks, institutional development and capacity building of relevant agencies and organizations.

In addition, the issue of land tenure and lack of secure land rights should be settled through explicit protocols for overseeing large-scale land acquisitions for agriculture to support local government, encourage affordable credit, involve women, respect the tenure of smallholder farmers, encourage responsible investment and support crop diversification and livelihood activities. Strategies for supplementary irrigation, groundwater access and rainwater gathering are necessary for sustainable agriculture productivity. Even in flood-prone areas, better water management is essential for smallholder production, food security, income growth and livelihood enhancement.

Much has to be done for Myanmar to meet its climate targets. While political stability is central, weak policy interactions, lack of political will, and limited technical, organizational and financial resources have restricted progress in the agricultural sector. To establish a meaningful institutional network, the government should support CSA practices and agricultural system resilience by rewarding farmers for their output and contribution to the environment.

The government should establish a conducive environment for better farming system resilience and adoption of CSA practices through policy and incentive initiatives. Better regional and national planning and management mechanisms should be implemented to ensure the efficient and effective implementation of CSA initiatives. Capacity-building, training and extension services are necessary to empower and strengthen the adaptive capacity of key stakeholders, policymakers, development planners, communities and farmers.

Improving climate change education and awareness is crucial to build public support for climate action. Community-based resilience planning can strengthen farmers' and communities' adaptive capacity and resilience. The government needs to speed up the early warning and response system (EWEAS) to educate the public on natural disasters and climate change-related occurrences. Enhancing climate data and monitoring systems is essential for effective climate change planning and decision-making. Development projects and programmes are needed to scale up and broaden the adoption of CSA initiatives (Tun Oo et al., 2023).

A profound evaluation of Myanmar's policies and initiatives is important to address climate change to promote sustainable development and create resilient communities that are heavily dependent on agriculture. Using innovative methods and technology, the CSA in Myanmar may be expanded by leveraging the agricultural research community comprised of NGOs, government and universities. However, the broad adoption of CSA will not be possible with just these technologies. The framework for the policy, including how it interacts with other policies and how people and agencies are involved, needs to be reinforced to guarantee success in ensuring broader CSA adaptation in Myanmar.

References

Çakmakçı, R., Salık, M. A., & Çakmakçı, S. (2023). Assessment and principles of environmentally sustainable food and agriculture systems. *Agriculture, 13*(1073), 1.

ADB. (2015). *Myanmar's agriculture sector: unlocking the potential for inclusive growth.* Asia Development Bank.

ADB. (2017). *The status of infrastructure services in East Asia and the Pacific.* Asian Development Bank.

Arvis, J., Wiederer, C., Ojala, L., Shepherd, B., Raj, A., Dairabayeva, K., & Kiiski, T. (2018). *Connecting to Compete 2018: Trade Logistics in the Global Economy—The Logistics Performance Index and its Indicators.* World Bank.

Barbon, W. J., Khaing, O., Thant, P., & Gonsalves, J. (2022). Exploring pathways for promoting and scaling up climate-smart agriculture in Myanmar. *Asian Journal of Agriculture and Development, 19*(1), 1.

Boori, M. S., Choudhary, K., Evers, M., & Paringer, R. (2017). A review of food security and flood risk dynamics in central dry zone area of Myanmar". *Procedia Engineering, 201*, 231–238.

Boughton, D., Haggblade, S., Kham, L., Kongabaugh, S., & Thaung, M. (2015). *Overcoming Isolation: An Exploration of the Rapid Growth in Pulse Exports from Myanmar.* Michigan State University, Research in Agricultural and Applied Economics.

Eckstein, D., Kunzel, V., & Schafer, L. (2021). *Global Climate Risk Index 2021, Who Suffers Most From Extreme Weather Events? Weather-related Loss Events in 2014 and 1995 to 2014.* Germanwatch e.V.

Eitzinger, A., Binder, C., & Meyer, M. (2018). Risk perception and decision-making: do farmers consider risks from climate change? *151.*

Eurocham. (2020). *Agriculture Guide 2020.* EuroCham.

Eurocham. (2020). *Myanmar Agriculture Guide 2020.* Eurocham Myanmar.

Fang, P., & Belton, B. (2020). *Maize and Pigeon Pea Production, Profitability, and Tied Credit in Southern Shan State.* Food Security Policy Project (FSPP).

Fang, P., & Belton, B. (2020, August). Maize Production, Farm Size, and Tied Credit in Southern Shan State, Myanmar. *IFPRI Discussion Paper 1961.*

FAO. (2019). *Agroecological and other innovative approaches for sustainable agriculture and food systems that enhance food security and nutrition. High-level panel of experts on food security and nutrition.* Food and Agriculture Organization.

FAO. (2020). *Follow Up Survey to Identify Drivers Affecting Monsoon Crops Plantation Area and Productivity of Small-Scale Farmers during COVID-19 Pandemic: Result of Telephone Interview.* Food and Agriculture Organization.

FAO. (2022, June 8). *Myanmar food security crisis: FAO and the Central Emergency Response Fund are enabling vulnerable farmers to feed themselves and their communities.* Retrieved from FAO website: https://www.fao.org/myanmar/news/detail-events/en/c/1538680/#:~:text=Myanmar%20is%20facing%20a%20rapidly,the%20entire%20national%20food%20system.

FAO. (2022b). *Myanmar: Response overview.* Food and Agriculture Organization.

FAO. (2023). *Agroecology Knowledge Hub.* Food and Agriculture Organization.

Flanders. (2020). *Agriculture in Myanmar.* Flanders Investment & Trade, Market Survey.

Hom, N., Htwe, N., Hein, Y., Than, S., Kywe, M., & Htut, T. (2015). *Myanmar Climate-Smart Agriculture Strategy.* Ministry of Agriculture and Irrigation (MOAI). CGIAR Research Program on Climate Change, Agriculture and Food Security (CCAFS), International Rice Research Institute (IRRI).

Horton, R., De Mel, M., Peters, D., Lesk, C., Bartlett, R., Helsingen, H., & Rosenzweig, C. (2016). *Assessing climate risk in Myanmar: Summary for policymakers and planners.* Center for Climate Systems Research at Columbia University. WWF-US and WWF-Myanmar, UN-Habitat Myanmar.

ILOSTAT. (2023). *Statistics on labour productivity, International Labour Organization.* Retrieved from https://ilostat.ilo.org/topics/labour-productivity/

Kyi, T. (2018). *Food Security and Nutrition in Myanmar: Policy Landscape.* Ministry of Agriculture and Irrigation (MOALI), Myanmar.

Kyi, T. (2019). *Policy Notes on Sustainable Land Use Management and Land Degradation in Myanmar.* Ministry of Agriculture and Irrigation, Myanmar.

Kyi, T. (2022). *Ensuring Access and Sustainable Consumption of Safe and Nutrition Foods: Trends in Consumer*

Behavior and Food Consumption in Myanmar. FFTC Agriculture Policy Platform.

Lambrecht, I., Belton, B., Fang, P., Minten, B., & and Naing, P. T. (2022). *Agricultural land and crop production in Myanmar.* Myanmar SSP Working Paper 24, International Food Policy Research Institute (IFPRI).

LIFT. (2015). *Uplands Programme: Scoping Assessment Report.* Livelihood and Food Security Trust Fund (LIFT).

MAPSA. (2021a). *Myanmar's poverty and food insecurity crisis: Support to agriculture and food assistance is urgently needed to preserve a foundation for recovery.* Myanmar Agriculture Policy Support Activity. International Food Policy Research Institute (IFPRI).

MAPSA. (2021b). *Monitoring the agri-food system in Myanmar: Agricultural input retailers – June 2021 survey round.* Myanmar Agriculture Policy Support Activity. International Food Policy Research Institute (IFPRI).

MAPSA. (2022a). *The State of Food Security and Nutrition in Myanmar: Findings from the Myanmar Household Welfare Survey 2021–2022.* Myanmar Agriculture Policy Support Activity. International Food Policy Research Institute (IFPRI).

MAPSA. (2022b). *Monitoring the agrifood system in Myanmar: Rice millers – April 2021 survey round.* Myanmar Agriculture Policy Support Activity. International Food Policy Research Institute (IFPRI).

MAPSA. (2022c). *Agricultural value chains in a fragile state: The case of rice in Myanmar.* Myanmar Agriculture Policy Support Activity. International Food Policy Research Institute (IFPRI).

MAPSA. (2023). *The state of food security and nutrition in Myanmar 2022: Findings from four rounds of the Myanmar Household Welfare Survey.* https://doi.org/10.2499/p15738. Myanmar Agriculture Policy Support Activity. International Food Policy Research Institute (IFPRI).

MAPSA. (2023b). *Livelihood Resilience and The Agrifood System in Myanmar: Implications for Agriculture and a Rural Development Strategy in a Time of Crisis".* Myanmar Agriculture Policy Support Activity. International Food Policy Research Institute (IFPRI).

MAPSA. (2023c). *Livelihood Resilience and The Agrifood System in Myanmar: Implications for Agriculture and a Rural Development Strategy in a Time of Crisis.* Myanmar Agriculture Policy Support Activity. International Food Policy Research Institute (IFPRI).

MAPSA. (2023d). *Livelihood resilience and the agrifood system in Myanmar: Implications for agriculture and a rural development strategy in a time of crisis.* Myanmar Agriculture Policy Support Activity. International Food Policy Research Institute (IFPRI).

MAPSA. (2023e). *Paddy rice productivity and profitability in Myanmar: Assessment of the 2022 monsoon season.* Myanmar Agriculture Policy Support Activity. International Food Policy Research Institute (IFPRI).

Maung, S. W. (2020). *Pulses Day in Myanmar.* Myanmar Pulses, Beans, Maize, and Sesame Seeds Merchants Association.

MIMU. (2023, August). *Who What Where When for Whom- Agriculture Sector, Projects under implementation as of 31 August 2023.* Retrieved from Myanmar Information Management Unit: https://themimu.info/sites/themimu.info/files/documents/5WMap_Agriculture_ALCB_IWR_CI_and_AD_Aug2023.pdf

MOALI. (2018a). *Myanmar Agriculture Development Strategy and Development Plan (2018–19~2022–23).* Ministry of Agriculture, Livestock and Irrigation.

MOALI. (2018b). *Myanmar Agriculture Development Strategy and Investment Plan 2018–19 ~ 2020–23.* Ministry of Agriculture Livestock and Irrigation, Nay Pyi Taw: MOALI.

MOALI. (2021). *National Pathway to Sustainable Food System of Myanmar by 2030 (Draft)",.* Ministry of Agriculture, Livestock and Irrigation (MOALI), Nay Pyi Taw, 15 September 2021.

MOC. (2023). *Myanmar Logistic Opportunities and Prospects.* Retrieved from Ministry of Commerce: Myanmar. https://www.commerce.gov.mm/my/content/myanmar-logistic-opportunities-and-prospects

MONREC. (2019). *Myanmar Climate Change Strategy (2018–2030).* Ministry of Natural Resources and Environmental Conservation (MONREC), Yangon, Myanmar.

MONREC. (2021). *Nationally Determined Contributions.* Ministry of Natural Resource and Environmental Conservation.

Myanmar National Portal. (2023). *Second Five-Year Short-Term Agriculture Policies and Strategic Thrusts.* Ministry of Agriculture, Livestock and Irrigation (MOALI).

NAPA. (2012). *Myanmar National Adaptation Programme of Action (NAPA) to Climate Change.* Ministry of Environmental Conservation and Forestry. Nay Pyi Taw, Myanmar: National Adaptation Programme of Action (NAPA).

Nortajuddin, A. (2020). *Myanmar Is at Risk from an Extreme Climate.* Retrieved 2023, from The ASEAN Post: https://www.preventionweb.net/quick/20269

Oberndorf, R. (2012). *Legal Review of Recently Enacted Farmland Law and Vacant, Fallow and Virgin Lands Management Law: Improving the Legal and Policy Frameworks Relating to Land Management in Myanmar.* Land Core Group, Food Security Working Group.

OCED. (2021). *Strengthening Women's Entrepreneurship in Agriculture in ASEAN Countries.*

Raitzer, D. A., Wong, L. C., & Samson, J. N. (2015). Myanmar's Agriculture Sector: Unlocking the Potential for Inclusive Growth. *ADB Economics Working Paper Series, no. 470.*

Rasiah, R. (2025). Building Sustainability and Food Security in Southeast Asia, In R. Rasiah., V. Anter. & N. I. Annizah (eds), *Sustainable Agriculture and Food Security in Southeast Asia: Towards Climate Resilience and Digitalization,* Singapore: Springer Nature.

Schneider, F., Feurer, M., Lundsgaard-Hansen, L. M., Myin, W., Nuam, C. D., Nydegger, K., & Messerli, P. (2020). Sustainable Development Under Competing

Claims on Land: Three Pathways Between Land-Use Changes. *Ecosystem Services and Human Well-Being, Eur J Dev Res, 32*, 316–337.

Statista. (2023a). *GDP contribution of total agriculture sector in Myanmar from financial year 2011 to 2020.* Retrieved from https://www.statista.com/statistics/1062444/myanmar-gdp-contribution-total-agriculture/

Statista. (2023b). *Share of GDP contribution from the total agriculture sector in Myanmar from financial year 2010 to 2019.* Retrieved from https://www.statista.com/statistics/1063691/myanmar-share-gdp-contribution-total-agriculture-sector/#:~:text=In%202019%2C%20the%20share%20of,towards%20the%20GDP%20in%20Myanmar

Statista. (2023c). *Production Volume of Main Crops in Myanmar*. Retrieved from https://www.statista.com/statistics/1060675/myanmar-main-crops-by-production-volume/

Sundram, P. (2023). *Food security in ASEAN: Progress, challenges and future.*

Swe, L. M., Shrestha, R. P., Ebbers, T., & Jourdain, D. (2015). *Farmers' perception of and adaptation to climate-change impacts in the Dry Zone of Myanmar.* Climate and Development.

Teng, P. P., Caballero-Anthony, M., & Montesclaros, J. M. (2021). *"ASEAN responses to COVID-19 for assuring food security" in Advances in food security and sustainability.*

Thant, P., Espino, A., Soria, G., Myae, C., Rodriguez, E., Barbon, W. J., & Gonsalves, J. (2022). Myanmar local food systems in a changing climate: Insights from multiple stakeholders. *Environmental and Sustainability Indicators, 14*, 1.

The Global Economy. (2023). *Burma (Myanmar): GDP share of agriculture.* Retrieved from The Global Economy.com. https://www.theglobaleconomy.com/Burma-Myanmar/share_of_agriculture/

Tun Oo, A., Boughton, D., & Aung, N. (2023). Climate change adaptation and the agriculture-food system in Myanmar. *Climate, 11*(6), 1.

Tun, A. (2022). *Agriculture in a State of Woe Following Myanmar's 2021 Military Coup.*

Union of Myanmar, MES, MGS, MIMU, ADPC, & DMH. (2009). *Hazard Profile of Myanmar (HPM), Institutional Arrangements for disaster management in Myanmar', Ministry of Social Welfare, Relief, and Resettlement, the Republic of the Union of Myanmar. 2009.* Ministry of Social Welfare, Relief, and Resettlement. Republic of the Union of Myanmar.

USDA. (2021). *Impact of Military Coup on Agriculture Sector and Trade.* United State Department of Agriculture, Foreign Agriculture Service.

Win, S. S. (2023). *Three innovative ways UNDP is helping farmers in Myanmar's Shan State.* Retrieved August 2023, from UNDP: https://www.undp.org/myanmar/news/three-innovative-ways-undp-helping-farmers-myanmars-shan-state

World Bank. (2007). *What Are the Links between Agricultural Production and Food Security?*

World Bank. (2016). *Myanmar: Analysis of Farm Production Economics.* Economic and Sector Work Report no. 100066-mm. World Bank Group, Livelihood and Food Security Trust Fund (LIFT).

World Bank. (2020). *World Development Indicators.*

World Bank. (2023a). *Myanmar: Natural Hazard Statistics.* Retrieved from World Bank Climate Change Knowledge Portal: https://climateknowledgeportal.worldbank.org/country/myanmar/vulnerability

World Bank. (2023b). *Employment in agriculture (% of total employment).* Retrieved from International labour Organization, "ILO modelled estimates database" ILOSTAT, World Bank Open Data. https://data.worldbank.org/indicator/SL.AGR.EMPL.ZS?locations=MM

World Bank. (2023c). *A Growing Crisis: Work, Workers, and Wellbeing in Myanmar.* World Bank Group.

World Bank. (2023d). *Agriculture, forestry, and fishing, value added (current US$)- Myanmar*. Retrieved from World Bank Open Data- Myanmar: https://data.worldbank.org/indicator/NV.AGR.TOTL.CD?locations=MM

World Bank. (2023e). *World Bank Open Data: Myanmar.* Retrieved from https://data.worldbank.org/country/myanmar

WTO. (2023). *Myanmar Trade Profile.* Retrieved from World Trade Organization: Myanmar and the WTO: https://www.wto.org/english/res_e/statis_e/daily_update_e/trade_profiles/MM_e.pdf

Zorya, S. (2016 May). *Unleashing Myanmar agriculture's potential.* Retrieved from World Bank Blogs: https://blogs.worldbank.org/eastasiapacific/unleashing-myanmar-agricultural-potential

From Subsistence to Sustainable Agriculture in Timor-Leste: A Challenging Transition

12

Jerry Courvisanos, Elsa Joaquina Araújo Pinto, and Apolinário Magno

Abstract

Two-thirds of households in Timor-Leste are engaged in farming, with more than a quarter of the working-age population dependent on subsistence agriculture. This large continuance of a very old form of agriculture creates an extremely challenging task for the country to make a strong and successful transition to sustainable agriculture. In this volume, Timor-Leste is unique in that it is the only country discussed that is not in the Association of Southeast Asian Nations (ASEAN), a political-economic union of the fastest growing economies in the world. Timor-Leste applied to join in March 2011 and is still waiting for a decision based on evaluations and monitoring of the country by ASEAN authorities. The subsistence nature of its agriculture and ASEAN demands to diversify away from the dominant gas-export basis of its economy is the major backdrop to this chapter's analysis. By analyzing available documents and secondary data, this chapter focusses on the country's ability to diversify its economy from subsistence to sustainable agriculture. The unchanging state of agriculture in the country is first examined, followed by charting past and present government policies on the economics of agriculture. Then, the difficulties in addressing a shift to sustainable agriculture are analyzed in the context of the need for infrastructure (including digitalization), climate resilience, self-sufficiency, and food security. Finally, Habermas' societal evolution framework is applied to address this transition into the future.

Keywords

Infrastructure · Technological capabilities · Labor productivity · Land reform · Organic farming · Habermas

Introduction: Perspectives Outlined

Timor-Leste occupies the eastern end of Timor Island (as well as the western enclave of Oecusse). The Democratic Republic of Timor-Leste came into existence on May 20, 2002. In the 24 years prior, it was occupied by Indonesia that includes the West Timor province. Timor-Leste is the only nation in this volume that is not in ASEAN but applied to join in March 2011 (with observer status in 2022). Accession to ASEAN would bind

J. Courvisanos (✉)
Federation University Australia, Melbourne, Australia
e-mail: j.courvisanos@federation.edu.au

E. J. A. Pinto · A. Magno
Universidade Nacional Timor Lorosa'e, Dili, Timor-Leste

R. Rasiah et al. (eds.), *Sustainable Agriculture and Food Security in Southeast Asia*, Sustainable Development Goals Series, https://doi.org/10.1007/978-981-95-0477-0_12

Timor-Leste to the ASEAN Economic Community (AEC)—a common market with a level playing field in the free flow of goods, investment, selected services, and collective infrastructure connectivity, but no free flow of labor (Seixas et al., 2023). The final AEC fact-finding mission to Timor-Leste was conducted on July 19–21, 2022. The ASEAN delegates acknowledged Timor-Leste's efforts in conducting economic reforms in diversification away from oil/gas as its overriding export earner and improving its human capacity, with encouragement to continue in this direction. However, the latest ASEAN Summit in May 2025 made no decision on accession of Timor-Leste and diversification has not yet become discernible.

Based on 2022 estimate Timor-Leste has 1.4 million (m) people with 56% of working age (15–64 years) and a massive 38% aged 0–14, with 12% below five years old. Only about a quarter of the 820,000 working-age people work in the formal economy (24%); the rest are made up of subsistence farmers or fisher people (27%, although they often sell a small amount of surplus), unpaid work (28%), and students (21%). The age of farmers is dominated by the 51–56 years cohort (30%) with relatively low formal education, noting particularly that 40% have only elementary school education (UNDESA, 2022). Roughly one-third of households have their main (usually only) income earner engaged in government, private sector, and micro family businesses occupations. This leaves two-thirds of households engaging in farming, even if they also do other work, while their agricultural holdings average only 1.5 hectares (Scheiner, 2021). As a result, about 80% of poor households who live in rural areas are dependent on mostly subsistence agriculture for their livelihood; the rest primarily depend on cash crops. Historically over two centuries, farmer households have sold surplus agricultural produce in local village markets for food items not grown, which formally makes them semi-subsistence but referred to in this chapter as "subsistence" as it is their primary endeavor. The role of the Timor-Leste Government in agriculture is addressed throughout the chapter in the context of sustainability issues raised.

In the Introduction chapter, Fig. 1.2 depicts the Gross Domestic Product (GDP) share of agriculture in Timor-Leste as rising from May 2002 (start of nation) to 2007, which is atypical compared to all the ASEAN nations with their declining agriculture shares (Rasiah, 2025). Then, following the 2006 internal crisis with 155,000 people fleeing their homes, the share of agriculture declined and continued to fall in line with all ASEAN nations, reaching below 10% of GDP in 2021.[1] Once adjustment for inflation and population growth is made, real per capita GDP from agriculture has declined by almost 20% between 2010 and 2020. Production of carbohydrate-rich foods (grains, seeds, nuts, root vegetables) has been the largest contributor to this decline by more than 40% in value. Over the same period, from a low base, there was a significant growth in meat-based protein-rich foods (chicken, buffalo, sheep), increasing by more than 30% in real per capita terms. Overall, agriculture is the largest non-government industry at just less than 20% of non-oil GDP, with public administration and government services (including education, health, and defense) comprising more than 30% of non-oil GDP (Centre for Development Economics & Sustainability, 2023).

The agriculture GDP share statistics provided are misleading as the calculation is based on assigning (or imputing) monetary value to subsistence agricultural production and land ownership (imputed rent), although unpaid work (volunteer non-government organizations [NGOs] and households) is not imputed (Scheiner, 2021). Applying GDP imputation to poor subsistence agricultural production overstates economic development, while neglecting social indicators like inequality, barter activity, and exchanges within extended families.

The decline in agricultural production in Timor-Leste reflects a shift toward the large infrastructure construction and burgeoning public sectors as the nation settles into peaceful political

[1] Note that fishery and forestry make up only 3% of all agricultural GDP, with 97% originating in crops, livestock, and small estates (Sabino and Harmadi, 2009).

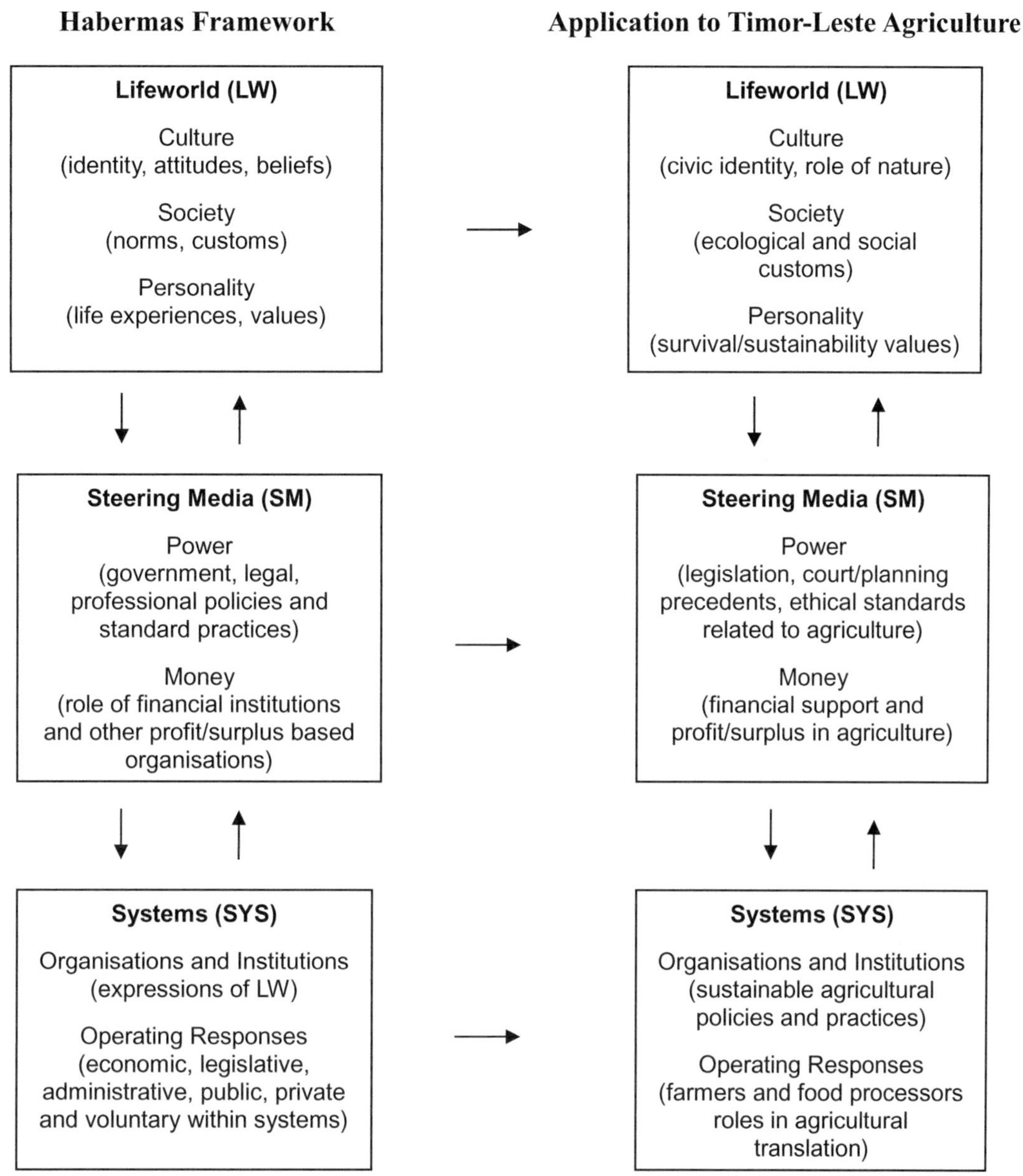

Fig. 12.1 Framework of analysis. *Source*: Adapted from Habermas (1987) and Zacharias (2007)

divisions and the United Nations ended its peacekeeping mission on 31 December 2012 (Shoesmith, 2020). Also, the large subsistence agricultural sector hides a great deal of underemployment when people go back to the family farm when they are not able to find work in construction or public administration. Unlike standard economic development in deindustrialization models (e.g., Rowthorn & Wells, 1987), decline in agriculture in Timor-Leste has not been accompanied by a rise in manufacturing. In fact, the minuscule manufacturing sector (as at 2021) produced less output than 20 years ago when it came into existence as a nation. The two growth sectors of construction or public administration are dependent "almost entirely from oil-funded state spending" (Scheiner, 2021, p.256). As a result, construction materials and daily food needs are from imports, including the staple rice. This has resulted since 2015 in a consistent pattern of deficit trade merchandise balance of about $500 m[2] per year imported goods (mostly from its neighbor Indonesia and China, both with mass produced cheap basic goods) and only $20 m per year non-oil exports, of which 90% is coffee (Banco Central de Timor-Leste, 2020, in US constant dollars).

On the sustainable development front, the Timor-Leste Government drove the United Nations Sustainable Development Goals (SDGs) agenda for adoption by developing nations (United Nations, 2015). At the agenda setting UN meetings in 2015 Timor-Leste led the group of fragile countries (called "g7+") in ensuring that the goal on peace, justice, and strong institutions (SDG #16) was fully included, offering detailed wording for targets related to developing countries. All Timor-Leste Governments have continued to advocate peaceful and inclusive societies for sustainable development, but their implementation into policy outcomes has seriously been questioned (Courvisanos & Boavida, 2018; Courvisanos & Doraisami, 2021).

For sustainable agriculture, Timor-Leste has an inherent natural advantage with Lonely Planet travel guide describing its physical environment: "From its ruggedly beautiful landscapes to its centuries-old traditions, Timor-Leste offers one of the world's last great off-the-beaten-track adventures" (Marx, 2016). Agriculture forms a significant part of both the landscape and its customs, with most of the agricultural systems operating in strong local communities under enduring traditional and organic farming, which does not use chemical fertilizers and pesticides. The soil in Timor-Leste has nutrition elements that do not damage quickly due to only 21% of the land being cultivated two or three times a year, while 79% of the rice paddy land having only one harvest per year (Sabino and Harmadi, 2009). However, the soils are sedimentary with high clay contents and pH that is generally alkaline, resulting in low concentrations of nitrogen that restrict crop yields (Williams et al., 2023). Due to customary history, as at 2023, 8.5% of Timor-Leste's total agricultural land is under organic farming (up from 7% in 2008), whereas all other Southeast Asian countries are below 1% (Kumar et al., 2023; Willer, 2008, p. 113). The one continuing customary ecological rural issue is the massive amount of wood burning that occurs for heat, waste, and clearing land (Tavarez et al., 2021).

There are many challenges with the agricultural system in Timor-Leste which results in low productivity. From a geographical perspective farming operates within a tropical forest ecosystem, but with a traditional dry land cultivation or shifting subsistence agriculture based on maize, sweet potatoes, cassava, taro, cabbage, spinach, and beans. This form of agriculture still dominates in the highland areas. Rice has become the preferred basic food with rice cultivation expanding significantly in the lower lands after "the introduction of improved varieties in the 1960s and the large increase in river-fed irrigation systems following the Indonesian invasion in 1975" (Asian Development Bank, 2021). Around 131,000 tons of rice are consumed annually by the local population, while local rice production in recent years has remained stable at about 86,000 tons per year, creating a shortfall of about 45,000 tons. Imported rice of $38 m occurred in 2021, primarily from (in constant US dollars) India ($20.4 m), China ($11.9 m), Thailand ($3.66 m), Pakistan ($1.78 m), and United States ($383 k) (OEC, 2023). Irrigation is required for rice growing due to the long dry season, since the "[m]onsoon in Timor-Leste that blows from Australia to the west and north tends to be hot and dry" (Tavarez et al., 2021, p. 49). This results in the "hungry season" from November to February with limited second rice crop due to dry conditions (Austin, 2016).

Two other perspectives amplify the geographical challenge. One is the agricultural production systems that operate are poor and inappropriate because traditional dry cultivation has been confronted with the need for large wet-based rice cultivation. This has resulted in a pattern of

[2] All Timor-Leste dollars ($) are equivalent to United States dollars (US$), as this is the legal tender for Timor-Leste.

production in which there are net imports of rice while maize is viewed as surplus and fed to livestock instead. There is also the reluctance to eat livestock (especially chicken) except for special events. The other is the "almost total lack of the agricultural technologies that are taken for granted in similar countries… [this is due to] many small-scale technical innovations that are proven to work in Timor-Leste, but there has been limited progress in disseminating these technologies to farmers" (Asian Development Bank, 2021). As noted by Mattson and Wood (2014, p. 2), "importing technology without adapting it to the specific developing world context is ineffective and unsustainable". Both perspectives are covered in more detail later in this chapter.

The three challenges identified are further exacerbated by independent commentators accusing Timor-Leste Governments across all different administrations in creating a situation such that "agriculture … [has] been neglected, and little has been done to strengthen the country's most valuable resource—its people" (Scheiner, 2021, p.254). Instead, public investment funded from the country's oil-based Sovereign Wealth Fund is centered on setting up the Greater Sunrise gas field and the Tasi Mane LNG export processing plants at an estimated capital cost of $18,050 m (La'o Hamutuk, 2019). Such a government commitment is a resource curse strategy by building "stranded assets" which end up as a liability before the end of their anticipated economic lifetime as a low-carbon global economy quickly emerges (Courvisanos & Doraisami, 2021). This new renewable energy global economy "rules out a lift in investment and employment in the coal and gas industries" (Garnaut, 2021, p. 4); however, Timor-Leste is unsustainably trying to achieve such a "lift" with Tasi Mane.

From a theoretical perspective, the analysis in this chapter is based on ecological limits set by global natural capital (or total renewable and non-renewable resources) and application of this principle in a specific country's socio-economic system. Applying this principle to Timor-Leste is in line with its strong commitment and support for the UN SDGs. This principle implies that market-driven economic growth is inconsistent with the sustainability principle; instead, markets need to serve the sustainability principle. From this perspective, sustainable agriculture for Timor-Leste requires urgent need to shift the dominant work activity in the country from subsistence agriculture to sustainable (green) agriculture. There are four elements to this perspective:

1. Investment in infrastructure that supports agriculture and has strong internal domestic multipliers and not controlled by monopoly capital (Kalecki, 1976).
2. Technological capabilities in agriculture developed for the local population which provides the basis for building workforce capacity for sustainable agriculture (Lall, 1996).
3. Keynesian demand management in support of small-scale agricultural markets and building exports of value-added agricultural products (Keynes, 1936).
4. Learning and innovation strategies for an agricultural industry policy that expedites structural change toward higher value-added activities based on farming (Nelson & Winter, 1982).

Together, these four elements set up the economic drivers to ensure Timor-Leste can make the long-term transition from subsistence to sustainable agriculture. However, the large continuance of a very old form of subsistence agriculture together with the more recent adoption of rice as a staple creates an extremely challenging task for this transition. This challenge is examined in detail below.

Methodology

The assessment of transition in this chapter is underpinned by a critical realist philosophy which resides between the two extremes of positivism (objective measure) and interpretivism (interpret experiences). Critical realism is based on the proposition that the problem under examination exists independently of what is currently known about it. In this chapter, the empirical data presented focuses on transition in agriculture, but this is not contingent on purely what is observable (Easton, 2010). The deep (and often unobservable) elements of the historical and social environment within which agriculture operates are critical to appreciating the challenges. Only a partial view of the problem and the circumstances around it can ever be known. After the state of play in the agriculture economy is examined, the chapter identifies the challenges and opportunities in transition. Finally, a framework for green transition in agriculture is proposed which enables a transitional analysis to be proposed with cultural foundations and effective government policies as a potential pathway toward transition. Such a framework and the analysis that it entails are never decisive or final (Sum & Jessop, 2013), but provide understanding of "the underlying mechanisms that have generated the phenomena of interest" (Bygstad et al., 2016, p. 84)—in this case agricultural transition.

Sustainable agriculture needs to be defined in terms of an approach (or paradigm) to the economic system that ensures the ecological dimension is addressed. Orthodox economics has a sustainability paradigm with market-based economic growth as its base and environmental concerns attended to by identifying market failures that need correction by appropriate scarcity pricing (e.g., carbon tax) and internalizing externalities (e.g., pollution fines). The growth in GDP inherent in this orthodox paradigm depends on the continuing exploitation of resources past ecological limits (e.g., global warming and ocean pollutants) and inequality of outcomes in which the exploitation of the populous less developed world meets the needs of the less populous developed world. The orthodox growth paradigm has been refuted by a deep body of research that exposes the limitations of this approach and proposes an alternative sustainability paradigm which requires humanity to live within ecological limits based on natural capital and social limits set by ethics-based moral capital (Schor & Jorgenson, 2019). This limits-based approach requires markets to operate a circular economy (where nothing is wasted) that serves "a mutually beneficial co-evolution of humans and natural systems in a partnered relationship" (Zhang, et al., 2015, p. 2). The potential transition pathway to sustainable agriculture for Timor-Leste is assessed using the three dimensions that underpin sustainability: circular economy (economic), natural capital (ecological), and moral capital (social).

Research design for this study is based on the perspectives outlined in the introduction. The next section sets out the current state of play in the Timor-Leste agricultural economy based on observation, stakeholder interviews, Timor-Leste literature discussing operation of sustainable agriculture, and secondary data accessed from a variety of sources—Timor-Leste Government documents, organizations' annual reports, NGO websites, World Bank reports, and various official databases.

State of Play in Agricultural Economy

The trade situation in Timor-Leste sets the agenda for the discussion. As a poor agrarian economy with a negligible manufacturing base, export of Liquefied Natural Gas (LNG) from offshore oil reserves dominates its trade income. The trade balance (goods and services) is consistently a large deficit, with significant reductions in the deficit in 2020 and 2021 due to abnormally high gas prices as gas accounts for 92% of all exports. Apart from coffee, there are small exports of nutmeg, konjac, coconut oil, and vanilla (WTO, 2023). The export of a fossil fuel has a very limited short time horizon, which means Timor-Leste needs to drastically alter its export base. This is the reason ASEAN is demanding diversification as a key element of any future accession. There are no tariffs set up by the Timor-Leste Government, as there is no nascent or existing industrial base to protect or support. Given this, Timor-Leste has only an observer status in the World Trade Organization (WTO).

Trade is a large part of Timor-Leste's economy, accounting for 41.5% of GDP (2020–2022). There is a consistent trade deficit pattern since 2015 of just under $1billion per year. This is due to about $500 m in merchandise exports (mainly gas) but double that in merchandise import value which is in roughly equal proportions for all three parts of the economy—basic agricultural foods, all forms of manufacturers, and fuel and mining products. A service trade deficit of about another $500 m is due to the very small export value in travel and the large import value for all forms of services not available domestically in the country (WTO, 2023).

The domestic agrarian sphere is much deeper than initially perceived. The main commodities produced for domestic markets are coffee, rice, corn, beans, carrots, cassava, cinnamon, clove, potatoes, taro, fruit, vegetable, cashew, and cocoa. Most are primarily produced for their own household consumption with any excess sold on small road stalls and village markets. This rich diversity of crops has the potential to create a strong alternative export market if there are market drivers. At this stage, rice is the main market driver with eating rice viewed as high status compared to low status of eating roots and vegetables. The inadequate production of rice leads to the government needing to provide short-term subsidy of imported rice during the hungry season. The new Timor-Leste Government (in 2023) is aiming to re-introduce root and grains crop production as a trial in a region of Timor-Leste (Lospalos), as well as implementing a program to mobilize young people to working on farms based on a South Korean model (Saemaul Undong). Both are small preliminary efforts to go back to local communities for developing agriculture. One longer-term effort is to increase rice production, by mobilizing a national plan. To this aim a team was sent to Israel with a view to understand how rice can be better produced in a dry land environment.

Agricultural labor productivity is below other small island developing states. This has resulted in Timor-Leste falling below the average for other low-income, food-deficit countries and the agricultural sector underperforms in the critical areas of adequate livelihood, food security, and overall economic growth. Because of this low productivity, many farm workers seek extra work on nearby construction projects (e.g., roads, bridges) or acting as market stall vendors. Such work reduces the potential for using labor to build a stronger sustainable agricultural base (World Bank, 2019).

Women contribute almost half of the total agricultural labor force in Timor-Leste and are highly empowered in terms of decision-making in crop farming. They have equal decision-making power when it comes to variety selection. Consequently, no significant difference was observed between male- and female-headed households in terms of adoption of improved varieties. However, as the production process moves up the level away from the farm through marketing, transport, and value added, the decision-making shifts to being dominated by male decision-makers (Akter et al., 2016). This shift removes the potential for women to drive stronger sustainability up the value-added chain in agriculture.

Farming systems are small scale and lack of enough information to bring about change from what is currently being implemented. Such systems mainly exist with low understanding of farming in relation to healthy household diet, nutrients required in the soil, damage done from slash and burn and waste burn-off, and role of technology and education. Opportunities exist for gaining knowledge through the support of Timor-Leste Ministry of Agriculture in forming regional clusters and cooperatives around farming communities. Examples of regional clustering are new attempts with neighboring West Timor farmers to gain knowledge of raising chicken and cows and also exporting candlenut, because West Timor farmers have much higher labor productivity. These cross-border arrangements have the potential to build stronger agriculture that extends past merely subsistence and limited cash cropping. Cooperatives is a form of community ownership that creates minimal scale necessary for producer market impact and mutual cluster of support among farmers with different aptitudes and skills (Moore, 2016). An example is the Kdadalak Sulimutuk Institute ["streams meet to become one river"] (KSI) which brings together farmer cooperatives in the Ermera region and provides them with information, legal advice, and regular meetings on effective access, entitlement, and use of land (da Silva, 2011). Another example is AI-Com which is a collaboration that aims to improve crop productivity by improving nutrients in soil (Williams et al., 2023).[3] Formal parastatals do not exist in Timor-Leste.

Digitalization is non-existent in Timor-Leste. This is due to intermittent electricity supply with some areas having none, poor coverage and power of the internet, and highly limited access to computers in the rural regions. Mobile phones with one provider are the extent of electronic communications and data analysis.

Subsistence farming exacerbates food insecurity in rural areas where 71.4% (as at 2020) of the population live (Timor-Leste National Institute of Statistics, 2023), with half of this population being below the poverty line (Statistics Timor-Leste, 2017). Based on the Food Insecurity Experience Scale (FIES), it is estimated around 20% of households in Timor-Leste suffer from moderate or severe food insecurity. Specifically, nearly 300,000 people experience "crisis" or "emergency" levels of food insecurity. The key long-term drivers are high food prices, reduced purchasing power, persistent impacts of the COVID-19 pandemic, and major flood events (National Directorate of Food Security and Cooperation, 2019). On a seasonal basis, the hungry season prior to maize and rice harvest in March is critical for survival. Farm households cope with this seasonality by diversifying crops with tuber crops (cassava, potato, and sweet potato) providing food when grain stocks dwindle. Short-term strategies include foraging for wild food, selling animals and other assets, and social networks (Erskine et al., 2015). Two fundamental concerns make food security persistent. On the supply side, it is low agricultural productivity as identified above. On the demand side, it is the lack of income to purchase food when stocks are depleted due to limited participation in markets as consumers of food due to subsistence agriculture (Islam et al, 2016).

Ceremonial observances on farms remain an integral part of the crop production cycle in Timor-Leste which can intensify food insecurity. Ceremonies have strong focus on storing food assets (e.g., chickens) and then eating in large quantities on major *lisan* (tribal) group events. There is an awareness by farmers of the need to successfully utilize ceremonies in a balanced approach, but cultural pressures to unbalance are highly prevalent (Browne et al., 2016).

Small, limited self-sufficiency agricultural-based activities are organized by small groups that recognize the importance of bringing about sustainable agriculture. At this niche level of innovation, activities are able to be developed, nurtured, and improved. Such niche innovation enables the bedding down of processes that actually work effectively (Geels et al., 2005). There are excellent examples of NGOs that are leading this niche-innovation phase in climate resilience sustainable agriculture with critical value adding for Timor-Leste. Notable examples are:

1. Permatil (Permaculture Timor-Leste) which aims to lead rebuilding and regenerating agriculture sustainable in culturally appropriate ways. This involves training and demonstration sites through techniques that use local resources, appropriate technology, and low-cost solutions. It has also developed permaculture education syllabuses for school gardens, youth camps, and permaculture guidebooks (Permatil, 2023).
2. PARCIC trains and mentors farmers to form their own producer cooperatives in the sub-district of Maubisse. They produce high quality of coffee "Cafe Timor" for the Japanese specialty market. They promote sustainable agriculture with technical guidance of organic farming and livestock breeding, effective use of firewood, and promotion of agroforestry (PARCIC, 2023).
3. The Timor-Leste Organic Fertilizer (TiLOFE) is a local organization based in Ermera focusing on producing organic fertilizers from local organic materials and water conservation. Volunteer youths sell organic fertilizers to provide capacity building.

[3] AI-Com is a collaborative research program improving agricultural productivity and profitability in Timor-Leste between the Ministry of Agriculture and Fisheries, the National University of Timor Lorosa'e, World Vision Timor-Leste and the University of Western Australia. It is supported by the Australian Centre for International Agricultural Research. See https://ai-com.tl.

4. Raebia work alongside communities in highland remote areas of Timor-Leste where shifting (slash and burn) agriculture exists to promote food and livelihood security and to ensure sustainable natural resource management. Raebia promotes principles of conservation agriculture, agro-biodiversity, and agroecology (Raebia, 2023).
5. Agora Food Studio is a restaurant in Dili which conducts food experimentation by young Timorese to collaborate and harness the power of nature through agriculture in the mountains. Techniques used are ones that local communities have long adapted to survive in remote and harsh environments, sharing vanishing knowledge which has evolved over generations (Board, 2021).
6. Dilicious Restaurant with Chef Cesar Trinito Gaio aims to serve only locally sourced food and runs gastronomy classes with native ingredients. The long-term objective is to educate people across the country with a traveling food truck about the benefits and taste of local food, exploring further possibilities of farm-to-table dining and help to build a strong agri-tourism sector (Board, 2021).
7. Distillery Lokal offers a boutique range of root-based sodas, wild ferment, raw honey meads, and traditional botany and bark-based vermouths with ingredients sourced directly from local producers (Board, 2021).

Two problems have emerged in efforts to effectively take off from such niche-level activities described in the above list through robust sustainability transitions into an embedded regime of change. One is on the demand side with the lack of scale in a very small domestic market. This prevents building significant financial resources, tools, and facilities needed for going to the next level of production (Board, 2021). The other is on the supply side with the lack in Timor-Leste of skilled experts in the fields of endeavor that these activities are involved, and sustainability education for such skills has not been developed.

Challenges and Opportunities

Land ownership and tenure is the primary area of government responsibility that needs attention. Secure land tenure and access to land is key to long-term development of agriculture. Colonial and invasion occupations distorted ownership from traditional claims, leaving great uncertainty as to who owns the land. This is a major disincentive to accessing bank loans and investing in the land. All international studies show that clear and unambiguous land tenure legislation (individual and/or collective) provides security and the right to use the land as a long-term economic asset. As the President of the National Parliament stated when opening it on September 20, 2016: "…land law is essential for the development of the country. It will create the necessary legal reliability for the property ownership regime to ensure investment and the integrated and sustainable development of the country" (da Costa, 2016). A land entitlement bill was passed in 2019, but its implementation has been stifled due to two major reasons. One is that a few large powerful families have much land spread all over Timor-Leste with strong commercial interests (note especially the Albano, Alkatiri, Alves, and Carrascalao families). The other is that many Timorese illegally occupy "abandoned" land of former Indonesian state and citizens (which the current state legally owns).

Some limited progress on resolving land tenure issues has been achieved. A computerized Land Registration and Information System using GIS mapping with the assistance of USAid has been developed to support decision-making on tenure, but poor internet access has limited its application. This system enables Timorese to check existing documents on land titles, ownership, and other relevant data through an online platform, but with no resolution of land title rights. Effective digitalization would contribute to securing land tenure by providing a reliable and transparent system for recording and managing land rights. Apart from poor internet access, successful digitalization requires capacity building of government officials and stakeholders. Some senior staff from the

Ministry of Agriculture were sent in 2023 to Israel to study agricultural and fishery technologies and policy implementations, including operating land tenure systems.

Coffee is the only non-subsistence crop. From Portuguese times, it has been an export crop, but the role of coffee in the economy of Timor-Leste is limited as it stands, with annual production of around 88,000 60 kg bags being country ranked 37th in the world. The Arabica coffee varieties produced are Catimor, Typica, and Timor Hybrid (Hibrido de Timor) in four key regions of Ermera, Ainaro, Aileu, and Same. Coffee is the largest agricultural export but well behind oil. The colonial Portuguese landowners introduced coffee into agriculture, but with the annexation of Timor-Leste by Indonesia in 1976, coffee production shrunk considerably as Indonesian administrators focused on wet-rice cultivation. With the expulsion of Indonesia, assistance through the United Nations and international aid made coffee the country's only cash crop as old Portuguese plantations were quickly brought back into production. Coffee now accounts for about 80% of total agricultural export revenue, with it being the primary source of income for a third of the population. However, around half of all coffee farms require intensive rehabilitation with plantation trees overgrown or long passed their productive life. The lack of land tenure certainty due to competing prior claims (ancestral, Portuguese, Indonesian, and post-independence) has prevented long-term investment in new plantation trees (Courvisanos & Boavida, 2018). Thus, coffee production "is based on relatively unmanaged plantations, with bean-gathering and processing by villagers, and scant attention paid to cleaning and weeding, pruning, pest and disease management or planting of new trees" (Ministry of Agriculture & Fisheries, 2019).

Timor-Leste has the potential for substantial expansion of its coffee exports. With little of the world's coffee grown wild at high altitudes of a tropical island based in small communities, Timor-Leste coffee production is unorthodox and product organic. The potential for substantial expansion of coffee exports is based on the new unique coffee variety known as "Hibrido de Timor" evolving from Timor-Leste's fertile mountain volcanic soils. Alf Kramer, a leading coffee expert, has described Timor's new variety coffee as "among the top 1% in quality and price worldwide" (Cafendo, 2023).

The challenges to develop the Timor-Leste coffee export market are many. Factors to deal with are the highly variable global price, deficient infrastructure, lack of capital investment in plantations, low nutrient-content in the soil, and arid weather with short rainy season that makes the coffee cherry difficult to grow. PARCIC has made significant strides in bringing coffee farmers together under cooperatives, aiding them to produce a high-quality sustainable crop, and marketing to the high-end Japanese coffee market. Increasing productivity of coffee farms by a national rehabilitation scheme is critical given the current farm average of 500 g of green coffee per tree (Timor-Leste Lulirema Natural, 2023). However, lack of budgeting by the new government in power has halted this scheme. Greater long-term planning and control of the Timor-Leste agricultural production system is required to address these challenges, which needs much Timor-Leste government administration and funding that has so far been lacking. Such government support also needs partnering with PARCIC and the Timor-Leste Coffee Association (Assosiasaun Café Timor-Leste–ACTL)—the industry association that brings together exporters, roasters, traders, farmers to exhibit the unique nature of Timor Coffee as a luxury coffee for western consumers. Associations in other food sectors need to be developed based on the needs, understandings, and input of the farmers who belong to such cooperatives.

Farm size is a challenge, but not in the way mainstream economists understand the term. Timor-Leste has everything small—population, plots of land (subsistence and supplying local markets), and rugged landmass. This means that scale-efficient economies are not achievable in reducing costs of production. Production will always be costlier in Timor-Leste. The aim is to increase the productivity of farms by more sustainable means of production, processing and organizing distribution and marketing. Land tenure, permaculture techniques, high-quality organic pesticides and fertilizers, better new seeds, cooperatives and producer associations to create effective clusters, and overall government coordination and planning. NGOs (like PARCIC and Raebia) have been the critical support to create niche-based sustainable agriculture and show transition can be successful. NGOs have provided climate resilience expertise on the farm itself or in building marketing strategies in countries around the world that are responsive to "quality at a price" organic, sustainably produced crops. However, NGOs do not have the ability to build a long-term productive sustainable overall farm system in Timor-Leste. This needs strong and well-funded government coordination.

In addition, there is the food security challenge of having access to seeds and the threat of extinction of local seeds in Timor-Leste within the next couple of years. The National Seed Producers and Trade Association has installed seed warehouses in all municipalities to collect locally grown seeds because Timor-Leste is "heavily dependent on imported seeds, especially the imported hybrid vegetable seeds from neighboring country Indonesia. In addition, nowadays, vegetable farmers could easily find and purchase any type of imported hybrid vegetable seeds from China at any Chinese store in Dili and municipalities" (Martins, 2022). For strong sustainable agriculture, the development of commercial seed trade is essential.

The new Timor-Leste Government elected in May 2023 has great ambitions in terms of extensive farming by broadening the variety of agricultural cash foods produced in the country and to double the production of rice and coffee. Central to this is to identify and encourage development of abundant unfarmed lands that are currently abandoned (no known owners) or unused (absent owners). However, a strategy for identifying and encouraging has not been determined. The best news is the significant GDP growth of protein-rich foods by farmers themselves, as this creates the opportunity for a healthier nutritional diet for a malnourished population, if encouraged and managed properly.

Intensive farming is even more of a problem. This requires a high level of input per unit of land, such as increased use of labor, machinery, fertilizers, and technology in order to maximize yield and raise productivity. Farming skills need substantial upgrading above mostly a subsistence type skill base with careful land management to avoid environmental degradation and resource depletion.

Framework for Green Transition in Timor-Leste Agriculture

A guide to an effectual green transition in Timor-Leste agriculture is presented in this section based on the framework of societal evolution and communicative action, developed by Jürgen Habermas (1979, 1987). This approach is underpinned by a critical realist understanding of how communities need to create, sustain, and develop social relationships through effective communication. The central assumption is that social relationships and actions take place through consensus and agreement attained by a common understanding of norms, meanings, and values (Habermas, 1979). Thus, this section is based on the Habermas (1987) framework that facilitates appropriate actions in the socio-cultural and political system of Timor-Leste that empowers all stakeholders to participate in agricultural green transition.

The Habermasian framework, presented in Fig. 12.1, brings into focus the Lifeworld, or socio-cultural and historical factors that provide an understanding of what sustainable agriculture actual means to the stakeholders, made up of farmers, processors, suppliers, consumers, nearby communities, and particularly to the politicians (and public servants) that need to guide implementation of such a transition. This Habermasian socio-political analytics model is adopted from Zacharias (2007) which examines causal mechanisms operating in a nation in relation to work-life balance government policies.[4]

[4] Brandenburg et al. (2022) adopts the same Zacharias (2007) Habermasian model to examine the causal mechanisms operating on stakeholders in a university environment (managers, administrators, research staff, and

Figure 12.1 has three integrating societal evolutionary stages shown on the left side: Lifeworld (LW), Steering Media (SM), and Systems (SYS), with arrows indicating influences from one stage to another can be both top-down and bottom-up. The left side shows the Zacharias (2007) exposition of the Habermas (1987) theory of societal action. The right side applies these three stages to Timor-Leste agriculture. LW has three factors which exist at the broad societal and cultural level, which in the case of Timor-Leste agriculture refers to the core cultural identity that two-thirds of households are directly linked to (and the rest indirectly linked to); the ecological (physical environment) and social customs that underpin life in Timor-Leste; and deep traditional sustainability for survival of individuals that defines Timorese personality. A strong cultural identity in Timor-Leste is associated with survival, resilience, and freedom. Leach (2017) shows how this identity has been forged through a long historical period of exploitative colonization and violent invasion. This distinctive LW stage underpins the operation of the next two stages.

The second stage (SM) reflects the activities of major actors in Timor-Leste who impact on the conduct of agricultural sector. Broadbent et al. (1991) show how government, professional, and financial institutions drive such actors. In this context, Fig. 12.1 shows two institutional factors that steer the operation of agriculture. One is the political and legal framework which sets laws, regulations, and standards (Power). The other is profit or surplus (Money) that drives financial institutions, investors, and supporting organizations like NGOs. The institutions of Power and Money play the conjoining role between the first and third stages to "…guarantee social cohesion" (Zacharias, 2007, p. 88). An example of Power and Money being integrated is the "Credito Suave" program of the VIII Constitutional Government (2018–2023) which aimed to provide a $50,000 loan for small and medium enterprises (including farms). The loan was allocated by BNCTL (Timor's national bank of commerce), but implementation was poor due to the complicated administration of the bank. As in this program, the conduct of agriculture is strongly shaped by the operations of actors with Power and Money in the way the agricultural sector plays out its economic, ecological, and social roles in Timor-Leste.

The third analytical stage in Fig. 12.1 is SYS, which represents individual actors (especially farmers and food processors) operating in systems that function at the community/local space in response to the SM. These agricultural operations systems are in response to the actions of the SM but follow LW imperatives, which Zacharias (2007, p. 88) sees as "expressions of the Lifeworld and which follow Lifeworld concerns". With the support of some major actors in Timor-Leste, especially NGOs, individual actors can cluster together as cooperatives and producer groups to propel back up the analytical framework from the bottom-up to influence the SM stage, and even address the critical LW perspectives so that sustainability becomes a dominant theme in agriculture going into the future. How the agricultural sector transitions from subsistence to sustainable activity is an empirical issue viewed from the lens of Fig. 12.1 and is discussed next in the context of past and future policy actions of major actors, especially government.

The evidence provided earlier in this chapter shows the neglect of agriculture by all government administrations since independence in 2002. In contrast, based on the study by Tavarez et al. (2021), the Indonesian occupational administration strongly supported agriculture through its top-down SM Power and Money in a transition from traditional dry land cultivation and shifting subsistence to wet-rice cultivation and large areas of shifting cultivation in the highlands at the SYS stage. However, the Indonesian administrators did not pay attention to the condition of local communities and farmers. This had the effect of maintaining subsistence with dependence on imported Indonesian rice for a new diet that aimed to tie Timorese to the Indonesian LW culture. Instead, East Timorese adapted their subsistence farming (based on existing LW) to supplementing imported rice for a rice-based diet,

industry partners) in order to encourage policies for effective interdisciplinary research.

resulting in regular hungry seasons and significant food insecurity.

From the SM stage, all Timor-Leste government administrations have regarded oil revenue as able to fund economic development through infrastructure construction supported by a large public administration, which will eventually trickle down to agriculture, changing the SYS to sustainability. The evidence shows that there has been no trickle down and of more concern is that such a top-down approach does not directly address the concerns of the vast farm-based household communities across the country (Courvisanos & Boavida, 2018). The new Timor-Leste Government (IX Constitutional) highlights the importance of diversification to agriculture. It has adopted the Blue and Green Economy for Sustainable Development Plan aiming to capitalize on both land and maritime resources sustainably by encouraging potential investors to focus on investment in agriculture and fishery oriented local food industries. This plan has a strategy to increase investment and improve the agriculture sector's market access and value chain so that current subsistence farming practices can progress toward commercial viability.[5] However, this current government has allocated only 2.2% of its 2024 state budget to the agriculture sector. This budget is 84% dependent on oil funds, but oil reserves are decreasing quickly with prospects of new oil revenues coming through remaining highly uncertain (Courvisanos & Doraisami, 2021). Thus, a top-down development strategy for sustainable agriculture like this Blue/Green Plan is severely hampered by lack of funds required in both infrastructure (roads, electricity, health, digitalization) and education (basic primary, nature-based, marketing) in the Timor-Leste rural regions, making it very difficult to attract private investment into the agriculture sector.

The alternative bottom-up approach begins with addressing SYS in the context of LW. The seven niche innovations in sustainable agriculture described above are such SYS examples of bottom-up development with the major assistance of NGOs. These sustainability alternatives need to take off in a significant commercial setting so that their activities will influence the broader SM stage in shifting the Power and Money factors away from fossil fuels and into sustainable agriculture. The two SYS factors that need strong development assistance are organizations/institutions (local-based community policies and practices), and operating responses by actors in the agriculture space (farmers, processors, nature-based food catering). This focus requires much less funding than top-down development and supports niche activities to become embedded so that they lead to transition. It does require institutional capacity building for effective policy implementation at the local level while ensuring that social inclusivity and environmental sustainability are integral components of the Timor-Leste agricultural SYS moving forward.

Collaborative small-scale sustainable farming is the basis of the first four niche-innovation activities described above. Small local producers should be the basis for good value-added food production SYS involving the processing raw agricultural products into higher-value products such as jams, sauces, or packaged goods with cassava, sweet potatoes, and beans. This adds value to a bottom-up agricultural sector and can lead to increased income for local farmers. Also, it can assist in establishing food processing industries which can create jobs in rural areas, contributing to rural development and poverty alleviation.

Formal education provides necessary learning devised from the top-down by centralized curriculums. There are many challenges and limitations to success in formal education in TL, see Taylor-Leech (2013), which has also made agriculture unattractive to youth (Courvisanos & Boavida, 2018). Social learning initiated by NGOs and small sustainability entrepreneurs need to be brought into the agriculture education space to address the limitations of learning through formal education. This latter learning type is appropriate for sustainable agriculture as it "...is not a search for the optimal solution to one problem but an

[5] The information for this example is based on the Timor-Leste's International Business Forum with the title: "Discovering Blue and Green Opportunities for Sustainable Investment", held in Dili, 23–25 November 2023. See http://timor-leste.gov.tl/?p=35243&lang=en&n=1.

ongoing learning and negotiation process where a high priority is given to questions of communication, perspective sharing and development of adaptive group strategies for problem solving" (Pahl-Wostl & Hare, 2004, 193–4). Social learning in sustainable education is required in agriculture. This form of community-oriented education has been successful with nurses and mothers in health and nutrition of Timor-Leste children (Walker & Kirk, 2013). As with health, there needs to be a serious commitment of education investment in social learning from a holistic ecological systems perspective incorporating community input in SYS factors, organizations, and small institutions and operating responses. The seven niche initiatives in Timor-Leste agriculture can form the basis of such ecosystems social learning. The experience with the KSI cooperative farmer's education institute in Ermera, where farmers gain greater control over their own means of production, shows the possibilities of social learning in sustainable agriculture (da Silva, 2011). Such an approach is consistent with the LW long-held culture, as can be noted by section 138 of the Timor-Leste Constitution under "Economic organization" which sets out the "co-existence" of cooperatives and social sector ownership in the economy together with the public and private sector means of production (RDTL, 2002). The reality through the existing SM of government is much more modest with 53% of producer cooperatives and 32% of credit unions being "inactive" (MCIA, 2013).

The World Bank (2019) provides a clue on how bottom-up sustainability development in agriculture can occur. It suggests the use of organic chemical supplements to increase food quality, promoting higher-value products, and technology transfers with knowledge exchange. The Bank states, "the following factors ought to be considered: facilitation of mechanization, improvement in the use of chemical (fertilizer, pesticides, and herbicides) inputs, enhancement in access to credit and extension, encouragement to farmers to join farmer groups, expansion of the commercialization of farm produce, and reduction in the gender gap in agricultural productivity" (World Bank, 2019, p. 3). In effect, what is required is exposure by farmers to new agricultural processes and practices that will alter the country's SYS from the bottom-up.

Strong opportunity exists for elements of a community-based agricultural sector in Timor-Leste to focus on activities like the three niche innovations described above by Board (2021)—Agora Food Studio, Dilicious Restaurant, and Distillery Lokal. The aim in this case is to build an agri-tourism sector that will address a global high-end tourism market that demands sustainable agriculture tourism experiences. This will encourage diversification of agricultural products, promoting traditional cuisines, and enhancing the overall quality of agricultural output.

Conclusion

The four elements of the theoretical perspective can now be brought back into this conclusion to provide an overall assessment of the Timor-Leste transition to sustainable agriculture. Firstly, the top-down investment to support agriculture has dealt with very basic infrastructure in electricity, roads, and internet. This builds the absolute necessary aspects of any economy, including agriculture. This is slow and ongoing, lacking strong internal domestic multipliers as all construction work is done by large overseas corporations with mostly non-Timorese workers. Secondly, technological capabilities have been developed through NGOs (sometimes partnering with government) with a strong focus on bottom-up but starting from a low base with sustainable agriculture concentrated on niche initiatives. Thirdly, demand management in support of small-scale agricultural markets is totally missing from the top-down and only with marginal impact through NGOs from bottom-up. Finally, learning and innovation strategies for higher value-added activities are not being planned. Implementation of such does occur, but in a piecemeal manner with little coordination toward transition to sustainable agriculture.

Sahlins (1972) shows how a subsistence economy per se is not bad. It can be the starting point for a strong viable development strategy built on reciprocity and altruism, and not aiming to copy commercial capitalist structures. As shown here, the Habermas framework to examine past and potential future transitions to sustainable agriculture can be an effective transitional analysis tool for such alternative sustainable agriculture. This transitional tool could guide greater coordination between top-down and bottom-up strategies in order to shift Timor-Leste towards a unique organic-focused sustainable agriculture sector.

References

Akter, S., Erskine, W., Branco, L.V., Agostinho, O.F., Imron, J., & Spyckerelle, L. (2016). 'Gender in crop production in Timor-Leste'. In H. Nesbitt, W. Erskine, C.J. da Cruz, & A. Moorhead (Eds.), *Food Security in Timor-Leste through Crop Production* (pp. 158–164), ACIAR Proceedings Series No. 146. Australian Centre for International Agricultural Research (ACIAR).

Asian Development Bank. (2021). Sector assessment (summary): Agriculture, natural resources, and rural development. *Country Partnership Strategy: Timor-Leste 2016–2020*. Asian Development Bank. https://www.adb.org/sites/default/files/linked-documents/cps-tim-2016-2020-ssa-07.pdf

Austin, N. (2016). "Forward". In H. Nesbitt, W. Erskine, C.J. da Cruz, & A. Moorhead (eds), *Food Security in Timor-Leste through Crop Production* (p. 3). Australian Centre for International Agricultural Research.

Banco Central de Timor-Leste. (2020). *Balance of Payments 2006–2020*. Central Bank of Timor-Leste. https://www.bancocentral.tl/en/go/balance-of-payments

Board, J. (2021). *Food everywhere, but nothing to eat: Timor-Leste is going hungry and the answer may be innovation*. CNA (Channel News Asia), 4 May. https://www.channelnewsasia.com/asia/china-belt-and-road-initiative-1-trillion-owed-debt-3902846

Brandenburg, R., Smith, J., Higgins, A., & Courvisanos, J. (2022). 'Breaking out': The Genesis, Development, and Implementation of an Interdisciplinary University Cross-School Research Group. *The Australian Educational Researcher, 49*(3), 489–510.

Broadbent, J., Laughlin, R., & Read, S. (1991). Recent financial and administrative changes in the NHS: A critical theory analysis. *Critical Perspectives on Accounting, 2*(1), 1–29.

Browne, M., Goncalo, L., Ximenes, A., Lopes, M., & Erskine, W. (2016). The place of ceremonial observances in sustainable livelihoods of crop producers in Timor-Leste. In H. Nesbitt, W. Erskine, C.J. da Cruz, A. Moorhead (Eds.), *Food Security in Timor-Leste through Crop Production* (pp. 128–134). ACIAR Proceedings Series No. 146. Australian Centre for International Agricultural Research (ACIAR).

Bygstad, B., Munkvold, B. E., & Volkoff, O. (2016). Identifying generative mechanisms through affordances: A framework for critical realist data analysis. *Journal of Information Technology, 31*, 83–96.

Cafendo (2023). Timor-Leste: Coffee Digital Catalog. https://www.cafendo.com/coffee/asia/timor-leste

Centre for Development Economics and Sustainability. (2023). *Agricultural GDP: Patterns and Trends, Research Briefs on Timor-Leste (RB-TL4)*. Monash University.

Courvisanos, J. (2012). *Cycles, Crises and Innovation: Path to Sustainable Development—A Kaleckian-Schumpeterian Synthesis*. Edward Elgar.

Courvisanos, J., & Boavida, M. (2018). *Review of the Roadmap for Sustainable Development in Timor-Leste: An Economic Policy Report*. Global Institute for Sustainable Prosperity, Policy Report No. 102.

Courvisanos, J., & Doraisami, A. (2021). The timor-leste petroleum fund: From buying peace to white elephants. In E. Okpanachi & R. E. Tremblay (Eds.), *The Political Economy of Natural Resource Funds* (pp. 233–260). Palgrave Macmillan.

da Costa, A. H. (2016). *Speech by the President of the National Parliament at the Opening of the Fifth Legislative Session*. 20 September, Dili.

da Silva, A. B. (2011). *FRETILIN Popular Education 1973–1978 and its Relevance to Timor-Leste Today*. PhD thesis, Armidale: University of New England, Australia.

Easton, G. (2010). Critical realism in case study research. *Industrial Marketing Management, 39*(1), 118–128.

Erskine, W., Ximenes, A., Glazebrook, D., da Costa, M., Lopes, M., Spyckerelle, L., Williams, R., & Nesbitt, H. (2015). The role of wild foods in food security: The example of Timor-Leste. *Food Security, 7*, 55–65.

Garnaut, R. (2021). *Reset: Restoring Australia after the Pandemic Recession*. La Trobe University Press.

Geels, F. W. (2005). *Technological Transitions and System Innovations: A Co-evolutionary and Socio-technical Analysis*. Edward Elgar.

Habermas, J. (1979). *Communication and the Evolution of Society (translated from German by Thomas McCarthy)*. Heinemann.

Habermas, J. (1987). *The Theory of Communicative Action: Vol. 2: The Critique of Functionalist Reason* (translated from German by Thomas McCarthy). Polity Press.

Islam, S., Saif, S. S., Maiden, J., & Raab, R. (2016). Strengthening agricultural markets to enhance food security in Timor-Leste. In H. Nesbitt, W. Erskine, C.J. da Cruz, & A. Moorhead (Eds.), *Food Security in Timor-Leste through Crop Production* (pp. 114–121). ACIAR Proceedings Series No. 146. Australian Centre for International Agricultural Research (ACIAR).

Kalecki, M. (1976). *Essays on Developing Economies*. Harvester Press.

Keynes, J. M. (1936). *The General Theory of Employment*. Cambridge University Press.

KONSSANTIL. (2023). Over 20% of the Timorese population suffering from food insecurity. *Tatoli*, 12

January. https://en.tatoli.tl/2023/01/12/konssantil-20-of-the-timorese-population-suffering-from-food-insecurity/15
Kumar, D., Willer, H., Ravisankar, N., Chandrabhanu, & Kumar, A. (2023). Current Scenario of Organic Farming Worldwide. In D. Kumar, N. Ravisankar, & A. Panghal (Eds.), *Transforming Organic Agri-Produce into Processed Food Products Post-COVID-19 Challenges and Opportunities*. CRC Press.
Lall, S. (1996). *Learning from the Asian Tigers*. Macmillan.
La'o Hamutuk. (2019). *South Coast Petroleum Infrastructure Project*. Updated 2 April 2019. https://www.laohamutuk.org/Oil/TasiMane/11TasiMane.htm#contracts
Leach, M. (2017). *Nation-building and national identity in Timor-Leste*. Routledge.
Mattson, C. A., & Wood, A. E. (2014). Nine principles for design for the developing world as derived from the engineering literature. *Journal of Mechanical Design, 136*(12), Article 121403.
Martins, F. (2022). ANAPROFIKO-Govt to increase local seed production. *Tatoli*, 14 March. https://en.tatoli.tl/2022/03/14/anaprofiko-govt-to-increase-local-seed-production/16/#:~:text=During%20the%20meeting%2C%20the%20Prime,seeds%20from%20neighboring%20country%20Indonesia.
Marx, S. (2016). *Timor-Leste's Non-Oil Economy Must Look to Tourism*. The Asia Foundation, 22 June. http://asiafoundation.org/2016/06/22/timor-lestes-non-oil-economy-must-look-tourism
MCIA. (2013). *Perfil Industria Ki'ik no Kooperativa [Profile of Small Industry and Cooperatives]*. MCIA (Ministério Comércio Indústria e Ambiente).
Ministry of Agriculture and Fisheries. (2019). *National Coffee Sector Development Plan 2019–2030*. Asian Development Bank. https://www.adb.org/sites/default/files/linked-documents/51396-001-sd-02.pdf
Moore, C. (2016). *Challenges of addressing and resolving post-colonial and post-conflict land issues and disputes*. Seminar presentation, America House, Liceu Campus, Universidade Nacional Timor Lorosa'e, 27 October, Dili.
National Directorate of Food Security and Cooperation. (2019). *The First IPC Analysis Report on the Chronic Food Insecurity Situation in Timor-Leste*. Ministry of Agriculture and Fisheries.
Nelson, R. R., & Winter, S. G. (1982). *An Evolutionary Theory of Economic Change*. Harvard University Press.
OEC. (2023). *Rice in Timor-Leste*. Observatory of Economic Complexity (OEC). https://oec.world/en/profile/bilateral-product/rice/reporter/tls
Pahl-Wostl, C., & Hare, M. (2004). Processes of social learning in integrated resources management. *Journal of Community and Applied Social Psychology, 14*(3), 193–206.
PARCIC. (2023). *PARCIC website*. https://www.parcic.org/english/timor-leste
Permatil. (2023). *A Short History*. Permatil website. https://permatilglobal.org/permatil-timor-leste
Raebia. (2023). *About Raebia*. Raebia website. https://www.raebia.org/about_us
RDTL. (2002). *Constitution of the Democratic Republic of Timor-Leste* [English version]. República Democrática de Timor-Leste (RDTL), 20 May. http://timor-leste.gov.tl/?cat=37&lang=en
Rasiah, R. (2025). Building Sustainability and Food Security in Southeast Asia, R. Rasiah, V. Anter & N. I. Annizah (Eds), *Sustainable Agriculture and Food Security in Southeast Asia: Towards Climate Resilience and Digitalization,* Singapore: Springer Nature.
Rowthorn, R., & Wells, J. (1987). *Deindustrialization and Foreign Trade*. Cambridge University Press.
Sabino, M. A., & Harmadi, S. H. B. (2009). *Agricultural policy strategies to address challenges facing Timor-Leste: Rural poverty alleviation and social transformation*. Paper presented in the World Civic Forum 2009, 5–8 May, Kyung Hee University, Seoul, South Korea.
Sahlins, M. (1972). *Stone age economics*. Aldine.
Scheiner, C. (2021). Timor-Leste economic survey: The end of petroleum income. *Asia & the Pacific Policy Studies, 8*(2), 253–279.
Schor, J. B., & Jorgenson, A. K. (2019). Is it Too Late for Growth? *Review of Radical Political Economics, 51*(2), 320–329.
Seixas, P. C., Mendes, N. C., & Lobner, N. (2023). Timor-Leste's Membership of ASEAN: The Political Process and Its Discontents. In P.C. Seixas, N.C. Mendes, N. Lobner (Eds.) *The Paradox of ASEAN Centrality: Timor-Leste Betwixt and Between* (pp. 25–46). Brill.
Shoesmith, D. (2020). Party Systems and Factionalism in Timor-Leste. *Journal of Current Southeast Asian Affairs, 39*(1), 167–186.
Statistics Timor-Leste. (2017). *Census Publications: Employment Sector*. General Directorate of Statistics. http://www.statistics.gov.tl/census-2
Sum, N. L., & Jessop, B. (2013). *Towards a cultural political economy: Putting culture in its place in political economy*. Edward Elgar.
Tavares, J., Rochwulaningsih, Y., & Rinardi, H. (2021). Implementation of modern agricultural food system in Timor-Leste, 1982–2007. *IHiS (Indonesian Historical Studies), 5*(1), 45–58.
Taylor-Leech, K. (2013). Finding space for non-dominant languages in education: Language policy and medium of instruction in Timor-Leste 2000–2012. *Current Issues in Language Planning, 14*(1), 109–126.
TiLOFe. (2023). *TiLOFe website*. https://en.tatoli.tl/tag/tilofe
Timor-Leste Lulirema Natural. (2023). *Passport Specialty Coffee*. https://passportcoffee.com.au/blogs/roastery-ramblings/timor-leste-lulirema-natural
Timor-Leste National Institute of Statistics [INTEL]. (2023). *Timor-Leste Population and Housing Census 2022: Main Report*. INTEL. https://www.inetl-ip.gov.tl
UNDESA. (2022). *World Population Prospects 2022 Demographic Indicators by Region, Subregion and Country: Annually for 1950–2100*. United Nations Department of Economic and Social Affairs (UNDESA), United Nations Population Division.

United Nations. (2015). *Transforming Our World: The 2030 Agenda for Sustainable Development*. Resolution A/RES/70/1.

Walker, C., & Kirk, J. M. (2013). From cooperation to capacitation: Cuban medical internationalism in the South Pacific. *International Journal of Cuban Studies, 5*(1), 10–25.

Willer, H., Yussefi-Menzler, M., & Sorensen, N. (2008). *The World of Organic Agriculture: Statistics and Emerging Trends*. Earthscan.

Williams, R., Belo, J. B., Lidia, J., Soares, S., Ribeiro, D., Moreira, C. L., Almeida, L., Barton, L., & Erskine, W. (2023). Productivity gains in vegetables from rice husk biochar application in nutrient-poor soils in Timor-Leste. *Scientific Reports, 13*, 10858.

World Bank. (2019). *Making Agriculture Work for the Poor in Timor-Leste*. World Bank. https://doi.org/10.1596/32017.

Zacharias, N. (2007). *Work/life Balance Through a Critical Gender Lens': A Cross-country Comparison of Parental Leave Provisions and Take-up in Australia and Sweden*. PhD thesis. University of Ballarat, Australia.

Zhang, X., Skitmore, M., De Jong, M., Huisingh, D., & Gray, M. (2015). Regenerative sustainability for the built environment–from vision to reality: An introductory chapter. *Journal of Cleaner Production, 109*, 1–10.

13 Summing up: Towards Strengthening Smartification and Climate Resilience

Rajah Rasiah

Abstract

This chapter presents a synthesis of the sustainable agricultural development experiences of the Southeast Asian countries. In doing so, the chapter focuses on initiatives to raise agricultural productivity and end hunger, and to quicken the proliferation of digitalization and strengthening of climate resilience. Despite the uneven nature of development and unequal capabilities differentiating the countries, the chapter argues that the prospects for establishing a climate smart sustainable agriculture system in Southeast Asia look bright. It calls for effective implementation and constant appraisals to strengthen the planning process with strong coordination between central, provincial, and local policies, the constant identification and implementation of best climate smart agricultural practices to enable the member countries to meet the relevant SDGs, even if it means that countries, such as the Philippines, Lao PDR, Myanmar, and Timor-Leste may defer their targeted deadlines. Much depends on political stability, continued cooperation among the countries, and strong persuasion to seek diplomacy to achieve political, economic, and social consensus.

Keywords

Smartification · Climate resilience · Sustainable agriculture · Southeast Asia

The pursuit of the United Nations' (UN) (2016) sustainable development goal (SDG) 2 to end hunger calls for a rethink of Ricardo's (2004) prescription to specialize on the basis of static comparative advantage,[1] as countries become vulnerable when they deepen their dependence on essential food imports. Although economists generally agree on the need to promote free trade owing to its overall benefits (Mankiw, 2014), exceptions to such a rule are often needed to avert potential vulnerabilities to national food security, as well as when nations target focused industrial policy interventions to stimulate industrial upgrading (Rasiah, 2025). The recent disruption on imports of wheat, and rape seed and sunflower seed oils following the outbreak of the

R. Rasiah (✉)
Asia-Europe Institute, Universiti Malaya, Kuala Lumpur, Malaysia
e-mail: rajah@um.edu.my

[1] Using real costs on the basis of commodities, Ricardo (2004[1921]: 135) had famously argued that England should specialize on clothing manufacturing and Portugal on wine manufacturing, and then to trade with each other.

R. Rasiah et al. (eds.), *Sustainable Agriculture and Food Security in Southeast Asia*, Sustainable Development Goals Series, https://doi.org/10.1007/978-981-95-0477-0_13

Russia-Ukraine war in 2022 is an example. Its consequences have been dire to several countries, including Britain and Germany. Although not associated directly with agriculture, economic sanctions on Russia since 2022 also attracted the still main source of energy, i.e. oil and gas from Russia, as well as, Russia and China restricting exports of fertilizers. In this context, most Southeast Asian nations earmarked staple food cultivation as the cornerstone of their agricultural as well as economic development policies. Since a massive share of the population in the poorer economies of Southeast Asia, such as Cambodia, Lao PDR, Myanmar, and Timor-Leste, and the lower middle-income economies of Indonesia, the Philippines, and Vietnam still work in farms and agriculture, employment with decent incomes still remains important SDG goals (especially SDG8).

Also critical to the SDGs is the calls to displace fossil fuels with clean and efficient energy (SDG7), sustainability of jobs and improvement in incomes (SDG8), and infrastructure to stimulate innovations (SDG9). While SDGs 7, 8, and 9 cut across all sectors in the whole economy of nations, governments have been encouraged to focus on these SDGs in their initiatives to raise agricultural productivity and reach to the masses.

Given the complex history, and wide differences in the period of integration with the capitalist world economy, it has to be expected that the market economies of Indonesia, Malaysia, Philippines, Thailand, and Brunei enjoyed a temporal advantage over Vietnam, Cambodia, Lao PDR, and Myanmar as these transition economies only started market reforms from the 1980s and 1990s. Meanwhile, although Timor-Leste gained independence from Indonesia in 2002, the country is still struggling to establish an orderly administrative framework of governance. Hence, it is to be expected that Southeast Asia will present a set of uneven inter-country developmental experiences. Also, the geographical differences, such as Singapore's status as a world-class entrepot centre but small in size, while Timor-Leste is also small but is located some distance away from the main trade routes. Meanwhile, Malaysia, the Philippines, Thailand, and Vietnam are medium-sized countries with strong historical links with trade. Indonesia is the largest country in Southeast Asia with a large population with several Islands significantly less developed than Java.[2]

This chapter seeks to offer a synthesis of the experiences the individual country authors have put together on the evolution of agriculture in these countries with a focus on government policies seeking to raise agricultural productivity in the light of initiatives to end hunger, and to promote the spread of digitalization and digitization, as well as to strengthen climate resilience.

Raising Productivity and Ending Hunger

Although all countries in Southeast Asia emerged from the Second World War, ravaged and gripped by poverty and underdevelopment, the different initial conditions and development policies have not only created uneven development between the countries but has also been characterized by highly contrasting capacities to address digitalization and climate change. Among the market economies, Singapore since its breakaway from Malaysia in 1965, Indonesia and the Philippines from the 1970s, and Malaysia since the 1980s began to focus more on industrialization than agriculture. Manufacturing expansion not only has increasingly reduced arable land for agriculture, farming technologies did not change aggressively to raise agricultural productivity in the face of rising populations. Consequently, not only has the contribution of agriculture in GDP fallen, gradually these countries have increasingly faced increasing dependency on food imports. It is only since the food trade imbalance became severe that Indonesia and Malaysia began introducing national agricultural policies again to reverse the

[2] The geographical classification of small, medium, and large is used to differentiate the countries in Southeast Asia. In doing so, the World Bank's use of low-, medium-, and high-income countries was dropped as the income differential does not effectively bring out the vast geographical differences between these countries. For example, Brunei, Singapore, and Timor-Leste are small but the last is the poorest in Southeast Asia while Singapore and Brunei are the two richest by per capita incomes among these countries.

scale of import dependence, though both countries had focused heavily on rice and other essential food cultivation in the 1970s, only to be replaced with greater emphasis on manufacturing since the 1980s.

A combination of overreliance on imports and the volatile fluctuations in food supply and prices drove Singapore and later Brunei to launch policies to reduce its dependence. Singapore implemented strong use of science and technology to engineer productivity rise and towards high value-added activities in safe and novel food products (Ludher & Teng, 2025). The launching of the sustainable development goals since 2016 simply intensified the emphasis by the countries to introduce smart farming and change the incentive framework to stimulate the displacement of conventional farming with climate-resilient farming.

Despite the recognition and the masterplans to rejuvenate agriculture, especially smart agriculture from the turn of the millennium, food import dependency has continued in Indonesian and Malaysia. Malaysia has since 1990 experienced food insecurity problems. The heavy focus on commercial plantation, industrialization, and urbanization, and the abandonment of farm subsidies for small farmers have driven a massive expansion in food imports in Malaysia so much so that country has faced a negative trade balance in food farming over the period 1990–2023 (Annizah & Rasiah, 2025). While Indonesia has successfully expanded commercial agriculture and fish exports, the country remains gripped by low productivity in paddy farming (Fadhil, Dianwecaksih & Fahcru, 2025). Despite the implementation of various poverty reduction programmes and social safety nets to support the vulnerable communities, including promoting structural change into manufacturing and service to create job opportunities, and a more stable environment to enhance peoples' well-being, both countries are still grappling with low wage and food insecurity problems. The masterplans to modernize agriculture require effective implementation with constant reviews to ensure that their goals are not missed.

The neo-liberal free trade policies did not help but undermine food cultivation in the Philippines over several decades. Ofreneo (2025) noted that the asymmetric relation between the small farmers and the large buyers undermined agricultural prices for the farmers. The monopolistic power enjoyed by the large buyers offered them high rents while keeping the small farmers poor, and hence, why he argues that hunger (SDG2) has remained entrenched in the country. Following a bloody war of independence, Timor-Leste is in need of serious build-up in basic infrastructure and its agriculture is still dominated by sedentary activities. Timor-Leste is in danger of following this path as economists driven by markets seem to discourage problem-solving interventions to build supply capability in the country (Courvisanos et al., 2025). There are also little resources that the government has at its disposal to develop the infrastructure to stimulate cultivation of crops. Timor-Leste requires the drawing of clear economic plans led by economists carrying tacit knowledge of the sectors and technological specificities associated with the type of agriculture suitable for the geography of the country.

Among the transition economies, Vietnam has shown remarkable advancement in the smartification of agriculture, as well as movement towards building climate resilience (Bach et al, 2025). However, Myanmar (Naw, 2025), Cambodia (Kum, 2025), Lao PDR (Songvilay & Kammerng, 2025), and Timor-Leste are characterized by poverty with little translation of agricultural master plans towards ending hunger and strengthen their capacity to quicken the displacement of fossil fuels with clean energy, driving full employment and improvement in wages, and digital infrastructure development to stimulate innovations in agriculture. While Myanmar showed the natural and historical endowments to stimulate rapid economic growth, the political crisis not only attracted economic sanctions from the United States in 2003, the 2021 political coup has seriously undermined the country's capacity to investment inflows and economic growth (Naw, 2025). Cambodia has introduced a number of masterplans to smartify agriculture while strengthening climate resilience to alleviate poverty; the country has shown little translation of these prescriptions into action (Kum, 2025). The deployment of robots manufactured by Huawei in potash mining in central Lao PDR by a Chinese firm shows promise towards raising agricultural productivity in the country (Songvilay, 2025). The Lao PDR government should work towards stimulating strong spillover towards national firms to spread the economic benefits that can be harnessed from the high-technology venture in the country.

Although Indonesian agriculture has seen the rapid expansion of palm oil production, especially in Sumatra and Kalimantan, and tuna fishing, the country is still gripped with low productivity in rice yields, which is why the country has sought technology transfer from China (Fadhil et al., 2025). The Philippines has also faced with a trend decline in food farming, which has not helped in ending hunger in the country (Ofreneo, 2025). Divya (2024) reported official estimates that 7 percent to 16 percent of Indonesia's population is vulnerable to starvation amid declining domestic rice production. Given the country's population of 281.6 million as of June 2024, that would translate to between 20 and 45 million people being at risk of hunger. The World Bank projected 2.5 percent of Malaysian population (938, 000 people) to experience hunger in 2019, while the Statistics Department of Malaysia reported 5.6 percent of the population to face poverty (1.83 million people) in 2020 using the new poverty line income (Intan Maizura, 2021).

Looking at the Southeast Asian countries where hunger (SDG 2) is still a problem, Thailand ranked 52nd among 127 countries in 2024 by the Global Hunger Index (GHI) with a score of 10.1. Among these countries, Thailand ranked ahead of Vietnam 56 (score of 11.3), Malaysia 61 (12.7), the Philippines 67 (14.4), Cambodia 68 (14.7), Myanmar 74 (15.7), Indonesia 77 (16.9), Lao PDR 87 (19.8), and Timor-Leste 104 (27). The poorest countries with the severest hunger index are in Africa (The Nation, 2024). Despite its higher per capita income, Malaysia ranked lower than Thailand and Vietnam in the higher index.

In addition, Cambodia's goal to reduce emissions by 41.7 percent by 2030 is contingent on accessing US$5.8 billion financing for mitigation and $2 billion for adaptation. However, Cambodia's cumulative external development finance assistance from 2015 to 2019 amounted to only around $630 million, falling far short of the required funding to meet their INDC targets. Similarly, Vietnam committed to increase its emission reduction target from 15.8 percent to 43.5 percent if provided with international aid, particularly from advanced economies through climate financing and technology transfers. Hence, collaborative efforts between developed and developing regions are imperative, as the impacts of climate change have cross-border consequences. Without adequate financing and cooperation, achieving the targets set by the Paris Agreement becomes highly challenging.

Myanmar's cause to attract foreign climate funds is not at all helped by the unnecessary political crisis that erupted in 2021. Timor-Leste lacks the capital resources to engender the conditions essential for the modernization of its economy, and hence, needs foreign assistance to enable the early development of smart climate agriculture.

Given that vegetables and livestock farming in Southeast East Asia are largely undertaken by small farmers, intensive farming methods must be promoted extensively to raise yields with common technical support centres, and with bulk purchase of organic fertilizers and insecticides, and price guarantees to the farmers. The bulk purchases can then be divided into smaller units for the small farmers. This practice is prevalent in Thailand and Vietnam and was also actively driven in Malaysia and Indonesia in the 1970s and early 1980s (Rasiah & Salih, 2019; Rasiah, 2022). While Singapore, Brunei, Thailand, and Vietnam are well positioned to continue the practice, the remaining Southeast Asian countries must take steps to implement it effectively. Indonesia, Malaysia, the Philippines, Cambodia, Lao PDR, Myanmar, and Timor-Leste should adapt such support systems that feed fertilizers, new production techniques (especially the use of drones and robots) for ploughing, weeding, injection of fertilizers and organic fertilizers, harvesting, constant renewal of irrigation and draining systems, and marketing support.

Ludher and Teng (2025) argue that Singapore should use its superior R&D support capability in Southeast Asia to ease the problems faced by their domestic producers, especially against cheap imports, but breaking down barriers they face at supermarkets and the normal markets, and to attract lower prices when approaching bulk purchaser markets, such as army camps and prisons. Brunei can also offer the same support given its high-income status. The governments of Indonesia, Malaysia, Philippines, Thailand, and Vietnam can also consider such approaches in their main metropolitan cities. Aqua fish farming and high-value vegetables can also be expanded in controlled environments.

Goh et al (2025) argue that post-harvest losses in crop cultivation is a big problem in Malaysia, which is also the case with the other countries other than Singapore. Governments should encourage the handling of post-harvest is done effectively. Associated with this problem is the need for promoting food recycling to raise valorisation, which is a problem that all Southeast Asian countries must address with its severity highest in Timor-Leste.

Despite its limited resources for agricultural expansion, Singapore has managed to keep hunger in control as it enjoyed rapid growth, and since 2000 has focused on creating self-sufficiency in essential food items (Ludher & Teng, 2025). The same can be said of Brunei. Despite strong reliance over oil and gas, Brunei has increasingly focused on raising self-sufficiency on food. Volatile food prices and serious supply disruptions following the outbreak of COVID-19 drove the government of Brunei Darussalam to offer incentives for the introduction of high yield food cultivation, including fishing (Ahmed & Rose, 2025). In doing so, Brunei is using Singapore's leveraging strategies to reduce dependence on food imports. Thailand followed by Vietnam appears the most advanced in addressing agricultural productivity with significant proportion of its population still reliant of the sector with strong initiatives to embrace digitalization and climate resilience followed by Vietnam. In fact, Thailand and Vietnam ranked higher than Malaysia in the World Hunger Index in 2024 (The Nation, 2024). Indonesia and Malaysia have also introduced policies to strengthen digitalization and climate reliance but more focus is essential to raise its outcomes. Philippines needs to show greater emphasis on digitalization and climate reliance to address its problems with poverty, underemployment, and climate change issues. Cambodia, Lao PDR, Myanmar, Indonesia, and Vietnam are still reliant strongly on foreign assistance to support their cause of digitalization and climate resilience. From 2018 to 2020, Vietnam, Myanmar, and Indonesia received massive blended climate financing, with Indonesia's SDG Indonesia One (SIO), which is managed by PT Sarana Multi Infrastructure (PT SMI); SIO-GFF employs a 20-year financial intermediation loan of $150 million from ADB to fund green and SDG-focused subprojects (Rasiah & Gopi Krishnan, 2025). This project supported PT SMI to partially finance up to 10 projects, totalling $423 million, with an average allocation of 35 percent per subproject. Furthermore, ADB's loan funds are expected to potentially unlock approximately US$1 billion for green projects to generate a leverage eight times the initial amount. The climate impact of SIO-GFF could lead to an annual reduction of around 480,700 tons of CO_2 emissions.[3]

Digitalization and Digitization

The spread of digital infrastructure and its facilitation of IR4.0 technologies, such as robots and drones through the amplification of innovations associated with SDG 9, has raised agricultural productivity through improved efficiency where such technologies are deployed. Workers equipped with digital competence have also enjoyed higher salaries. However, Indonesia, Malaysia, the Philippines, Thailand, Myanmar, Lao PDR, Cambodia, and Vietnam have been struggling to train through the diffusion digital

[3] The calculation of the impact of SIO_GFF is reported in: https://www.adb.org/sites/default/files/publication/806411/sdg-indonesia-one.pdf.

competence in technical and vocational education training institutes. Some of these countries have even sought support from Germany, South Korea, China, Taiwan, and Japan to upgrade their national TVET organizations.

Except for Timor-Leste (which is still dominated by sedentary agriculture and extremely poor physical infrastructure) (Courvisanos et al., 2025), the remaining Southeast Asian countries have not only launched master plans to widen and intensify the spread of both digitalization and digitization, their proliferation has produced varying effects in the economies of each of these countries. Some countries have also enlisted the help of foreign companies to install robots and drones equipped with smart technologies. An example is the role played by Huawei in central Lao PDR to mine and process potash as fertilizer (Songvilay & Khammerng, 2025), and the spread of IR4.0 technologies in Indonesia in rice farming through help from China (Fadhil et al., 2025).

Singapore is arguably the most advanced in the deployment of IR4.0 technologies in agriculture, though along with Brunei, it is the most import-dependent country in Southeast Asia when it comes to agriculture. Singapore is not only powered by its huge capital endowments, but is also driven heavily by the need to establish a minimum of 30 percent self-sufficiency in critical food items by 2030 (Ludher & Teng, 2025). Although the developments have taken place only since 2018, as the Brunei chapter in this book has shown, it is pursuing an ambitious plan to reduce dependency on agricultural food imports and on raising the value added through the introduction of IR4.0 technologies. Brunei also shows semblance of Singapore's leveraging strategy here as it has tied incentives and grants to agricultural value-added upgrading.

IR4.0 technologies are also proliferating strongly in Malaysia, Indonesia, Thailand, and Vietnam but it appears that there are lower levels of decentralization of these artificial intelligence-driven technologies in Indonesia and Malaysia than in Thailand and Vietnam, which may be a major reason why the latter two countries face lower level of hunger than Indonesia and Malaysia. The integration of advanced technologies like precision farming and IoT, coupled with a commitment to support R&D focused on sustainable practices, can optimize yields and ensure long-term viability. As the chapters show comprehensive TVET education, training, and financial support for farmers contribute to the development of a knowledgeable and empowered agricultural community. There is a clear need for raising digital competence and integrating small farmers with technical assistance providers in Indonesia and Malaysia to alleviate hardcore poverty (SDG 1). Pro-active strategies are necessary to reach the farmers to integrate them into an IT and digital network connecting members of the value chain, providers of technical training, markets, and government agencies tasked to allocate subsidies and technical assistance. The recommendations Naguib et al. (2024) and Saffa (2023) offer interesting applications of drone technology to support agriculture in Malaysia and Indonesia that should be enhanced to raise productivity and farmers income.

The Philippines, Cambodia, Lao PDR, Myanmar, and Timor-Leste hold different histories but are gripped by low agricultural productivity and significant levels of hunger. Significant strategies are necessary to end hunger (SDG 2) in these countries. Whereas the government has been able to address poverty, employment, and wage issues in Cambodia and Lao PDR, these countries still lack the STI infrastructure to quicken digitalization, and hence, still seek external support to strengthen the process. Now that the second Trump administration has started to withdraw USAID allocations that have been important in addressing hunger issues in Cambodia, Lao PDR, and Timor-Leste, alternatives are needed to address assistance related to ending hunger. Not only that Myanmar shares this predicament, it is also riddled in an unnecessary political crisis that has led to its isolation by many countries. Timor-Leste has been thrust into the world economy at such an underdeveloped level and little prepared to seek positive synergies from the globalization process let alone appropriate synergies from digitalization. Yet, digitalization and IR4.0 are critical for all economies to compete in the global economy. The country needs significant help from the international organizations and its neighbours in Asia to embrace digitalization to pursue poverty alleviation and hunger-ending strategies (The Straits Times, 2025).

Climate Resilience

Traditional farming methods have long been used in Southeast Asia, which for econological and biodiversity reasons have been the earliest sustainable and climate-resilient agricultural methods deployed in Southeast Asia (FAO et al., 2018, p. 102), especially integrated farming of wet paddy with fish in Malaysia, Indonesia, Vietnam, the Philippines, Laos, and Thailand (Frei & Becker, 2005). The same can also be said of small-scale multiple crop farming with animal grazing in Thailand, Lao PDR, and Vietnam (Prapimphan, 2022). A significant portion of such traditional sources of knowledge is indeed indigenous in nature. While such small-scale farming methods provide strong ecological justification for expansion, much of Southeast Asia's farming has moved towards modern farming methods.

It is also important to note that greening policies have taken on strong pursuit of organic farming far more extensive than the traditional farming methods that often use indigenous sources of knowledge in Southeast Asia. Singapore and Brunei in particular have embarked strongly on organic farming (Ludher & Teng, 2025; Ahmed & Rose, 2025). Timor-Leste is still faced with serious problems of poverty with the economy still very much immersed in early-stage primary processing. Singapore and Brunei lack natural non-fossil endowments, such as hydropower but have moved swiftly to raise the share of renewable energy in total energy consumption, suggesting that these countries are poised to meet SDG 7 with increasing use of clean energy, including in agriculture while intensifying organic farming. While hydropower resources have ensured that Thailand, Vietnam, Malaysia, Indonesia, and the Philippines enjoy relatively high share of renewable energy in overall energy consumption, the renewable shares show a trend fall owing to increasing use of fossil fuels and rising carbon emissions. Cambodia, Lao PDR, and Myanmar show high use of hydropower but these countries should quickly reverse the reliance on fossil fuels. Meanwhile, with exception of Timor-Leste and the Philippines, the remaining countries have introduced master plans to focus on sustainable agriculture through the smartification of agriculture and building climate resilience. It is interesting to note that Singapore, Brunei Darussalam, Vietnam, Malaysia, Indonesia, and Thailand have started to emphasize the smartification of agriculture to strengthen climate resilience. Also, Singapore, Brunei, and Malaysia have resorted to IR4.0 technologies to reverse food import dependence. While these developments augur well for the region, it is pertinent for these countries to strengthen their review process to ensure that climate change targets remain contemporaneously upgraded to meet the challenges that emerge.

In addition to emphasis on organic farming, Singapore's strong R&D capability has enabled the country to increase its emphasis on climate-resilient food safety and novel foods since the turn of the millennium (Ludher & Teng, 2025). If as expected Singapore achieves local production of 30 percent of nutritional demand domestically by 2030, the country still be heavily reliant on global supply chains and trade. Nevertheless, Singapore's high GDP per capita of US$84.7 thousand in 2023 and wage share of 40 percent of GDP in 2022 offer the country a significant lead in assisting the country in facing hunger.[4] Brunei appears to show effort to imitate such climate-resilient practices in agricultural promotion in Southeast Asia with initiatives taken to raise the R&D capability in the country (Ahmed & Rose, 2025). Hence, despite the low starting positions owing to the lack of hydropower, Singapore and Brunei show strong implementation of climate mitigation measures to focus on renewable energy. Singapore has already introduced carbon taxes to mitigate climate change (Rasiah & Gopi Krishnan, 2025). Brunei's policies go beyond Singapore to achieve self-sufficiency in

[4] The data was obtained from World Bank (2024).

cattle, goat, sheep, buffalo, vegetables, and paddy (Ahmed & Rose, 2025) through the adoption of smart technologies to raise labour productivity through upgrading human capital, resource optimization, reduction in post-harvest losses, and agrifood processing.

Indonesia, Malaysia, Thailand, and Vietnam have begun efforts to check climate change since the 1990s but especially from 2016 (Fadhil et al., 2025; Annizah & Rasiah, 2025; Mot & Teerawat, 2025; Bach et al, 2025). The achievements recorded in Thailand and Vietnam are noteworthy, which is why both countries rank higher than Malaysia despite having a lower GDP per capita than Malaysia. The mechanisms for carbon capture, carbon taxes, and carbon trading have already been worked out to mitigate climate change. The big problem involves reaching the small farmers in all these countries. There are also problems facing vulnerable areas, such as mangroves, inflows of sea water, heavy uncontrollable rains that lead to serious floods, and landslides. While Indonesia and Malaysia have progressed considerably over the period since the 1960s, but these countries are still grappling with food security issues in critical items, such as vegetables, mutton, and beef. For these countries to break out from the middle-income trap to reach developed status, among other things these countries should quicken the transition to high value added through the diffusion of digitalization, digitization, and IR4.0 technologies into agriculture. Efforts must also be taken that need to strengthen CSA diffusion into Malaysian agriculture accompanied by a serious use of an appraisal mechanism that ensures stringent targets are met. Since food farming is significantly influenced by intensive small-scale farming, governmental emphasis on ecosystem focus should stimulate collective support that can synergize the small farmers. In addition, government policies should be complemented by ground-level execution of digitalization and IR4.0 technologies, and strengthen climate resilience in the country. However, to achieve zero poverty (SDG (1) and zero hunger (SDG (2), climate-friendly development emphasis also requires Indonesia and Malaysia should also meet SDG 7 (provision of quality jobs with decent incomes), SDG 8 (building climate resilience), and SDG 9 (infrastructure and innovation) (UNESCO, 2021).

Indonesia's exemplary performance in tuna fishing where the country has become the world's leading producer of that fish can act as a model for Malaysia, the Philippines, and Thailand. However, Brunei and Thailand may be better examples to follow in aqua farming of fish and animal husbandry for Malaysia, the Philippines, and Indonesia. Given its tiny size and lack of resources, Timor-Leste should consider applying to join the Parties to the Nauru Agreement (PNA) and the Vessel Day Scheme (VDS) in which the Pacific Islands cooperate to generate significant income from tuna fishing, including licensing foreign fisherman to fish in its waters to support government expenditure (Rasiah, 2024).[5] The Philippines in addition should consider the critique levelled at government policy by Ofreneo (2025) to address market failures to ensure national capabilities are upgraded to compete with imports. In making this point, it will be interesting for Indonesia, Malaysia, Philippines, Lao PDR, Myanmar, Cambodia, and Timor-Leste to consider Corv et al.'s call to revisit the Habermas framework to examine past and potential future transitions to sustainable agriculture to establish effective transitional analysis to strengthen sustainable agriculture, which should be predicated on greater coordination between top-down and bottom-up strategies towards organic-focused climate-resilient sustainable agriculture sector.

The carbon net-zero target years submitted to the United Nations Convention for Climate Change UNFCCC) for Thailand (2065) (Mot & Teerawat, 2025) and Indonesia (2060) (Fadhil et al., 2025) are the highest in Southeast Asia followed by Singapore, Malaysia, Vietnam,

[5] The Parties to the Nauru Agreement (PNA) was begun established in 1982 to establish an exclusive economic zone (EEZ) covering 10,000 square miles with the Federated States of Micronesia, Kiribati, Marshall Islands, Nauru, Palau, Papua New Guinea, Solomon Islands, Tuvalu, and Tokelau as members as of 2024 (Rasiah, 2024). The Vessel Day Scheme (VDS), which was introduced in 2007, enhanced the effectiveness of the PNA.

Cambodia, and Lao PDR by 2050 (Rasiah & Gopi Krishnan, 2025), and then, Myanmar using the LULUF measure by 2040 (Naw, 2025) and the common measure by 2050 (Rasiah & Gopi Krishnan, 2025); the last involving Myanmar is questionable, particularly owing to the political crisis facing the country. The Philippines and Timor-Leste have yet to submit their carbon net-zero targets with both countries subjected to severe weather issues.

Building Cooperation Within ASEAN

With especially the United States and the European Union raising tariffs to render the World Trade Organization (WTO somewhat powerless as geo-political tensions with China and the trade and investment sanctions on Russia that followed the invasion of Ukraine by the latter, it will help ASEAN members to use the ASEAN Economic Community (AEC) to raise cooperation among members to specialize on particular technologies and products to raise competitiveness of their agricultural exports while pursuing digitalization and climate resilience. Singapore and Brunei are positioning to specialize on high value-added, robotized organic crops and fisheries beyond just meeting their domestic demand (Ludher & Teng, 2025; Ahmed & Rose, 2025). Indonesia, Lao PDR, Vietnam, Cambodia, and Timor-Leste should deepen their harvesting of civet-based organic coffee for the ASEAN and global market (see also Rasiah & Zhang, 2015).

With Timor-Leste likely to join the ASEAN Economic Community in 2025, the Southeast Asian countries can work towards strengthening trade and investment relations within the community a *la* the European Union. ASEAN members should strengthen relationship among member nations to use the 40 percent rule of origin value-added condition to be recorded within ASEAN to earn tariff exemption.

The margin of preference (MOP) is one of the instruments used with preferential trading arrangements under the ASEAN preferential trading arrangement (PTA) since its introduction in 1977 (ASEAN, 2024). All products originating from ASEAN which are under the PTA will be accorded preferential tariffs which will be a margin lower than the MFN tariff rates of the ASEAN member countries. MOPS are usually defined in terms of a percentage applied to the rate of an ASEAN country, and in 2024, it was 50 percent for all PTA items. The 40 percent rule of origin constitutes a cumulative total of value added recorded in the whole of ASEAN. The use of the common effective preferential tariffs (CEPT) within ASEAN should also allow for stronger exports from the successful agricultural economies of Thailand and Vietnam to the rest of the member nations.

In ASEAN, sustainable bond issuance has significantly increased, with consolidated sustainable bonds outstanding recording a 79 percent annual average growth rate per annum, rising from $1.7 billion in 2015 to $30.5 billion in 2021 (Rasiah & Gopi Krishnan, 2025). The development of ASEAN Green Bond Standards has contributed to this growth, providing guidelines for regional regulators to create their own national green bond guidelines. The ASEAN Capital Market Forum (ACMF), comprising capital market regulators from ASEAN countries, developed these standards based on the International Capital Market Association's (ICMA) Green Bond Principles (GBP). The ACMF also proposes the establishment of a "Green Lane" to facilitate cross-border issuance and distribution of labelled bonds without compromising regulatory standards. Furthermore, the ACMF emphasizes the importance of the ASEAN Collective Investment Schemes (CIS) framework and streamlining the review framework for the ASEAN Common Prospectus to enhance cross-border capital flows and inclusive financing opportunities.

Meanwhile, collaborations among the ASEAN countries on decarbonization efforts have evolved with the ASEAN joint statement at COP 27, which achieved a 23.8 percent reduction in energy intensity (based on 2005 levels) and a 14.2 percent share of renewable energy in primary energy supply. Indeed, the renewable energy contribution in installed power capacity through the Lao

PDR–Thailand–Malaysia–Singapore Power Integration Project (LTMS-PIP) increased to 33.5 percent in 2020, which demonstrates the collaborative commitment of these countries to achieve their overall carbon net-zero targets. In this regard, Indonesia, Brunei, Cambodia, the Philippines, and Vietnam are all expected to establish similar collaborative initiatives (Rasiah & Gopi Krishnan, 2025). While the internal political crisis in Myanmar has slowed down smartification initiatives, signs show that all ten of the ASEAN members and the prospective new member of Timor-Leste will work towards strengthening climate resilience.

While Singapore has already developed experience in developing regional satellite farms, for example in Jilin in China and Bulan Island in Indonesia (Ludher & Teng, 2025), Brunei should also include such as focus to solve its land scarcity problems. Sarawak and Sabah in East Malaysia should be considered for rice and food crop cultivation. Similar farming contracted by Japanese agencies in Cambodia, Lao PDR, and Malaysia (Sarawak) exists to strengthen food security in Japan. In doing so, the Lao-China Economic Corridor (LCEC) programme should be revisited for its extension to integrate the whole of Southeast Asia, especially so with Indonesia receiving technology transfer in rice cultivation from China.

Overall, despite massive differences in initial endowments, the Southeast Asian experience shows that the nature of policy execution has played a critical role in the productivity. Besides, the prospects for the establishment of a climate smart sustainable agriculture system in Southeast Asia look bright. In addition to drawing more through effective implementation and constant appraisals to strengthen the planning process with strong coordination between central, provincial, and local policies, the constant identification and implementation of best climate smart agricultural practices should enable the countries of Southeast Asia to meet the relevant SDGs, even if it means that countries, such as the Philippines, Lao PDR, Myanmar, and Timor-Leste, may defer their targeted deadlines. Much depends on political stability, continued cooperation among the countries, and strong persuasion to seek diplomacy to achieve political, economic, and social consensus. In the spirit of appropriating collective synergies through cooperation among ASEAN members, the secretariat should strengthen collaboration among these countries to share farming methods, including traditional and organic.

References

Annizah, N. I. & Rasiah, R. (2025). Agriculture, Food Security and Sustainability in Malaysia. In R. Rasiah, V. Anter, N. I. Annizah (Eds.), *Sustainable Agriculture and Food Security in Southeast Asia: Towards Climate Resilience and Digitalization,* Singapore: Springer Nature.

ASEAN. (2024). ASEAN Joint Statement on Climate Change to the 29th Session of the Conference of the Parties to the United Nations Framework Convention on Climate Change (UNFCCC COP-29), October 12, 2024, Downloaded on January 2, 2025 from https://asean.org/asean-joint-statement-on-climate-change-to-the-29th-session-of-the-conference-of-the-parties-to-the-united-nations-framework-convention-on-climate-change-unfccc-cop-29/

Bach, S.T., Dang, K.K., Nguyen, T.B., Le, V.T.T., Nguyen, H.K. (2025). Agricultural Development and Food Security in Vietnam. In R. Rasiah, V. Anter, N.I. Annizah, (Eds.) *Sustainable Agriculture and Food Security in Southeast Asia* (forthcoming).

Courvisanos, J., Pinto, E. J. A., & Apolinário, M. (2025). From subsistence to sustainable agriculture in Timor-Leste: A challenging transition. In R. Rasiah, V. Anter, N.I. Annizah (Eds.) *Sustainable Agriculture and Food Security in Southeast Asia: Towards Climate Resilience and Digitalization,* Singapore: Springer Nature..

Divya, K. (2024). Up to 45 million Indonesians at risk of hunger: Home Ministry, *Jakarta Post*, August 17. Accessed on December 2027 from https://www.thestar.com.my/aseanplus/aseanplus-news/2024/08/17/up-to-45-million-indonesians-at-risk-of-hunger-home-ministry

Fadhil, M.H., Dianwicaksih, A. & Fachru, N. (2025). Agriculture, Food Security and Sustainability in Indonesia. In R. Rasiah, V. Anter, N.I. Annizah (Eds.), *Sustainable Agriculture and Food Security in Southeast Asia: Towards Climate Resilience and Digitalization,* Singapore: Springer Nature.

FAO, IFAD, UNICEF, WFP, and WHO. (11 September 2018). *The State of Food Security and Nutrition in the World 2018. Building climate resilience for food security and nutrition.* FAO.

Frei, M., & Becker, K. (2005). Integrated rice-fish culture: Coupled production saves resources. *Natural Resources Forum, 29*(2), 135–143.

Goh, L.T, Yong, S.K., Santha, C. & Ong, S. Li (2025). *Food for Thought: Addressing ASEAN-5's Food Security Challenges* (forthcoming).

Intan Maizura, A. K. (2021*). Hunger no more: How data sharing can help tackle Malaysia's hunger issue, New Straits Times, July 18.* Accessed on December 27, 2024 from https://www.nst.com.my/lifestyle/sunday-vibes/2021/07/709478/hunger-no-more-how-data-sharing-can-help-tackle-malaysias

Kum. K. (2025). Agricultural Development and Food Security in Cambodia, R. Rasiah, V. Anter & NI. Annizah (Eds), *Sustainable Agriculture and Food Security in Southeast Asia: Towards Climate Resilience and Digitalization,* Singapore: Springer Nature.

Ludher, E.L. & Teng, P. (2025). Agricultural Development and Food Security in Singapore, In R. Rasiah, R., V. Anter & N.I. Annizah (eds), *Sustainable Agriculture and Food Security in Southeast Asia: Towards Climate Resilience and Digitalization,* Singapore: Springer Nature.

Mankiw, N. G. (2014). *Principles of Economics* (7th Edn.). Cengage.

Naguib, M. N., Rasiah, R., & Kamarul A. M. (2024). Smartifying Agriculture to Strengthen Food Security. In R. Rasiah, K. Salih, B. Hafriza, M.I.A.R. Tawfik, (Eds.), *Tun Abdul Razak: Imagining Malaysia's Future through his Lenses.* Universiti Malaya Press.

Naw, S. (2025). Sustainable Agriculture in Myanmar: Towards Climate Resilience and Food Security. In R. Rasiah, V. Anter, N.I. Annizah, (Eds.), *Sustainable Agriculture and Food Security in Southeast Asia: Towards Climate Resilience and Digitalization,* Singapore: Springer Nature.

Ofreneo, R. (2025). Sustainable Development Challenges in the Agricultural Sector in the Philippines. In R. Rasiah, V. Anter, N.I. Annizah, (Eds.), *Sustainable Agriculture and Food Security in Southeast Asia*: Towards Climte Resilience and Digitalizaton, Singapore: Springer Nature.

Ahmed, O.H & Rose, A. (2025). Smartifying Agriculture to achieve Food Security and Sustainability in Brunei Darussalam. In R. Rasiah, V. Anter, N.I. Annizah (Eds.), *Sustainable Agriculture and Food Security in Southeast Asia: Towards Climate Resilience and Digitalizaton, Singapore: Springer Nature.*

Prapimphan, C. (2022). A ground-up approach to climate-resilient agriculture in Southeast Asia. *Perspective,* October 25, No. 103.

Rasiah, R., & Zhang, M. (2015). Reducing poverty in Timor-Leste through stimulating growth and structural change. *Institutions and Economies, 7*(3), 40–68.

Rasiah, R., & Salih, K. (Eds.). (2019). *Driving development: Revisiting Razak's role in Malaysia's economic development.* Kuala Lumpur: Universiti Malaya Press.

Rasiah, R. (2020). Industrial policy and industrialization in Southeast Asia. In O. Arkebe, C. Cramer, H. J. Chang, & R. Kozul-Wright (Eds.), *Oxford handbook of industrial policy.* Oxford University Press.

Rasiah, R., Low, S. W. Y., & Nurliana, K. (2022). *Digitalization and Development: Ecosystem for Promoting Industrial Revolution 4.0 Technologies in Malaysia.* Routdledge.

Rasiah, R., & Gopi Krishnan, V. (2025). Green Southeast Asia. In J. Mathews, O. Arkebe (Eds.), *Oxford Handbook of Green Industrial Policy.* Oxford University Press.

Rasiah, R. (2024). *The industrial policy landscape of Asia and the Pacific Islands.* Background report prepared for the UNIDO Industrial Development Report 2024. United Nations Industrial Development Organization.

Rasiah, R. (2025). Strengthening Agricultural Sustainability and Food Security in Southeast Asia. In R. Rasiah, V. Anter, N.I. Annizah, (Eds.), *Sustainable Agriculture and Food Security in Southeast Asia: Towards Climate Resilience and Digitalization, Singapore: Springer Nature.*

Ricardo, D. (2004[1821]). On the Principles of Political Economy and Taxation. In *Volume I of The Works and Correspondence of David Ricardo, edited by Pierro Sraffa with the Collaboration of Maurice Dobb.* Liberty Fund Press

Saffa, A. (2023). *AI, Drones Enhance Crop Management, October 3, OpenGov.* Accessed on December 27, 2024 from https://opengovasia.com/2023/10/03/indonesia-ai-drones-enhance-crop-management/

Songvilay, L. & Kammerng, B. (2025). Agricultural Development in Lao PDR: Towards Raising Productivity, Climate Resilience and Sustainability. In R. Rasiah, V. Anter, N.I. Annizah, (Eds.), *Sustainable Agriculture and Food Security in Southeast* Asia: T*owards Climate Resilience and Digitalization,* Singapore: Springer Nature.

The Nation. (2024). *With 5.6% of the population undernourished, Thailand ranked 52nd on hunger index.* Accessed on December 27, 2024 from https://www.nationthailand.com/sustaination/40042449

The Straits Times. (2025). *The growing pains of Asia's newest country.* February 16, downloaded on February 18, 2025 from https://www.straitstimes.com/asia/the-growing-pains-of-asias-newest-country

UN. (2016). *Sustainable Development Goals*, downloaded on January 2, 2025 from https://sdgs.un.org/2030agenda

World Bank. (2024). *World Development Indicators.* World Bank Institute.

Yot, A. & Teerawat, C. (2025). Agriculture, Food Security and Economic Development. In R. Rasiah, V. Anter, N.I. Annizah, (Eds.), *Sustainable Agriculture and Food Security in Southeast Asia: Towards Climate Resilience and Digitalizaton, Singapore: Springer Press.*

Index

R. Rasiah et al. (eds.), *Sustainable Agriculture and Food Security in Southeast Asia*, Sustainable Development Goals Series, https://doi.org/10.1007/978-981-95-0477-0

MIX
Papier aus verantwortungsvollen Quellen
Paper from responsible sources
FSC® C105338

If you have any concerns about our products,
you can contact us on
ProductSafety@springernature.com

In case Publisher is established outside the EU,
the EU authorized representative is:
Springer Nature Customer Service Center GmbH
Europaplatz 3, 69115 Heidelberg, Germany

Printed by Libri Plureos GmbH
in Hamburg, Germany